Irrigation and Agricultural Development in Asia

Also edited by E. Walter Coward, Jr.
(with George M. Beal and Ronald C. Powers):

Sociological Perspectives of Domestic Development

Irrigation and Agricultural Development in Asia

PERSPECTIVES FROM THE SOCIAL SCIENCES

Edited by
E. Walter Coward, Jr.

CORNELL UNIVERSITY PRESS
ITHACA AND LONDON

First published 1980 by Cornell University Press.
Published in the United Kingdom by Cornell University Press Ltd., 2–4 Brook Street, London W1Y 1AA.

International Standard Book Number (cloth) 0-8014-1132-7
International Standard Book Number (paper) 0-8014-9871-6
Library of Congress Catalog Card Number 79-24319
Printed in the United States of America.
Librarians: Library of Congress cataloging information appears on the last page of the book.

Contents

Preface

This book is positioned at the confluence of two interests—the scholarly concern with the study of organizational arrangements by which people enter into relations with each other to manage important natural resources and the practical concern with creating appropriate irrigation organizations that will enhance effective systems operations.

Many workers, acting through both national bureaucracies and international aid agencies, have significantly expanded irrigable cropland. Since most of this irrigation development has been very capital-intensive, vast financial investments have been made. The outcomes of those investments, however, often differed distressingly from those that had been expected: less land was served than originally planned; cropping patterns were not modified as intended; water users failed to complete local irrigation and drainage facilities.

When these ubiquitous disappointments were assessed, one major difficulty emerged: the need for improved farmer organization in the local units of irrigation systems. Policy makers began to formulate strategies for the organization of irrigation groups, at various levels in the irrigation systems, which were intended to assume such responsibilities as completing and maintaining the local infrastructure, collecting irrigation fees from group members, settling disputes regarding water distribution, and coordinating the timing of planting and harvesting.

As the essays in this book show, social scientists have been interested in the problems of irrigation organization for some time, and their interest has been paralleled by the concern of irrigation planners and operators with creating, or sustaining, viable irrigation groups and associations. This is not to say, of course, that the two parties define the issues and problems similarly—only that they have interests in common phenomena.

Lamentably, this collection of readings provides a satisfactory response to neither of these concerns. The scholar of organizations will not find a well-articulated theory of irrigation organization to be further tested and refined, nor will the administrator find a handbook on the organization of water users. What both will find, and I hope will find useful, is an array of experiences with irrigation organization and tentative propositions that suggest both what is known and what requires further inquiry.

The book is divided into three sections. Part I is composed of three chapters that provide introductory comments regarding the nature of irrigation organization and its relationships to irrigation management.

In Part II a group of chapters reports on the organizational patterns and arrangements that have been observed in several community-based irrigation systems in Asia, and two chapters illustrate potential relationships between community irrigation experience and broader irrigation development policy. The importance of community-based irrigation systems has been highlighted in the last few years as such nations as Indonesia, Malaysia, and the Philippines have given community systems a more prominent position in overall irrigation policy.

In Part III attention turns to studies of bureaucratically operated systems. Systems of this type, of course, have been the usual focus of irrigation development. As is demonstrated in the chapters in this section, major problems often arise with respect to coordination of actions by the agency and actions by the water users. Geographical, cultural, and social distances frequently contribute to ineffective interaction between these two groups.

In selecting these readings I have tried to choose those that emphasize the patterns of interaction within irrigation systems. Sometimes the writer has related these internal dynamics to external causes or consequences, but the emphasis is not on internal–external relationships. In general, the collection emphasizes the patterned social relations among water users and, when appropriate, between water users and the irrigation agency. While in no sense assuming that this emphasis represents a comprehensive social science view of irrigation, I am confident that it represents one significant perspective.

I should note that writings on irrigation organization in general, and these readings in particular, are deficient in two important ways. First, with a few important exceptions, the social science literature does not adequately reflect an understanding of the natural and engineered setting in which the irrigation organizations operate. A more ecological perspective is being incorporated into current research efforts and should further enhance our understanding. Second, relatively little cooperative work has been done by researchers and agency personnel to design and test organizational alternatives. Currently a few such "action research" activities are under way, and the insights and understandings derived from these experiences will be reported in the future. Much can be learned from this emerging research mode.

I would like to express my appreciation to those people who have significantly assisted me in thinking about these topics, though in no manner are they to be implicated in the faulty or incomplete thinking that remains. For the past several years I have been a co-leader of a graduate seminar orginally entitled "Peasants, Water, and Development." Many of the readings included in this book, and some not included, were used in these seminars. I have greatly benefited from the questions, comments, and criticisms of the seminar participants, many of whom brought to the discussions considerable experience with irrigation development efforts in a variety of settings.

More specifically, I am grateful to two colleagues with whom I have collaborated both in that seminar and on an array of irrigation research and policy topics: Milton Barnett and Gilbert Levine. I have greatly benefited from their considerable familiarity with and inquisitiveness about irrigated agriculture.

And finally, I appreciate the support that my wife, Rosalyn, and our children, Jeff, Doug, and Dana, have provided me over the past several years of work on irrigation. My pursuit of these interests has required frequent travel and sometimes prolonged absences from home. During those times, which have often been exacerbated by abundant snowfalls, they have assumed the extra responsibilities of caring for one another and managing the practical tasks of each day.

E. W. C.

Ithaca, New York

PART I

Introductory Concepts

Introduction

Irrigation development requires the successful joining of irrigation technology and appropriate institutions and organizations to govern that technology. Chapter 1 discusses the meaning of institutions and organizations in the context of irrigation by identifying three critical tasks to be organized: water allocation, system maintenance, and conflict management. It also emphasizes the importance of an ecological perspective in analyzing patterns of irrigation behavior.

Chapter 1 highlights articulation between water users and the irrigation bureaucracy as a central organizational issue. This articulation problem can be particularly acute in systems operated by a specialized irrigation staff. The performance of irrigation institutions and organizations can be judged by their effectiveness in creating patterns that coordinate the actions of these two groups.

Robert Chambers's discussion in Chapter 2 is a useful blend of field-based information on several irrigation systems in Tamil Nadu (South India) and Sri Lanka, with provocative typologies, concepts, and propositions. I shall leave to the reader the considerable detail that Chambers provides on the operation of specific irrigation systems. Here I wish to note two important points in his discussion.

First is his provisional typology of irrigation systems based on the nature of the involvement of communities and bureaucracies in the distribution of water. His overall classification includes five categories, of which the last three seem the most important: systems of community allocation, systems of bureaucracy allocation, and systems of joint community-bureaucracy allocation.

The second part of Chambers's discussion which I wish to highlight is his stimulating thought on the division of responsibilities that should occur in joint bureaucracy-community systems. He identifies four critical activities. Both communities and the bureaucracy, he suggests, should be involved in decisions on general policy, including decisions about the timing of water delivery, the units of land to be served, and perhaps the cropping patterns to be followed. Responsibility for the implementation of these policies should reside with the bureaucracy whenever more than one irrigation community is concerned, but should be the responsibility of the community itself when implementation occurs within the zone of a particular community. And finally, responsibility for the enforcement of policies and prosecution of infringements should remain with the bureaucracy. A sensible arrangement, I think, one that seems to be followed in the actual operation of successful joint systems (see Canute VanderMeer, Chapter 11).

In Chapter 3, Gilbert Levine is concerned with the problems of appropriate design and management of irrigation systems, particularly in the context of the developing nations. Of concern to him is the fundamental concept of water requirement. While the quantification of this concept is a basic step in current design procedures, Levine points out the relatively narrow way in which it is defined and measured. As used in system design, water requirement is simply the amount of water required for crop production in a given setting when two important physical features are considered: evapotranspiration rate and the amount of rainfall. The difference between these two elements defines the water requirement, which is then used to determine the critical characteristics of the engineered systems.

Thus, the parameter with which the designer operates is crop water requirement, not field water requirement, and certainly not farmer water requirement. Designing on such a limited parameter would seem inevitably to lead to problems.

The difficulties associated with many irrigation development schemes have led to a widespread belief that "irrigation modernization can take place only with radical departures from traditional practice" (presumably the traditional practice of the water users and not the traditional practices of the designers and systems managers). Levine is skeptical of this approach and suggests that more attention be given to the possibilities of management adaptations "based upon indigenous social relationships." Even though these indigenous arrangements may have their limitations, Levine suggests that they have a greater chance of success than do imported forms of social organizations.

1

Irrigation Development: Institutional and Organizational Issues

E. Walter Coward, Jr.
CORNELL UNIVERSITY

In the pursuit of agricultural production, human beings rearrange their microenvironment to favor the growth of their cultigens and inhibit the growth of other plants and animals. On those parts of the earth's surface where rainfall is either scarce or poorly distributed during the agricultural year, an important rearrangement is the creation of a water environment suitable for the manipulation of plant growth. Often management of the water environment depends on the coordinated action of two or more potential water users. Impounding the water, conveying it to the desired location, disposing of the excess are activities that may require the assistance or permission of one's neighbors, and sometimes of one's government. Irrigated agriculture therefore characteristically requires coordination of activities. While such coordination is not totally absent in other modes of agricultural production, it is of prime consideration in most irrigated agriculture.

A system of irrigated agriculture can be defined as a landscape to which is added physical structures that impound, divert, channel, or otherwise move water from a source to some desired location. These structures are operated cooperatively for the purpose of producing food or fiber.[1]

Irrigation units may be viewed in a variety of perspectives: as hydrologic entities, as engineering networks, or as farming systems, for example. It also is possible, and frequently useful, to view irrigation units as organizational entities: units in which a collectivity of individuals establishes actual patterns

1. The basic elements of this definition are contained in Canute VanderMeer, "Changing Water Control in a Taiwanese Rice-Field Irrigation System," *Annals of the Association of American Geographers* 58 (1968): 720–47 (Chapter 11 this volume).

of behavior that are more or less related to a set of ideas and rules regarding acceptable action and subject to modification as conditions and consequences change. In irrigation organizations, patterns of action and rules of behavior are often closely related to significant features of the physical habitat, the crops being grown, and the engineering apparatus.

In actuality, of course, an irrigation unit is simultaneously a hydrologic, engineering, farming, and organizational system, and irrigation development is the task of creating these complex units. Consequently, it is a task that is concerned with the development not only of landscapes and physical structures but also of patterns of interaction and organization. It is complicated by the fact that these various components are interdependent; for example, patterns of organization and layout of the physical design may need to be selected conjunctively, since not all combinations of layout and organization will be effective.

In brief, irrigation development must confront the issues of governance and enlist human and other resources and procedures to arrange appropriate institutions and organizations in addition to appropriate irrigation technologies. These resources can often be mobilized within the communities and among the individual farm operators who will benefit from the irrigation development activities. But they cannot be assumed to emerge spontaneously or inevitably, regardless of the bureaucratic and physically structured environment in which they are situated. Some forms of bureaucratic arrangements and some bureaucratic procedures will facilitate this local mobilization whereas others will severely retard it. The same can be said of the technological apparatus of the engineering system.

Irrigation Development and Agriculture

The development of these complex irrigation units is an important element in the agricultural development strategy of many of the world's nations and an important part of their current and projected budgets. A recent discussion of the magnitude of irrigation development possibilities was presented by A. H. Boerma.[2] He outlines the need for massive efforts both to construct new irrigation facilities and to improve the operation of existing systems: "The returns obtainable through an overhaul of irrigation systems are enormous. A reasonable target for achievement by 1985 would be the renovation of 46 million hectares of such land at a cost of about $21 billion, $6.5 billion of which would be foreign exchange."[3] In addition to these renovation efforts, Boerma projects the construction of 23 million hectares of new irrigation. The

2. A. H. Boerma, "The World Could Be Fed," *Journal of Soil and Water Conservation* 30 (January/February 1975):4.
3. Ibid., p. 6.

majority of this new construction is expected to occur in Asia (15 million), with smaller areas in Latin America (4 million), the Near East (3 million), and Africa (1 million).

Irrigation Institutions and Organizations

There is increased awareness of the critical role that institutions and organizations play in the process of irrigation development. Most frequently institutional and organizational difficulties are considered the problems of water users. The individual farm operators are viewed as unorganized or only poorly organized, and thus inefficient in the use of water, unpredictable in their behavior. or simply "uncooperative" with the water agency. The call for farmer organization is, of course, usually stated more positively, as in the following:

> The success of an irrigation project depends largely on the active participation and cooperation of individual farmers. Therefore, a group such as a farmers' association should be organized, preferably at the farmers' initiative or if necessary, with initial government assistance, to help in attaining the objectives of the irrigation project. Irrigation technicians alone cannot satisfactorily operate and maintain the system.[4]

While this recognition of the role of the water users in the management of irrigation systems is laudable, particularly in an era of increased concern for the participation of various beneficiaries in the projects designed to serve them, it is worrisome to suggest that the success of a project depends largely on the participation and cooperation of the water users (and by inference only to a small degree on the participation and cooperation of the irrigation agency). In large-scale irrigation systems under a specialized irrigation authority, institutional and organizational problems may arise with the irrigation agency, the water users, or both. Thus, there is need for more discussion of the point made by P. R. Crosson: ". . . the management of large irrigation projects is in the hands of public officials who are too far removed from the on-farm situation to know the conditions of efficient use, who lack economic incentives to achieve it even if they knew how, and who typically are bound by inflexible operating rules of water allocation impeding their response to economic incentives even if they had them."[5] Partial views of the institutional and organizational problems of irrigation may be related to the confusion that exists regarding the meanings of these terms, their applicability to particular irrigation situations, and their relevancy to policy formulation.

"Institution" is a concept associated with ideal behavior and expectations

4. Asian Development Bank, *Regional Workshop on Irrigation Water Management* (Manila, 1973), p. 50.

5. P. R. Crosson, "Institutional Obstacles to Expansion of World Food Production," *Science* 188 (May 9, 1975):522.

and can be used as a generic concept for the variety of rules that help to pattern social behavior: norms, folkways, mores, customs, convention, fashion, etiquette, law.[6] Some economists have used a similar notion of institutions; T. W. Schultz defines an institution as a behavioral rule.[7] In this sense, the rule of continuous irrigation, the custom of performing a ritual ceremony at the headworks of a community irrigation system, and the law requiring payment of an irrigation fee are examples of irrigation institutions.

In addition to these existing expectations, one observes in any human group actual patterns of social behavior and interaction that are referred to as social structure. These patterns of behavior sometimes are formal, purposive, and enduring enough to warrant the use of a group name: the Cruz family, the Royal Irrigation Department, the San Lorenzo Farmer's Irrigation Cooperative Association, Inc., or the Muda Area Development Authority. Of course, social structure also is composed of patterns and groups less formal, purposive, and enduring: an evening meeting between irrigation authorities and an assembly of water users, a partnership between two farmers that allows one to move water across the fields of the other, or a temporary band of farmers from a common lateral working to clean a canal.

Institutions and social structure are fused through the basic concept of role, which in turn is composed of two elements: role expectations (the institutional dimension) and role performance (the structural dimension). In part, a role can be thought of as the cluster of expectations associated with a given function. In addition, a role has associated with it actual patterns of action. Roles help one to predict the actions and reactions of others and thus enable social patterns and social organization to emerge. In an irrigation system a cluster of institutions associated with the function of water allocation is found in the role of ditch tender and another cluster is found in the role of water user. The presence of these two roles allows for the emergence of social organization in an irrigation system in the form of patterned relationships between and among the ditch tenders and water users. It is important to note that this perspective of institutions and organization allows us to consider patterns of social organization in a particular irrigation system even though no formal irrigation association may exist.

Finally, an understanding of the basic relationship between institutions and social structure requires recognition of the frequent inconsistency between what people believe should occur (the institutional element) and what actually occurs (the structural element). The basic "lack of close correspondence between the 'ideal' and the 'actual' in many and pervasive contexts of social behavior"[8] is one important force for change in either the institutional or

6. See Ely Chinoy, *Society* (New York: Random House, 1961).

7. T. W. Schultz, "Institutions and the Economic Value of Man," *American Journal of Agricultural Economics* 50 (1968):1113–22.

8. Wilbert E. Moore, *Social Change* (Englewood Cliffs, N.J.: Prentice-Hall, 1965).

Table 1.1.
Irrigation system activities, by institutional and organizational elements

Institutional and organizational elements	System activities		
	Water allocation	System maintenance	Conflict management
Key rules	Rules for allocation of water between subunits of system, farms, etc.	Rules for mobilizing human and other resources to maintain and repair physical apparatus	Rules for resolving disputes between subunits of system and between individuals
Important roles	Roles involved in establishment of water allocation policies and implementation of water distribution	Roles involved in identification of maintenance needs, mobilization of resources, supervision of repairs	Roles involved in mediating disputes, making judgments, enforcing sanctions
Significant social groups	Groups that influence establishment of water allocation policies; groups involved in actual implementation of water distribution	Groups that exist to provide routine or emergency repairs to system	Groups that participate in settling disputes and in enforcement of sanctions

social structural pattern. A major reason for this inconsistency is that changes in the social or nonsocial environment either make it difficult or impossible to act in certain established ways or make it easy or possible to act in certain new ways. Change in either the institutional or the social structural component creates demand for change in the other.

The general concepts of rules, roles, and groups are useful in understanding human behavior as it occurs in many contexts and for a variety of purposes. To be useful in the understanding of irrigation, however, these concepts need to be matched with critical tasks or actions that occur in irrigation systems. While many tasks must be organized to sustain the operation of an irrigation system, three are of fundamental importance: (1) the organization of water allocation, (2) the organization of physical maintenance activities, and (3) the organization of conflict management. Understanding the institutional and organizational aspects of an existing irrigation system or attempting to design appropriate institutional and organizational arrangements for a new irrigation system is facilitated by a matching of the basic concepts of rules, roles, and groups with the fundamental tasks of allocation, maintenance, and conflict management (see Table 1.1).[9]

Following this analytic scheme, one would, for example, seek to identify

9. The organization of these tasks in a specific irrigation system is discussed in E. Walter Coward, Jr., "Principles of Social Organization in an Indigenous Irrigation System," *Human Organization* 38 (1979): 28–36.

the rules that exist in a system regarding maintenance activities, the major roles that are used to organize maintenance activities, and the important groupings (formal and informal) that are involved in implementing needed repairs and improvements. A similar approach would be employed in examining the tasks of water allocation and conflict management.[10]

Thus, the institutional and organizational milieux of all irrigation systems can be seen to refer to the rules and patterned behavior of both the water users and the water authorities. And, as we shall note later, the critical test of the performance of these institutions and organizations is their effectiveness in articulating the various users and authorities—establishing rules and creating patterns of behavior that coordinate the actions of water authorities and water users, for example, so that authorities provide water when and where farmers need it and are prepared to use it effectively.

While the authority–users linkage is important in all group irrigation systems, it is most important in those in which the water authority is a formal agency or office of the government. In this situation the independent centripetal forces of the agent's bureaucracy and the user's community can greatly inhibit successful interaction.

The Ecological Perspective of Institutions and Organization

Having identified particular rules, roles, social units, and other regularities in the social behavior of a group, one may wish to try to understand the reasons for these ordered actions. One useful approach to regularities in irrigation behavior is the ecological perspective (sometimes referred to as the perspective of human ecology, cultural ecology, or more recently ecological anthropology).

The distinguishing element of the ecological perspective is its emphasis on "the role of the physical-environmental factors in shaping, limiting, or determining various forms of group-shared behavior and the regularities which lie behind them."[11] Most social science, in contrast, views institutional and organizational patterns as relatively independent of the physical setting in which they occur. The limitations of this nonecological approach have been discussed by E. R. Leach:

> Of course every social anthropologist recognizes that societies exist within a material context which is primarily natural—terrain, climate, natural resources—and partly man-made—houses, roads, fields, water supply, capital

10. Each of the chapters in this book provides information for one or more of the cells identified in Table 1.1, though few would enable the reader to complete all of the cells. The analytic scheme can provide one useful basis for the comparison of systems as well as a guide for the collection of new data.

11. John W. Berry, *Human Ecology and Cognitive Style* (New York: Halsted Press, 1976), p. 10.

> assets—but too many authors treat such things as nothing more than context, useful only for an introductory chapter before getting down to the main job of analyzing the social structure. But such context is not merely a passive backcloth to social life; the context itself is a social product and is itself "structured"; the people who live in it must conform to a wide range of rules and limitations simply to live there at all. Every anthropologist needs to start out by considering just how much of the culture with which he is faced can most readily be understood as a direct adaptation to the environmental context, including that part of the context which is man-made.[12]

It is not assumed that the physical and natural habitat are the only factors that influence institutional and organizational patterns, but that these factors need to be given consideration along with the social science factors that are more commonly considered. In fact, as many have noted, the relevant environment of human populations includes not only the physical and natural habitat but also other groups and the variety of activities in which they engage.[13]

A major premise of the ecological perspective is that human groups relate to the environments in which they operate—both the physical and natural habitat and the sociopolitical milieu—through the mediation of "socially organized activities aimed at satisfying the requisites of collective survival."[14] It is this socially organized relationship between the group and its environment that ecologists refer to as the group's adaptation.[15] As Yehudi Cohen notes, "The concept of adaptation is historical: when we say a population is adapting we mean that it is altering its relationship to its habitat to make that habitat a more fit place in which to live or to make itself more fit to live in that milieu."[16]

When the ecological perspective is applied to the context of irrigation, a number of interesting implications arise. At one level, irrigation as a mode of agricultural production can be viewed as a sociotechnical adaptation to particular habitat features and population characteristics (see, for example, the work of Julian Steward and Robert McAdams).[17] At a more micro level the ecological perspective suggests that apparently irrational, unarticulated, or random activities in a particular irrigation system may be unraveled and made

12. E. R. Leach, *Pul Eliya, a Village in Ceylon: A Study of Land Tenure and Kinship* (Cambridge: University Press, 1961), p. 306 (Chapter 5 in this volume).

13. John W. Bennett, *Northern Plainsmen: Adaptive Strategy and Agrarian Life* (Chicago: Aldine, 1969), p. 12.

14. Michael Micklin, ed., *Population, Environment, and Social Organization: Current Issues in Human Ecology* (Hinsdale, Ill.: Dryden Press, 1973), p. 5.

15. Yehudi A. Cohen, ed., *Man in Adaptation: The Cultural Present* (Chicago: Aldine, 1974).

16. Ibid., p. 3.

17. Julian Steward, *Irrigation Civilization: A Comparative Study,* Social Science Monograph no. 1 (Washington, D.C.: Pan-American Union, 1955); Robert McAdams, *The Evolution of Urban Society: Early Mesopotamia and Prehispanic Mexico* (Chicago: Aldine, 1966).

more intelligible if they are examined in relation to the habitat and sociopolitical context in which they occur.[18] Irrigation behaviors may be patterned, in part, as a response to rainfall patterns, canal layouts, bureaucratic procedures, or a variety of other environmental features.

To return for a moment to Table 1.1, having identified the various rules, roles, and social groups associated with the process of water allocation, the analyst is then prepared to move toward an explanation of these social patterns. In doing so, he may find that he requires important nonsocial information including data on soils, the water requirements of crops being grown, the degree of variability in local rainfall patterns, and the nature of the physical structures available for the transport and direction of water flows—all in addition to the information that he may have on the nature of kinship arrangements, patterns of social stratification, and critical themes in the group's ideological system.

If one operates with the assumption that irrigation institutions and organization are in part a response to the physical and natural habitats in which they occur, one is alerted to the notion that planned modifications in the relative water supply available in a system or rearrangements of the canal layout or turnout locations can have important implications for the organization of social relations. In effect, most so-called technical decisions are simultaneously technical and organizational decisions: as soon as a particular technical arrangement has been selected, organizational arrangements are partially prefabricated. Thus, one can speak fruitfully of the ecology of irrigation institutions and organizations and use this analytic approach to compare various forms and arrangements for system operation, to understand the absence of irrigation institutions and organizations in a given context, or to anticipate the likely organizational consequences of an irrigation arrangement being planned.[19]

Irrigation Development: Two Contexts

Irrigation development typically occurs in one of two situations: a setting in which irrigated agriculture is already practiced and some form of irrigation system exists, or a setting in which some other mode of agriculture is practiced (such as pastoralism, dry-land farming, or rain-fed production). Clearly, these two general contexts have different implications for public policies and development decisions regarding local irrigation organizations.

18. A good example of this approach is found in Theodore E. Downing, "Irrigation and Moisture-Sensitive Periods: A Zapotec Case," in *Irrigation's Impact on Society*, ed. Theodore E. Downing and McGurie Gibson (Tucson: University of Arizona Press, 1974), pp. 113–22.

19. Few of the chapters in this book have explicitly used this ecological perspective, though many identify important ecological-organizational interactions.

If irrigated agriculture is an established mode of agricultural production in a location selected for irrigation development activities, one can assume that there exists, in addition to the physical structures for capturing, conveying, and distributing water, some form of local governance for the operation and maintenance of these systems. Thus, an early procedure to be followed in this context should be the gathering of information required to describe and understand the realities of irrigation institutions and organizations in the location.

When this ethnographic study of the local scene has been concluded, a fundamental policy decision needs to be made regarding the existing institutional and organizational milieux. Are the existing arrangements to be maintained? Are they to be modified, and if so, in what ways? Or are the existing patterns so inappropriate as to require replacement? Obviously, those are not easy choices, and they will be influenced by a variety of factors, including national political commitments to indigenous social organization and the relative political power of the existing local groups. Whatever the actual influence on those choices, decisions will be enhanced if they are made on the basis of a good understanding of what exists rather than an assumption that nothing of importance is present or that the existing patterns are simple and rigid and therefore inappropriate for "modern" irrigation management.

An additional point is that when a decision has been made regarding the use, modification, or replacement of the existing irrigation institutions and organizations, decisions regarding the physical design of the irrigation system should be made consistent with this fundamental policy decision. For example, if traditional irrigation organization is based on canal groups of fifty farmers operating under the direction of a locally selected canal leader, and if the basic policy is to encourage continuation of those canal groups, the design of a reticulation system that creates new canal groups of a hundred farmers will be inappropriate. Alternatively, if a decision has been made to create a new form of local irrigation organization in which the residential community leaders play a large role, then design choices should be made that will create irrigation zones congruent with village boundaries so as to interlink village and irrigation affairs.

If an irrigation project is to be located in a setting in which irrigated agriculture is not being practiced, a modified procedure needs to be followed. As in the first case, a critical policy decision to be made very early concerns the preferred nature of local irrigation responsibility. That is, to what extent are local people to be involved in the design, construction, operation, and maintenance of the irrigation system? (We may delay for the moment questions about the forms of local organization appropriate for obtaining the desired participation.) As in the previous context, choices in the design of the physical layout of the system should be made consistent with the fundamental choice regarding local responsibility. The matrix of canals and turnouts should serve to create a pattern of human zonation that facilitates the creation of local groups capable of performing critical operation and maintenance tasks, as well as the formation of a decentralized system staff.

Again, the fundamental point is that the technological and organizational elements interact, and thus preferred organizational patterns (old or new) will be affected by the technological arrangements that are selected and constructed.

Irrigation Development: Alternative Organizational Modes

In general, past irrigation development has occurred in one of two modes: through the development of community irrigation systems or through the development of agency-operated systems. It seems highly likely that those patterns will continue into the future.

It is, of course, easy to associate irrigation development with large-scale activities—the Gezira scheme in the Sudan, the major projects in northwest Mexico, the extensive systems of the Indian subcontinent. The size of those projects requires the involvement of large-scale public agencies that are able to recruit the technical personnel required to plan and manage complex undertakings and to obtain the enormous financial resources required for construction. It is less obvious that the public agency needs to be involved in the full range of management activities required to operate the system. Some have maintained a very elaborate administrative network that extends from the water source to the individual farmer's fields. The extreme case has been the Gezira scheme, in which nearly all management activities are in the hands of the irrigation agency.[20]

Large-scale irrigation systems often have organizational patterns that may be characterized as joint.[21] In such systems some management functions are the responsibility of the agency and others are lodged with farmer groups within the command area. For example, the maintenance of all main and secondary canals and the allocation of water to them may be the responsibility of the irrigation agency, whereas user communities are in charge of maintaining and allocating water along specific tertiary channels. Since this organizational form is discussed by Robert Chambers in Chapter 2, no further detail need be provided here.

Not all irrigation development, however, takes the form of large-scale,

20. But see the interesting report by Tony Barnett, "The Gezira Scheme: Production of Cotton and the Reproduction of Underdevelopment," in *Beyond the Sociology of Development*, eds. Ivar Oxaal, Tony Barnett, and David Booth, (London: Routledge & Kegan Paul, 1975), pp. 183–208, indicating that farmers may have a kind of "invisible" involvement in actual system operations.

21. What Robert Chambers has referred to as a bureaucratic-communal system (see his Chapter 2 in this book; see also Chapter 11 by Canute VanderMeer and my Chapter 15).

bureaucratically managed systems. In many countries, extensive portions of irrigated land are served by small-scale community systems. Such countries (for example, Indonesia, the Philippines, and Thailand) are interested in improving the facilities already in place, and sometimes in the creation of new community systems.

Small-scale community systems have demonstrated a remarkable persistence even in the face of enormous changes in the political economy in which they have operated and in the context of significant technological changes. Community irrigation systems remain important in the highly industrialized context of the United States,[22] have endured in the Philippines through the changing irrigation policies of two colonial powers (Spain and the United States) and the present government, and have demonstrated their suitability for managing deep-well pumps in Bangladesh.[23]

The Articulation of Local Organization and System Bureaucracy

Perhaps the most serious institutional and organizational issues in irrigation development are those related to the articulation between water users and the irrigation bureaucracy. Articulation problems are important in systems managed by communities as well as in agency systems.

In general, one can distinguish two important organizational configurations in an irrigation system. One element is that composed of the role of water user, the significant social groups of which this role is a part, and the key rules that orient the water-user groups and persons occupying the water-user role. A second configuration is composed of the various water-authority roles, the significant groups of which they are a part, and the key rules that orient the behavior of water-authority groups and persons occupying water-authority roles.

In community irrigation systems these two organizational components are highly integrated because they are guided by similar sociocultural rules and because of the temporal circulation of individuals between the two configurations or the simultaneous occupancy of roles in both configurations. Thus, in many small community irrigation systems the ditch tender may be simultaneously producing his own rice crop, and an individual who this year is serving in the role of canal chief may next year occupy only the role of water user.

In contrast, of course, are the many irrigation systems in which the water authorities are more specialized and differentiated from the water users. The

22. Arthur Maass and Raymond L. Anderson, *. . . And the Desert Shall Rejoice: Conflict, Growth, and Justice in Arid Environments* (Cambridge: M.I.T. Press, 1978).

23. E. Walter Coward, Jr., and Badaruddin Ahmed, "Village Technology and Development: Patterns of Irrigation Organization in Comilla District, Bangladesh," *Journal of Developing Areas* (forthcoming).

shifting of roles or the dual occupancy of roles is unlikely to occur in such systems because of the specialized processes of role recruitment that exist for the water authorities. Furthermore, the key rules that orient the behavior of the water authorities are often dissimilar to those that guide the action of water users.[24]

Thus, in irrigation situations with a high degree of differentiation, a major organizational feature influencing system performance is the degree of articulation and integration of water users and water authorities. From an ecological perspective, we can suggest that this integration will be influenced by both the institutional and organizational arrangements that are devised to link the two elements and the nature of the physical and engineered habitat that defines the need for certain irrigation tasks.

In agency-managed systems the problems of interaction may involve a full range of issues, from design and construction procedures through water allocation and the adjudication of farmer disputes. Since community systems often are relatively independent in organizing their operation and management activities, however, community–agency interactions more commonly occur when outside assistance is being provided to construct or rehabilitate the system, or when disputes occur between two or more community systems.

There is increasing recognition that the style of articulation needs to be modified from a highly agency-directed pattern of interaction to one in which water users play active information-providing and decision-making roles. Clearly, the operation and maintenance of irrigation systems has been improved when irrigation agencies have succeeded in eliciting and using local information, local workers, and local management skills.

Some interesting experiences with the linking of water users with the irrigation agency at the time of system design and construction have occurred recently. One such case is the Laur Project of the National Irrigation Administration (NIA) in the Philippines.[25] In this project the NIA is assisting in the rehabilitation of an existing community system. The water users to be benefited, through their existing irrigation association, have participated rather fully in the rehabilitation project: they assisted in the preconstruction survey work of the engineers, commented on the initial construction designs, and have maintained a careful record of construction expenditures.

The recording of expenditures is important to the local group because the NIA assistance is being provided to them on a loan basis. The serious involvement of the local water users' group can result in some uncomfortable

24. For a good discussion of this point see Maass and Anderson, . . . *And the Desert Shall Rejoice*.

25. Carlos Isles, "Sociological Considerations in Water Management: Strengthening Irrigation Associations through People's Participation," in Philippine Council for Agriculture and Resources Research, *Seminar-Workshop on Determinants of Developing Country Irrigation Problems* (Los Baños, Laguna, 1978), pp. 43–52.

scrutiny of agency procedures. For example, the irrigation association questioned the action of a particular equipment operator who drove his equipment from the work site back to the village each day for his noon meal. Since the cost of gasoline was being charged to their account, they suggested that he either take his lunch to the work site or arrange alternative transportation for the noon break.

While this critical review of routine behavior doubtless has been annoying to the agency staff detailed to the project, early assessments of the program suggest that the agency will gain in at least two ways: (1) its design and construction activities will be enhanced by the location-specific information regarding the current and historical characteristics of the physical habitat and (2) a local commitment to the project has been created and is likely to continue into the operational period of the system.

Summary

For much of the developing world, investment in irrigation development is considered a fundamental element in agricultural development. Irrigation development depends on the adequate provision of water management technology as well as appropriate institutions and organizations for the governance of the irrigation apparatus.

When irrigation institutions and organizations are considered, an ecological perspective is useful because it recognizes the important interplay between rules and patterns of behavior and the natural and engineered environments in which these rules and behaviors occur. Management procedures and administrative rules that are not in harmony with such natural features as rainfall patterns or such engineered features as turnout locations will result in poor "institutional" performance.

In general, existing irrigation systems approximate one of three types: they may be *community systems,* operated and maintained by the water users themselves and/or their representatives; they may be *bureaucratically managed systems,* fully administered by an agency of the government; or they may be *jointly managed systems,* in which some functions are performed by the irrigation agency while others are the responsibility of one or more water-user communities. In any of those situations a major purpose of the irrigation institutions and organization is to structure effective linkages between the water users and the system authorities.

Irrigation development occurs sometimes in settings where irrigated agriculture is already practiced and sometimes in areas in which it is absent. Of course, both situations require attention to issues of governance. The former situation offers the unique opportunity to build on existing institutional and organizational arrangements—an opportunity that can be explored only if an effort is made to identify and understand those structures.

2

Basic Concepts in the Organization of Irrigation

Robert Chambers
UNIVERSITY OF SUSSEX

To a remarkable degree, many writers on irrigation ignore and even appear unaware of the relationships between people and irrigation water. Attention is usually fixed on hydrological, engineering, agricultural, and economic aspects. Especially in official documents it is rare to find described, let alone analyzed, the human side of the organization and operation of irrigation systems—the management of those who manage the water, the procedures for irrigation control, the processes of allocation of water to groups or individuals, the distribution of water within groups. There may be almost as many instances of these omissions as there are reports on irrigation.

Thus the report of the working group for the formulation of the Indian Fourth Five-Year Plan proposals on soil and water management under irrigated conditions (ICAR, 1966) is entirely technically oriented, has no place for any social scientist on any research station, and proposes no research on organizational aspects of irrigation or on the management of the staff who manage the water. A report of an irrigation program review in Sri Lanka (part of an IBRD/FAO cooperative program: MPEA, 1968) is overwhelmingly oriented toward capital works and their planning and execution, and while recommending that there should be many more extension staff and stating the need for coordination at the field level, does not go into any detail about the procedures for achieving this. This was despite the terms of reference which included instructions to review and recommend institutional, organizational, manage-

Previously published as "Men and Water: The Organization and Operation of Irrigation," in *Green Revolution?: Technology and Change in Rice-Growing Areas of Tamil Nadu and Sri Lanka*, ed. B. H. Farmer (Boulder, Colo.: Westview Press, 1977), pp. 340–63. Reprinted by permission of Westview Press and Macmillan, London and Blasingstoke.

rial, and technical measures to ensure successful execution *and operation* of existing and future projects (my italics). Nor were the operational and organizational aspects of water management and their economic and social implications a concern of an international seminar on economic and social aspects of agricultural development in irrigated areas, held in Berlin in 1967 (German Foundation for Developing Countries, 1967). Finally, a recent publication of the National Commission on Agriculture in India dealing with modernizing irrigation systems and integrated development of commanded areas shows much the same blind spot: it embodies a top-down view of irrigation and omits operational detail (Government of India, 1973).

There are several reasons for this neglect (see also Chambers, 1975a, pp. 2–6): first, the common preoccupation with capital investment, construction, and settlement processes at the cost of the vital operating processes which follow; second, cramped vision from within narrow disciplinary boundaries, including mutual ignorance between social scientists and technologists and a reluctance to explore a no-man's-land between disciplines; third, the intensity of research required to examine what happens at the lower levels of administration, and difficulties in generalizing from one or a few cases, which are all that one researcher may hope to study; fourth, the maddening nature of water itself, with its tendency to flow, seep, evaporate, condense, and transpire, and the problems it presents in measurement—problems which tie down natural and physical scientists to research-intensive tasks, denying them time, even if they had inclination, to branch out and examine wider aspects such as the people who manage the water and how they behave.

These tendencies have left several gaps in comparative knowledge. Where water is administered to communities, there is a gap geographically between the last point at which it is officially controlled or measured and the point at which it enters a farmer's field. Organizationally there is a gap between what happens at the level of senior officials and what happens in the community which receives the water. Politically there is ignorance of the processes of decision making and allocation which influence the timing and quantity of water which farmers receive. In terms of political economy there has been little analysis of who gets what, how, when, and why, and with what costs and benefits. In terms of human management there is a widespread failure to perceive the problems and opportunities of managing those who manage the water, the men in organizations and communities.

This chapter uses the comparison of irrigation systems in the study areas in India and Sri Lanka as a basis for some preliminary steps among the minefields of interdisciplinary no-man's-land which these gaps represent. Much of the evidence is used with misgiving, being based on one-off interviews on day visits to villages and cultivation committee areas. For Sri Lanka, additional sources have been visits to and studies of the records of two major irrigation systems—Gal Oya and Uda Walawe. For India and elsewhere, some secondary sources have also been drawn upon. The purpose is to open up some comparative, analytical, and practical aspects of the organization, operation, and political economy of irrigation.

A basic point is that water is usually a scarce resource for which men and groups of men compete and the benefits from which should be optimized in relation to other scarce resources. In the dry zone of Sri Lanka there is much evidence that water is more limiting than land (Chambers, 1975a, pp. 19ff.), although scarcities of draught power and labour are also constraining (Harriss, 1976) and may at some times and places be more limiting than water. In parts of North Arcot the scarcity of water is even clearer and more acute. Surface irrigation water from tanks is often inadequate for a second crop and the groundwater level is undergoing a serious secular decline as numbers of wells grow, more pump sets are installed, and groundwater extractions increase (Madduma Bandara, 1976). As population presses more and more on the resources available for food production in these and other environments, so understanding the relations between people and water for irrigation becomes a more and more vital priority.

Typologies of Irrigation

A first step is to try to identify useful categories. The descriptive terms used by engineers and agriculturalists dominate discussion of irrigation systems. This is partly because they themselves have such key roles in irrigation, partly because their categories refer to physically observable phenomena such as structures, field layouts, and methods of water application. These categories may not be the most useful ones for an analysis of the organization and operation of irrigation. But classifying the irrigation systems encountered in Sri Lanka and India in terms of their more obvious physical characteristics does provide a starting point.

In the study area in Sri Lanka almost all irrigation is by surface gravity flow, most from storage tanks. Tank water is received from various combinations of catchment run-off and river diversion. Scarcely any wells are used for irrigation. The commonly used classification of gravity-flow irrigation into "major" and "minor" corresponds with differences in scale and organization, not with differences in physical type of source, conveyance, or storage of water. The management of water under major irrigation is the responsibility of the Territorial Civil Engineering Organization (TCEO), which distributes it down to the field channel level. Water management on minor irrigation is the responsibility of village communities, which organize their own distribution systems. Under a major irrigation project there are usually several cultivation committees, whereas under minor irrigation there is usually only one.

From the more obvious characteristics of scale and type of water source and storage, the cultivation committees in our sample can be classified as in

Table 2.1. In all cases, distribution from the tank or from the main canal is by gravity through channels of diminishing size to farmers' fields. There is only one well and pump known under any of these systems (under Tissawewa tank) and that is not in one of the survey cultivation committee areas.

In the study area in India there is a greater variety and mixture of irrigation systems. The most common form of gravity irrigation consists of canals from anicuts from rivers which are dry for most of the year, and which supply chains of village tanks in series. In our sample, large tanks are represented only by Dusi, which is one of 18 villages served by the large Dusi-Mamandur tank. In addition, in all villages there are wells used for lift irrigation. Three forms of lift are used—etram (human power), kavalai (ox power), and pump sets (oil or, much more commonly, electric power). These wells are usually found both on the dry land (land which is not under command for tank or channel irrigation) and on the wet land (under command for tank or channel irrigation). The villages in the sample can be classified as in Table 2.2.

The categories in the table follow the necessary but well-worn discipline-bound criteria of engineers and hydrologists. They are much concerned with the acquisition, transport, and storage of water and less with its distribution. An engineer talks and thinks in terms of diversion channel, tank, dam, gravity, well, pump, major and minor irrigation, with type of structure and scale of operation as his main criteria. A hydrologist thinks and talks in terms of water cycles and sources of water—shallow or deep well, spring, surface run-off storage, and river diversion irrigation, for example. But other disci-

Table 2.1.

Water source and storage categories

Cultivation committee	Minor/ Major	Water source	Storage system
Kachchigala	Minor	Small catchment run-off	Small tanks
Metigatwala	Minor	Small catchment run-off (now supplemented by major irrigation)	Small tanks
Kataragama	Minor	Small catchment run-off	Small tank
Tenagama	Minor	Small catchment run-off and spills of higher tanks with small area sometimes supplemented by major irrigation	Small tanks in series, close together
Wellawaya	Minor	Anicut and channel from permanent stream	Nil
Hanganwagura Jansagama Rotawala	Major (WRB)	Anicut and long channel from Walawe river with perennial flow	Nil
Jayawickremayaya	Major (KOLB)	Anicut and channel to tank from Kirindi river (water not always available)	Tank (Debarawewa)
Kachcherigama	Major (KOLB)	Anicut and channel to tank from Kirindi river (water not always available)	Tank (Tissawewa)
Udasgama	Major (KOLB)	Anicut and channel to tank from Kirindi river (water not always available)	Tank (Tissawewa)
Companniwatta	Major (KORB)	Anicut and channel to tank from Kirindi river (water not always available)	Tank (Wirawila)

WRB = Walawe right bank; KOLB = Kirindi Oya left bank; KORB = Kirindi Oya right bank.

Table 2.2.
Village classification

Village	Non-well water source	Tank storage	Wells in wet land	Wells in dry land
Kalpattu	Nil	Nil	No wet land	Yes
Vegamangalam	Excavated springs near river, permanent flow	Nil	Nil	Yes
Dusi	Channels leading from large seasonal rivers	Large tank	Negligible	Few
Meppathurai, Vinayagapuram	Channel leading from seasonal large river direct to village tank	Village tank	Yes	Yes
Amudur, Duli, Randam, Sirungathur, Vayalur, Veerasambanur, Vengodu	Combinations of natural drainage lines and channels from seasonal rivers leading through chains of tanks to village tank	Village tank	Yes	Yes

Note: Some villages have additional small tanks which are fed by catchment run-off. All tanks receive some water from their catchments in addition to amounts received from the source named.

plines would classify irrigation systems quite differently: for an agriculturalist the field application of water is central and includes flood, border strip, check basin, furrow, underground, and sprinkler irrigation. In the social sciences the only large-scale attempt at comparative analysis of the organization and operation of irrigation has apparently been Wittfogel's eccentric polemic on oriental despotism (1957), although a recent start in classification has been made by Thornton (1976). After considering the physical acquisition and transport of water, Thornton points out that it is with distribution that "the largest number of organizational alternatives occur." Distribution is also a potential focus for classification since it corresponds with much of the unexplored no-man's-land in irrigation.

Categories depend both on the subject matter and on the orientation of the observer. Classifications of irrigation organization can themselves be classified as top down, bottom up, or middle outward, depending on the focus of concern and the stance of the typologist. Thornton's types derive from a top-down view, using formal organization and the distribution of responsibilities within the organization to separate out categories, with a major division into private and public organizations. A bottom-up view of irrigation, starting with the farmer and his preoccupations, might differentiate between irrigation systems according to the cost, adequacy, convenience, and reliability of the supply of irrigation water to the farm. A middle-outward view of irrigation organization would start geographically and organizationally in the middle of the distribution system. It might differentiate systems according to the decisions, communication, and allocations which affect distribution, looking both upward toward the source from which the water derives and downward to the farmer. All three views—top down, bottom up, and middle outward—deserve to be developed. Here we will start in the relatively unexplored middle ground and move outward from there, paying particular attention to the organization and operation of communities and bureaucracies in the distribution of water.

A central and universal issue in the distribution of irrigation water is who gets what, when, and where. This is the very stuff of politics and it is surprising that political scientists, political anthropologists, and those who study political economy have not devoted more attention to it. Where water is scarce and often constraining and when individual farmers and communities of farmers compete for it, the focus is on the processes of allocation and acquisition which determine the access of users to water. These processes can be classified as:

Direct appropriation — The user acquires water directly from a natural source such as a private dam or well.

Acquisition through contract — The user acquires water through agreement with a supplier in exchange for goods or services.

Community allocation — A communal source of water is allocated among a community of users.

Bureaucratic allocation — Water is allocated by bureaucratic organization direct to individual users.

Bureaucratic-communal allocation — Water is allocated by a bureaucratic organization to one or more communities of users, each of which manages distribution to its members.

These types are represented in the examples available as in Table 2.3. The categories adopted must be treated warily. They are designed for convenience

Table 2.3.

Water allocation

Type of allocation/ acquisition	Sri Lanka	India
Direct	Negligible	Very common (individual wells)
Contract	Negligible (except where tenancy carries water rights)	Negligible (except where tenancy carries water rights)
Community	All minor irrigation (Kataragama, Wellawaya, Tenagama, Metigatwala, Kachchigala)	Amudur, Duli, Meppathurai, Randam, Sirungathur, Vayalur, Veerasambanur, Vegamangalam, Vengodu, Vinayagapuram
Bureaucratic	Uda Walawe	Nil
Bureaucratic-communal	All major irrigation (Hanganwagura, Jansagama, Rotawala, Jayawickremayaya, Kachcherigama, Udasgama, Companniwatta)	Dusi

without necessarily implying that they have great explanatory power. As with many other distinctions in the social sciences, the edges blur and overlap in practice. Thus Dusi is immediately a bad fit in bureaucratic-communal irrigation, since the size of the paddy tracts under the large Dusi-Mamandur tank would lead anyone familiar with irrigation in Sri Lanka to look for a bureaucracy which distributes the water; but in the strict sense of bureaucracy—an organization with its own norms, roles, terms of service, and so on—there is none. The PWD only controls issues from the tank sluices, leaving the rest to the traditional officers of the villages. Again, Amudur, though having a community system of allocation and acquisition, has something verging on its own ''bureaucracy'' in the form of three Harijan thoddis who distribute the water to individual farmers. These two examples are cited not to undermine the classification, but to discourage any tendency to think that words refer to classes of entities which are more consistent and distinct than they really are.

Analysis will concentrate on those types of which there are numerous examples: direct acquisition, almost entirely through wells in India; community allocation, widely represented in both Sri Lanka and India; and bureaucratic-communal allocation, mainly in Sri Lanka. The focus will be further narrowed by concentrating on the levels at which decisions and actions affecting allocation and acquisition are taken; for these three irrigation types they are as shown in Table 2.4.

The main attention will be at the community and system levels. ''Community'' here refers to users with an interest in a common source of supply, the water from which is distributed among themselves. This usually refers to what in Sri Lanka is called minor irrigation, to what in India is village tank irrigation, and in both countries to groups of users on larger irrigation projects who depend upon the same feeder. ''Systems'' refers to whatever organization or arrangement exists above the community level for the management and allocation of water.

The discussion which follows is in two sections: the first deals with the organization and operation of community irrigation, examining allocation and appropriation of water, equity and productivity, enforcement and arbitration, and action by irrigation communities; the second deals with the organization and operation of bureaucratic-communal irrigation.

Table 2.4.

Levels of decisions actions

	Farmer level (within fields)	Community level (within community area)	System level (within irrigation system area)
Direct	Yes	No	No
Community	Yes	Yes	No
Bureaucratic-communal	Yes	Yes	Yes

Community Organization and Operation

The Allocation and Appropriation of Water

The allocation and appropriation of water can be described in terms of two stages: decisions about areas to be irrigated and about timing; and actual allocations and appropriations.

In the first stage a decision may have to be taken as to which areas under command to irrigate. Leach has described for Pul Eliya in Ceylon the nice decision which has to be taken with a village tank:

> The issue is a subtle problem of economic choice since, if the water resources of the irrigation system are overextended, the outcome may be total crop failure. The village meeting makes its collective decision on the basis of the level of water in the tank and a gambling estimate of rain in the weeks to come. [Leach, 1961, p. 53; Chapter 5 below, pp. 103–4]

This type of decision is not limited to village tanks. Wellawaya depends on diversion from a small perennial stream which is not always sufficient for all of its six blocks of asweddumized land; similar decisions have to be taken about which and how many of the blocks to cultivate in Yala. The only Indian village in the sample known to have a similar system is Duli, where, when water is short, a decision is taken to allow the same fixed acreage to each holder of wet land and to supply water only for that. Under the other Indian villages with tanks there appears to be no formal decision about the acreage to be cultivated: the decision is left to individuals, who must rely on their own judgment of the water likely to be available and their chances of obtaining enough of it, through whatever system of allocation and appropriation operates and subject to the physical layout of the irrigation system and of their fields. Where, as in Vegamangalam, there is a perennial supply of water adequate for more or less continuous cropping, the question of which land to irrigate or not to irrigate does not arise in the same form but depends on the timing and phasing of cultivation operations.

The second stage of decision is the allocation and appropriation of water within an irrigation community, affecting those areas which it has been decided to irrigate. There are at least four forms this can take:

1. A physical division of waterflows between channels. The *karahankota* described by Leach for Pul Eliya (1961, pp. 160–66 [Chapter 5 below, pp. 119–22]) is an example. Water was divided by a wooden weir into which flat-bottomed grooves of various widths had been cut, the water allocations being the amounts of water which flowed through different grooves into different channels. The physical system (though not the proportional allocations) had fallen into disuse in Pul Eliya even in 1954 and no case of any similar system was found in our survey either in Sri Lanka or in India.

2. Rotational rationing on a roster basis. This is widespread throughout

the world. The warabandi system in Haryana (Vander Velde, 1971, p. 132) and the waqt (sunrise to sunset or sunset to sunrise) system in Iraq (Fernea, 1970, pp. 124–25) are examples. In our survey we found that time had been estimated in various ways in the past including judging by the sun during the day and by the stars at night, measuring the lengthening shadow of a stick either in finger breadths or paces (Amudur), and taking the time a leaking pot took to empty (the murai palla system in Vengodu). These methods have, however, fallen into disuse and have been replaced by the wristwatch, sometimes in Sri Lanka combined with paper chits (tundu) as in Wellawaya and Companniwatta (where four-hour spells have been used in periods of scarcity). In several Indian villages in the sample there was a karai system in which a sequence of turns was taken by family groups, the duration of the turns being a matter of tradition. But given the dispersal of family lands and the complication of pump sets, what happens in practice must be an open question. A principle often stated, however, was that the duration of water was related to the acreage owned or to the acreage actually cultivated in the season in question.

3. Allocation by restricted acreage. The rationing system at Duli is based on the principle that each cultivator should restrict his acreage to a fixed amount and then, in rotation, be supplied with the water needed. This has some similarities with the bethma system in some purāna villages in Sri Lanka (Farmer, 1957; Leach, 1961) in which, in a season when acreage had to be restricted, all holders of wet land were able to cultivate a portion of the irrigated field.

4. "Anarchy." Water may be not so much allocated as appropriated, as described by John Harriss for part of Kirindi Oya right bank: "I have found . . . the suggestion of a kind of anarchy in which in time of scarcity water supplies depend upon the strength of a man's right arm" (1976, p. 16). The apparent disintegration of traditional allocation systems under Indian village tanks may also sometimes verge on this situation.

Equity and Productivity

These two sets of actions—deciding the location and timing of irrigation, and then the allocation and appropriation of water to those lands which are being cultivated—raise acute questions of equity. Rural inequity is often associated with differing sizes of landholdings. But this misleads when a man with a secure water supply is able to crop his land three times a year while a man who has to rely on only one irrigation takes but one crop. The physical position of fields relative to channels is critical. Those near the top of channels have an immense physical advantage of access which it can be very difficult for those farther down to control. In the absence of countervailing custom, social sanction, or physical force, the privileged top-enders satisfy their own needs first before allowing water to flow on down a channel to their less fortunate neighbors below. The tail-enders often receive less water less reliably and in a less timely fashion than those near the top. There is a striking variation in the extent to which the communities studied in India and Sri Lanka moderate these inequities and in the methods they use.

In India the most common systems for distribution under tanks favor those at the top end. In Meppathurai, Randam, Sirungathur, Vayalur, and Veerasambanur, top-enders are said to take water first. Moreover, the karai system, and any other system of time rationing, is liable to deliver less water to tail-enders because of seepage and evaporation losses en route (see Vander Velde, 1971, *passim*). However, informants from Vinayagapuram, Amudur, and Vengodu all claimed to have systems which made special provision for tail-enders in time of water scarcity: in Vinayagapuram, the first issue was said to be from the top downward with the second issue in reverse from the tail end upward back toward the top; in Amudur since about 1955 it was said that water had been issued to tail-enders first (this was part of a major reform in which the supervision of water allocation was also changed); and in Vengodu, where tail-enders had been suffering, a partially effective convention was said to discourage those with pump sets in the wet land from using tank water so that it could be supplied to those less fortunate cultivators who did not have pump sets. It is, however, Duli's system, allowing adequate water to equal plots of land, which scores highest for equality. In Sri Lanka the systems also varied but information on them is incomplete. On major irrigation, however, the practices appeared to follow the principle of "the devil take the hindmost."

Questions of equity are linked with questions of productivity. With food production a major objective and water a critically scarce resource, measures which might be more equitable have to be weighed also in terms of productivity. The main issue is that the conveyance of water involves losses through percolation and evaporation. Duli scores highly for equity but the water losses in distributing water as in Navarai 1972 to small plots of 0.3 acres each scattered over the ayacut must have been substantial. Had it been possible to adopt an equivalent of the bethma system, in which all cultivators participated but in which the water was applied to one block of land near the tank, then the productivity of water and the total output of the land should have been higher. Similarly the supply of water to tail-enders first is wasteful, not only in conveyance losses but also in the loss of opportunity to reuse drainage water and to raise the water table; for when top-enders in an ayacut take water first, seepage in their fields may raise the water table lower down and thereby reduce subsequent water duties there, and surface run-off into drains may be reused by cultivators nearer the tail end, as at Kataragama and under Tissawewa in Sri Lanka.

The questions are complex and interlinked with the patterns of wealth and power in irrigation communities. Any government may hesitate to intervene in such a difficult policy area; but several of the Indian villages had themselves within living memory changed their water allocation systems, in one case at least (Amudur) in the direction of greater equity in distribution. The systems used are by no means a sacred part of the social fabric, to be tampered with only at the risk of severe disruption. The evidence suggests that water distribution under tanks was usually both inequitable and inefficient in terms of productivity. A particular example is the tendency for those with water available from wells and pump sets none the less to take tank water (since they do not have to pay for it), denying it to their less fortunate neighbors who may not have wells. The result may often be that a village cultivates a much smaller area than it could if the pump-set owners were to use only well water. Could those with pump sets be persuaded or forced to forgo tank water? The suggestion was greeted with laughter in Randam and Vayalur, but informants in Vengodu suggested that some such idea was at large there and might even be partially implemented. If, with the introduction of pump sets in wet land and the progressive fragmentation and dispersal of family lands, the distribution systems under tanks in North Arcot are looser and less effective than in the past, this may be a time when an official initiative to increase both equity and productivity is feasible. Differential taxation to provide an incentive to pump-set owners in the wet land to abstain from using tank water might be considered.

Enforcement and Arbitration

An intriguing set of questions arises over infringements and disputes and their adjudication. There is a sharp contrast between Sri Lanka and India. John Harriss has described (1976) the work of the *vel vidanes* who were appointed by government under the colonial regime in Sri Lanka, armed with authoritarian powers, and remunerated with a share of the crop; and the subsequent system of enforcement through the elected administrative secretaries (Govimandala Sewaka) of the cultivation committees, who received 40 percent of an acreage tax. It seems to be widely accepted that the vel vidane system could be quick-acting and technically efficient, whereas the cultivation committee system has always been slow-acting and permissive. Cultivators canvassed in our survey gave responses which can be interpreted as preference for a system, whether vel vidane or other, which was authoritative, quick, and effective (Chambers, 1976b, text and Appendix A). It would be easy, if no other system were known, to conclude from this that a more authoritative and more efficient system is needed at the irrigation community level; that a committee cannot perform this function; and that a man whose reward is unrelated to the value of the crop is unlikely to perform it well.

The contrast with the Indian villages is then striking. Under the South

Indian tanks there is no equivalent of the vel vidane. There is no tradition of a government servant being concerned with allocations within the paddy tract under small tanks. The system is radically different. Whereas the vel vidane was usually an influential and prosperous local person, those responsible for the execution of water control in the South Indian villages are Harijans, the thoddis or neer thoddis. Their responsibilities vary considerably, as does their remuneration. In some villages they are responsible only for closing and opening the sluice. In Amudur, however, they have extensive responsibilities in executing the allocations in the paddy tract. One of the three Amudur thoddis said (1974) that he would never allow anyone else to move water and if they did there would be an ur panchayat meeting and the miscreant would be fined; but this had never happened. Evidently, if our informants were correct, rights and allocations in Amudur are clearly understood and the thoddis have clear guidelines to follow. One Amudur farmer went so far as to say that under the system practiced before 1955 there were many disputes, but now he did not even bother to go to his fields when water was due as he had complete trust in the fair operation of the system by the thoddis.

The extent to which an arbitration role is demanded must depend on the extent to which there are infringements or, in the absence of clear rules, the extent to which there are acts which cause serious resentment. No doubt cultural differences and different developmental experiences profoundly influence attitudes toward different forms of arbitration. But appeals to outside authorities are common. On the basis of a comparison of fifteen irrigation systems in the Philippines, Ongkingco has written, "It is striking to note the satisfaction of farmers when somebody in authority, like a policeman or a major, attends to water distribution problems. Under these circumstances, farmers even seem to be satisfied with reduced water supplies" (Ongkingco, 1973, p. 242). In Sri Lanka, one administrator has lamented the volume of cases and appeals presented to him over water matters, deflecting him from the main task of stimulating agricultural production (Weerakoon, 1973, p. 7). Performing these arbitration functions, whether the arbitrator is a government servant or a local person, is not easy. Administrative secretaries interviewed in Sri Lanka were generally unenthusiastic about their work, several of them complaining about the arduous duties involved. In the Philippines again, Ongkingco found one hereditary water master (whose duties were roughly similar to those of an administrative secretary) who wanted to relinquish his position because he got no benefit from it, but felt he could not do so because of community tradition (Ongkingco, 1973, p. 240).

One objective of government policy may be to improve equality and productivity while avoiding involvement in administrative costs. Once government intervenes, there is a danger of an endless series of cases and appeals, and of a need to provide more staff to deal with them. There is also a danger of inducing attitudes of dependence among communities. To secure a "fair" distribution of water within irrigation communities may often be difficult (and in any case there are problems with the connotations and interpretations of "fair"). But cultivators do appear generally to agree that they value quick action. And even where governments cannot institute "fairer" distribution of water, there may be opportunities for them to enable crucial decisions and judgments to be made more promptly.

Action by Irrigation Communities

Governments benefit if they can rely on action by irrigation communities for the operation and maintenance of irrigation works. The survey villages are of interest because they present four cases in India where considerable communal labor is called for to maintain an irrigation system, one of which has collapsed; and one case in Sri Lanka of partial collapse.

The four cases in India all involve work required to acquire and transport a communal water supply. They are Dusi, Vegamangalam, Meppathurai, and Vinayagapuram.

The Dusi case involved collaboration between the eighteen villages served by the Dusi-Mamandur tank. On 16 August 1971 the Dusi-Mamandur irrigation board, consisting of one representative of each of the villages, a secretary, and a president, met to decide how to secure the flow in the channel from the anicut to the tank. This, they maintained, was the responsibility of the PWD, but as the PWD could not be relied on to act swifty enough, the villages themselves had to take action. They decided that each village should send labor at the rate of one man to every ten acres irrigated, in order to divert the Palar River into the channel. The work was apparently successful.

The Vegamangalam case is a continuing and customary activity. When the long channel bringing the spring water to the pangu lands of the village requires a cleaning out, every family with a share provides labor at the rate of one man per anna of land (1.6 acres of wet plus 0.74 acres of dry). The system apparently works well.

The Meppathurai case is an example of a practice abandoned: of what it was there were several differing accounts. What was agreed is that the run-off flow into the Meppathurai tank had for many years been supplemented by a channel from the Cheyyar River. When the river flooded, villagers dug in the river and in the channel to divert water into the channel and along it to the tank. Much work was involved in removing silt from the channel. In about 1967 there was a heavy flood and the channel seriously silted up. According to some, the task of clearing was too great for the village and appeals for government assistance failed. Others state that there were political differences between the larger, older farmers (who were Congress supporters and stood to benefit more from clearing) and the smaller, younger farmers (who were DMK supporters and stood to benefit less). Yet another contributory factor may have been a high degree of absentee ownership of wet land. It is also possible that the farmers were not unduly concerned because they could anyway rely on their pump sets in the wet land. But whatever the cause, Meppathurai failed either to obtain government assistance or to carry out the clearing itself. In 1974, some six years later, the situation was even less remediable; the two miles of channel were heavily overgrown with bush and the poorer people who used it as a source of firewood for sale were opposed to any clearing.

The Vinayagapuram case is an interesting contrast. The main water supply for the tank comes from a five-mile channel taking off from the Cheyyar River. This requires extensive and heavy work to clear off sand during the period from the beginning of January until the end of April. All those cultivators who benefit from the channel have an obligation to clear three feet per day for every acre of wet land they hold. The work is closely administered and arduous, but the second (Navarai) crop depends on it. There is a long history of conflict with Konaiyur, a village which lies astride the channel above Vinayagapuram but which has no rights to the water. Twenty years ago, when the channel silted very badly and Vinayagapuram was appealing for government help to clear it, Konaiyur people said they would clear it and take it over. However, Vinayagapuram obtained government assistance and managed to continue maintenance. More recently, theft of water by people from Konaiyur has led to violence and court cases. When the channel is running, Vinayagapuram posts night guards where it runs through Konaiyur. Since the main crisis twenty years ago the system of communal labor appears to have been continuously effective.

The final case, from Sri Lanka, raises the issue of the division of maintenance responsibilities between communities and bureaucracy. In one instance, a long canal was heavily silted and overgrown. Partly because of this, water only reached the tail end four to six weeks after it began to flow at the top. It was in the interests of the tail-enders but not of the top-enders that the canal should be cleaned and maintained. The maintenance responsibility lay with the Territorial Civil Engineering Organization, which was unable to carry it out. The TCEO suggested that the communities themselves should clean the canal. The tail-enders, in whose interest it was that the canal should be cleaned, might have done the work, but by then the top-enders already wanted water. The result was no maintenance and continuing inefficiency and inequity in water distribution.

These examples support common-sense conclusions about communal labor. First, communal labor is most likely to be effective where the community will benefit directly and where labor obligations are proportional to expected benefits. Thus Dusi and the other seventeen villages could mobilize labor to divert the river into the tank, and Vegamangalam and Vinayagapuram maintained their channels. In all these cases the labor obligation was related to irrigated acreage. Conversely, where there is no direct link between the work done and the benefits gained, communal maintenance will be much more difficult. One of the reasons given for the abandonment of the Meppathurai channel was that the young men and small farmers felt that they were being required to do more than their share in relation to the benefits they might expect. Even more so, it is unrealistic to expect maintenance to be undertaken by people who will not benefit at all, as with clearing of silt at the top of channels by top-enders, which only helps those farther down.

A second conclusion concerns the role of government. Intervention to help a community may be critical in sustaining a system of communal maintenance when it is under exceptional stress. Vinayagapuram's system survived after a successful appeal for government help; Meppathurai's collapsed after a similar appeal failed. The judgments involved are nice since too much help too easily given generates attitudes of dependence which in turn may lead to collapse. One error to avoid is uncertainty about the physical boundaries of responsibility for maintenance. Such uncertainty arose in Sri Lanka following an instruction to the TCEO (which was not well received by staff at the local level) that they should extend their maintenance work farther down some channels. The outcome of such a situation is liable to be that neither government nor the community maintains the works. In general, government should unambiguously avoid doing what communities can do for themselves in their own interests, but should intervene when exceptional problems are beyond a community's power to overcome.

A third conclusion is that those who design irrigation systems in countries where labor is abundant and government poor should consider designs which encourage community action. These require that the maintenance work shall be within the capacity of the numbers of cultivators anticipated, and that they shall benefit from the work being done. The recurrent costs to government of the irrigation system should then be less than if government itself were obliged to provide maintenance. Higher capital costs, for example with more separate channels to communities which would then maintain them, might be justified by reducing the recurrent costs of maintenance by government.

Bureaucratic-Communal Organization and Operation

Perhaps the most interesting, important, and difficult questions concern the organization and operation of bureaucratic-communal irrigation, in which water is controlled first by a bureaucracy and then by a community or communities. The issues which arise within irrigation communities also arise now within the bureaucracy, between the bureaucracy and the communities, and between communities. The problems of water allocations between competitors, the questions of productivity and equity, and the difficulties over enforcement and adjudication which all occur within communities are now replicated but on a bigger, more visible, and sometimes more dangerous scale on the larger irrigation system.

Although the variations are legion, a recurrent concern and source of intercommunity conflict on bureaucratic-communal irrigation arises over the allocation and appropriation of water. With community irrigation, without a bureaucracy, we have already seen how the poaching of Vinayagapuram's water by farmers from Konaiyur, higher up the channel, led to violence and litigation, with the difference that there is a mediating bureaucracy. Common practices include constructing illegal outlets, breaking padlocks, drawing off water at night, and bribing, threatening, or otherwise in some way inducing officials to issue more water. Typically those at the top end get their water first and get most of it, while those at the tail end suffer. Many examples could be given. On Kirindi Oya right bank canal in Sri Lanka, there are several extra pipes off the main canal which were not part of the original irrigation design (personal communication, John Harriss) extracting water higher up, often to the detriment of those lower down. In North India the tension between villages may erupt into serious threats to law and order. Vander Velde reports an intervillage dispute in which ten cuts were made in an embankment in less than twenty-four hours and major violence between villages threatened (1971, p. 154). Both in the allocation of water and in the execution of the allocations the competition between communities is an inescapable problem.

Productivity and equity are involved here, as they are in intracommunity distribution. Other things being equal, water is less productive after conveyance losses to the tail end of a channel than if it can be applied at the top end. Moreover, when a canal is long, conveyance losses are high, and delays in the arrival of water at the tail end run into weeks or even months, as they do with the seventeen miles of the Walawe right bank in Sri Lanka, then planting at the tail end becomes untimely, either forcing cultivators to grow lower-yielding, shorter-duration varieties, or involving them in risks of inadequate water at critical periods in the growth of the crop, or condemning the crop to climatically suboptimal conditions, or some combination of these. Excessive extractions higher up commonly contribute to these delays and inadequacies of supply to the tail end. On much major irrigation in Sri Lanka it is notorious that top-end farmers flood their fields more than is necessary for the growth of paddy and substitute water for labor in weeding, with little or no regard for their neighbors waiting dry farther down the channel. Their behavior is rational, given their interests; but it is also antisocial, both in denying their less fortunate neighbors timely and adequate water and in denying the country the additional paddy which their neighbors might be producing. The same is true with water issues on the two largest schemes in Sri Lanka—Gal Oya and Uda Walawe—where the acreage cultivated is much less than it might be because of permissive and excessive water issues. In the one Indian example of bureaucratic-communal irrigation (Dusi-Mamandur) the problem may be less acute, but even there tail-enders complained that they could grow fewer crops in the year than top-enders.

The challenge here is to be inventive in devising institutions and relationships which will moderate intercommunity strife and be both equitable and productive in the allocation and application of irrigation water. There are four clusters of functions to be performed:

1. Strategic decisions about water use, including timing, amounts, allocations to communities, which lands to be irrigated, what crops to grow, and the maintenance of channels.
2. The execution of those decisions.
3. Allocation of water and arbitration *within* communities.
4. Policing, and prosecution of infringements.

The question is how officials on the one hand and communities of users or their representatives on the other should be combined or separated in order best to perform these functions. A problem here is the word "best." The criteria for evaluating solutions already include the productivity of water and the equity of its distribution. To this some, democrats, would add maximizing participation by the users, while others, technocrats, would add its antithesis, maximizing the decision making and control by tehnical staff.

In deciding the balance to strike between the democratic and technocratic views it is chastening to reflect on the side differences which can be observed. At one extreme is the system operated under the Dusi-Mamandur tank in India, with its ayacut supporting eighteen villages. Intercommunity water allocation decisions are made by the president of the irrigation board elected by the villages. Villages send their traditional functionaries to him with requests for water which he then forwards, after whatever amendment he judges necessary, to the section officer of the PWD, who instructs one of his staff to open or close the sluice from the dam accordingly. In Sri Lanka, on this size of irrigation system, the distribution from the channels below the tank would be the responsibility of government staff, but according to the evidence given, all water movement below the Dusi-Mamandur tank is the responsibility of an irrigation board of village representatives. Among the examples available, this is an extreme version of user participation in strategic decisions and their execution. At the other extreme are projects where the bureaucracy controls water issues right down to the level of the farmer (as on Uda Walawe in Sri Lanka) or even to his individual field (as on the Mwea irrigation settlement in Kenya [Chambers and Moris, 1973]).

Both extremes have disadvantages. The Dusi-Mamandur system is probably inefficient in water use: certainly there is an irrigation engineering opinion that water use would be much less wasteful if the bureaucracy controlled water issues from the main canals to the irrigation communities; certainly too, the tail-enders only manage one or at best two crops a year while those at the top end regularly have two or even three. With tighter management the distribution of water might be both more productive and more equitable. On the other hand, the bureaucratic extreme, as on the Mwea irrgation settlement, is expensive in government staff and in the associated loss of community self-management and communal labor for maintenance. Government is liable to be doing for communities what they could and would otherwise do for themselves. Some middle course between these two extremes may combine greater productivity and equity without forgoing communal labor and without the need to maintain a large bureaucracy.

Taking this point of view, we can examine the four clusters of functions and see how they might be allocated.

First, there is a good case for strategic decisions being taken jointly by representatives of users and by government officials. Where representatives of users take decisions alone, they are likely to lack some of the technical knowledge needed, as probably on Dusi-Mamandur. Where administrators or technocrats take decisions on their own they are liable to ignore some needs of users, leading to later difficulties. Moreover, as the assistant government agent Hambantota wrote in 1922, "The proprietors are more likely to adhere to dates which they have agreed to than to regulations imposed from without" (letter to Government Agent, Southern Province, 8 November 1922; Hambantota Kachcheri file E85). Better decisions are likely where they result from discussion which benefits from an engineer's knowledge of water availability, an agriculturalist's appreciation of the cropping position, farmers' knowledge of their resources and problems, and a presiding administrator's appreciation of all of these. This is very much the system practiced in water meetings in Sri Lanka, presided over by government agents. In that form it has both strength and weakness in the openness of the meeting to all farmers affected and who may or may not fairly represent all the interests involved. Given the large attendances, it is not surprising that they decide on dates for operations (such as opening the sluices from a tank, starting cultivation, and completing water issues) but do not decide the detail of rotational issues. Were there a more representative but smaller body, elected by "irrigation constituencies" which would ensure that tail-enders were included, then it might be possible for such meetings or a succession of them to decide in more detail what system of water issues to communities, with what volumes of water, should be adopted.

Second, with the execution of these decisions the question is how far the bureaucracy should extend down the irrigation system. On Dusi-Mamandur it is restricted to the sluice itself. On major irrigation in the dry zone of Sri Lanka it extends down the main channels to the points at which water is issued into field channels to communities. Communities are unlikely to agree among themselves that those higher up will take less in order that those lower down may benefit. More usually, an independent and impartial organization is needed and this is mostly some form of bureaucracy. The need of such bureaucracy is underlined by the experience of the elected Thannimurrippu Paripalana Sabai, reported by Ellman and Ratnaweera, who state that while strategic decisions were satisfactorily taken, the problem was implementation and enforcement in which the elected body was not interested (Ellman and Ratnaweera, 1973, pp. 10, 15). There were difficulties over the blurred division of responsibilities between the elected body and the government officers and "depersonalizing the process of rule enforcement" was needed (ibid., pp. 8–9, 27). A crucial link is, it seems, between the strategic decisions and those who implement them. A degree of impartial independence is required, with willingness and ability to carry out instructions earlier arrived at without bowing to particularistic local pressures. For this, a bureaucracy loyal to the decisions, but with its discipline partly deriving from a larger national or regional department, seems the most promising solution.

Third, allocation and arbitration within communities can usually be left to those communities, with perhaps some provision for appeal and for intervention by the bureaucracy in emergency. If water has to be rationed on a rotational basis, the difficulties of allocation within the community irrigation tract may be lessened if, as suggested by Levine et al. (1973, p. 11), the intermittent issues of water are large.

Fourth, there is a persistent need for policing and the prosecution of infringements above the community level. These are sometimes carried out by communities themselves. Vinayapuram's night guards on its canal where it passes through Konaiyur, and the observation of the Dusi-Mamandur president (interview, May 1974) that, if government were to be responsible for distribution below the tank, it would be continuously necessary to call in the police, are reminders of the power of community organization. But it is also noteworthy that under Dusi-Mamandur there was ten years of conflict between two villages, Pallavaram and Kanikillupai, over the height of a weir alleged to be diverting too much water to one village to the detriment of the other, a dispute which provoked intermittent damage and repair to the offending structure. Wherever water is scarce, communities resent extraction of water from higher up on their own supplies, whether apparently legal, as with a rubber company upstream from Wellawaya and with two pumps in the river above Vinayagapuram, or evidently illegal, as with the surreptitious raising of diversion weirs, the use of pumps at night to lift water from channels, the digging or breaching of canal banks, and the like. For these, if not a police force, then something like one is needed.

Police are anyway quite often called in to intervene with both allocation and enforcement. During the crisis of water shortage on Kirindi Oya right bank in *yala* 1922, police helped with the allocation of water (letter, Divisional Engineer SD to the Director of Irrigation, 25 August 1922, Hambantota Kachcheri file E85). In the intervillage conflict in Haryana cited by

Vander Velde, "the resulting inter-village acrimony required the intervention of the police on a major scale to prevent serious violence" (Vander Velde, 1971, p. 154). In India the Irrigation Commission of 1972 drew attention to the need for efficient policing and prosecution and to "the success which has been achieved in Haryana through extensive patrolling and inspection of canals and channels by flying-squads of officers, adequately armed. These flying squads carry out surprise night inspections and whenever offenders are caught, heavy penalties are imposed on them. The essence of the system is surprise, and prompt and condign punishment. A similar system of inspection by flying-squads could be adopted with advantage elsewhere" (MIP, 1972, p. 300). A widespread complaint in Sri Lanka was precisely the lack of "prompt and condign punishment." Within communities, administrative secretaries rarely bothered to file cases which they knew would be subject to long delays; and at a bureaucratic level many cases filed by government servants were not heard for months, or even years.

A careful mix of relationships may be best: with user participation in strategic decisions and with management by communities of their own water supplies once allocated, but with a disciplined organization responsible for executing decisions, policing the system, and prosecuting delinquencies. It has to be made rational for the staff involved to deny resources to people who want them, in particular to issue less water to top-enders than they would like to receive. To achieve this the bureaucracy needs, first, high-level political support, and second, an internal style and supervision and incentive system which supports and rewards such unpopular actions (Chambers, 1975b and 1976).

Comparisons, Theory and Practice

These various comparisons help toward some theoretical and practical conclusions.

At the theoretical level, irrigation presents social scientists with tantalizing invitations, too rarely taken up, to speculate. Expressions like "irrigation society" and "hydraulic organization" hint that there may be strong causal links between irrigation systems and technology and social and economic relations. Irrigation organization has an appearance of inevitability which lends itself to deterministic interpretations. Wittfogel (1957) succumbed to the temptations presented by the apparent imperatives of large-scale irrigation, requiring, as he saw it, totalitarian organization in order to muster the labor forces necessary for the maintenance of hugh flood-control works and irrigation systems. This is not the place to discuss the validity of his thesis. The importance of Wittfogel here is that he illustrates the tendency to see the forms of irrigation organization as unavoidable, as generated and required by imperatives of the physical system and its technology.

There are perhaps two main reasons for this tendency. First, on all irrigation systems which are larger than "community" and in which water is controlled and allocated by a bureaucracy, that bureaucracy has to be fitted geographically to the permanent physical irrigation network. Certain tasks have to be carried out and staff are thought to be needed to perform them. Second, many statements about irrigation are based on detailed analysis of only one example, from which generalizations are extrapolated. The rather superficial information gathered in South India and Sri Lanka has provided an opportunity to see what variations in organization there may be over a wider range of examples than is usually available.

The outcome is surprising. It presents alternatives to the authoritarian, disciplinary, and totalitarian organizations postulated by Wittfogel, and shows considerable variance in the discretion of the bureaucracy on major irrigation. It is sobering to think how much simpler the conclusions might have been had only Sri Lanka's irrigation been considered. As it is, with the corrective of the system of community management under Dusi-Mamandur tank in Tamil Nadu, there seems nothing inevitable in the Sri Lanka pattern of a bureaucracy controlling issues down to the feeder level. The culture in which an irrigation system exists appears a major determinant of the form of organization: thus in the Sri Lanka examples, where the society is more egalitarian and more anarchic, bureaucracy extends farther down the physical system and the case for tighter bureaucratic controls seem clear; but in India, where the controls already exist in the hierarchical structure of the society, it has not been necessary for bureaucracy to extend so far down the system and the need for stricter bureaucratic controls is less obvious.

The technologies used for water acquisition, storage, and distribution, and for the maintenance of works, also underly the organization and political economy of irrigation. Direct individual appropriation from wells is sensitive to technology and an innovation like pump sets can radically differentiate access to water in a community and also deplete a communal resource. When larger-scale technology is used, there arise multifarious problems of allocation and appropriation, some of which have been discussed above. As Wittfogel argued, the requirements of construction and maintenance are powerful influences on social organization. Again, however, the technology used has a bearing. Wittfogel assumed that human labor was the main means used to build and maintain ancient irrigation works. It is at least possible that this was not the case with the ancient tanks of the dry zone of Sri Lanka, and that elephants were used as the bulldozers of that day. If so, then the form of organization may well have been closer to a modern PWD or military engineering unit than to a totalitarian bureaucracy exacting forced labor from peasants. Moreover, with present-day irrigation it is only at the lower levels, as at Dusi, Meppathurai, and Vinayagapuram, that communal labor and not machinery has to be mustered.

At a practical level, both organization and technology are manipulable and subject to choice. The objectives of irrigation can be variously stated, but in the conditions of South Asia a list might include:

productivity (of water)
equity (in its distribution to users)
stability (in maintaining the water supply over the years)
continuity (in water use throughout the year)
carrying capacity (in sustaining population at acceptable levels of living)

In achieving these objectives, and subject to trade-offs, the prescriptions vary by type of irrigation. For bureaucractic-communal and communal irrigation in the examples analyzed, the key lies in the reform of organization and operation—in short, in improved management of men. For direct-acquisition irrigation, the key lies in the design of appropriate technology for the acquisition process. In both organization and technology we are only at the beginning of appreciating the potential. In view of the rapidly increasing pressure of population on water supplies—especially in parts of India but elsewhere also—exploring and exploiting that potential is a high priority. On the organizational side it requires more and better research, especially by social scientists, combined with and supporting management consultancy and staff training. On the technological side it requires imaginative and vigorous research and development to create technologies appropriate for future rural life.

References

CHAMBERS, ROBERT. 1975a. *Water Management and Paddy Production in the Dry Zone of Sri Lanka*. Occasional Series no. 8, Agrarian Research and Training Institute, Colombo.

———. 1975b. *Two Frontiers in Rural Management: Agricultural Extension and Managing the Exploitation of Communal Natural Resources*. Communication Series no. 113, Institute of Development Studies, University of Sussex.

———. 1976. "On Substituting Political and Administrative Will for Foreign Exchange: The Potential of Water Management in the Dry Zone of Sri Lanka," in *Agriculture in the Economic Development of Sri Lanka*, ed. S. W. R. de A. Samarasinghe (Peradeniya: Ceylon Studies Seminar).

——— and MORIS, JON, eds. 1973. *Mwea: An Irrigated Rice Settlement in Kenya*. Munich: Weltforum Verlag.

ELLMAN, A. O., and RATNAWEERA, D. de S. 1973. *Thannimurrippu Paripalana Sabai: The Transfer of Administration of an Irrigated Settlement Scheme from Government Officials to a People's Organisation*. Occasional Series no. 1, Agrarian Research and Training Institute, Colombo.

FARMER, B. H. 1957. *Pioneer Peasant Colonization in Ceylon*. London: Oxford University Press.

FERNEA, R. A. 1970. *Shaykh and Effendi: Changing Patterns of Authority among the El Shabana of Southern Iraq*. Cambridge: Harvard University Press.

German Foundation for Developing Countries. 1967. *Report on the Seminar on Economic and Social Aspects of Agricultural Development in Irrigated Areas*. Berlin: Reiher-Werder.

Government of India. 1973. *Interim Report on Modernising Irrigation Systems and Integrated Development of Commanded Areas*. New Delhi: National Committee for Agriculture.

HARRISS, J. C. 1976. *Aspects of Rural Society in the Dry Zone Relating to the Problem of Intensifying Paddy Production*, in *Agriculture in the Peasant Sector of Sri Lanka*, ed. S. W. R. de A. Samarasinghe. Peradeniya: Ceylon Studies Seminar.

ICAR. 1966. *Report of the Working Group for the Formulation of Fourth Five-Year Plan Proposals on Soil and Water Management under Irrigated Conditions*. New Delhi: Indian Council for Agricultural Research.

IRRI. 1973. *Water Management in Philippine Irrigation Systems: Research and Operations*. Los Baños: International Rice Research Institute.

LEACH, E. R. 1961. *Pul Eliya: A Village in Ceylon*. Cambridge: Cambridge University Press. [Chapter 5 in this volume.]

LEVINE, G.; CARPENER, H.; and GORE, P. 1973. *The Management of Irrigation Systems for the Farm*. Research and Training Network reprint no. 2. New York: Agricultural Development Council.

MADDUMA BANDARA, C. M. 1976. *The Prospects of Recycling Subsurface Water for Supplementary Irrigation in the Dry Zone*, in *Agriculture in the Peasant Sector of Sri Lanka*, ed. S. W. R. de A. Samarasinghe. Peradeniya: Ceylon Studies Seminar.

MIP. 1972. *Report of the Irrigation Commission, 1972*, vols. i–iii. New Delhi: Ministry of Irrigation and Power.

MPEA. 1968. *Report of the Irrigation Program Review—Ceylon*. FAO/IBRD Cooperative Programme, Ministry of Planning and Economic Affairs, Colombo.

ONGKINGCO, P. S. 1973. *Organization and Operation of 15 Communal Irrigation Systems in the Philippines*, in IRRI, 1973, pp. 235–42.

THORNTON, D. S. 1976. "The Organisation of Irrigated Areas," in *Policy and Practice in Rural Development*, ed. Guy Hunter, A. H. Bunting, and A. Bottrall. London: Croom Helm.

VANDER VELDE, E. J., JR. 1971. "The Distribution of Irrigation Benefits: A Study in Haryana, Inda," Ph.D thesis, University of Michigan. [Chapter 14 in this volume.]

WEERAKOON, B. 1973. *Role of Administrators in the Context of a Changing Agrarian Situation—A District Point of View*, Seminar on Economic and Social Consequences of the Improved Seeds, Kandy, mimeo.

WITTFOGEL, K. A. 1957. *Oriental Despotism*. New Haven: Yale University Press.

3

The Relationship of Design, Operation, and Management

Gilbert Levine
CORNELL UNIVERSITY

Introduction

The design and operation of many irrigation systems in the tropics, especially in developing countries, are often inefficient because the importance of the management component and of social constraints has been, or is, underestimated. This paper considers these inefficiencies primarily in terms of the seasonal (wet–dry) tropics of monsoon Asia.

The argument of this paper is that, although every irrigation scheme is location-specific, there are some important management constraints on irrigation efficiency which are of general application and which are not under the control of the irrigation engineer, either in the design or operational stages, but which greatly influence the efficiency of water use in practice. Thus:

1. Our knowledge of the interrelationships between water and plant growth far exceeds our knowledge of the interrelations between water and the human element in delivery and utilization: in other words, irrigation engineers face the same social problems as, say, veterinary surgeons.
2. The efficiency concepts used in irrigation system design tend to understress the human component as a factor in water-use crop production.
3. Irrigation systems, on the one hand, and the farmers they serve, on the other, have criteria of optimal efficiencies of water use which may not coincide. When they are far apart there is friction between the system and the farmers and/or among the farmers.

Previously published as "Management Components in Irrigation System Design and Operation," *Agricultural Administration* 4, no. 1 (January 1977): 37–48. Reprinted by permission.

4. Within the resources available to the farmers and to the system, the operational optima for both parties can be brought closer together by effective liaison, e.g., feedback and response mechanisms.

5. As a result of 1 to 4 above, it is usually better for the irrigation engineer to recognize probabilities initially and strive, through reasonably acceptable change, toward possibilities.

Water Use, Crop Growth, and the Human Element

These five points can be illustrated by considering water delivery, water use, and crop growth.

A key decision in designing an irrigation system is to establish its water requirement. The design water requirement determines *either* the maximum area that can be served *or*, if the area is specified, the amount of the required water supply. The water requirement is the basic parameter dictating channel water-carrying capacity and has implications for the physical, biological, economic, and social environment locally. The estimated water requirement frequently is based on potential evapotranspiration (PET) (sometimes termed T). Potential evapotranspiration is the water potentially evaporated from the leaves of a crop and from the land or water it is growing in. When PET is satisfied, plant growth is near or at its maximum. This is well shown in Figures 3.1 and 3.2. Potential evapotranspiration has been, or can be, calculated for most parts of the world from published meterorological data. The water requirement for irrigation is expressed as acre-inches (hectare-

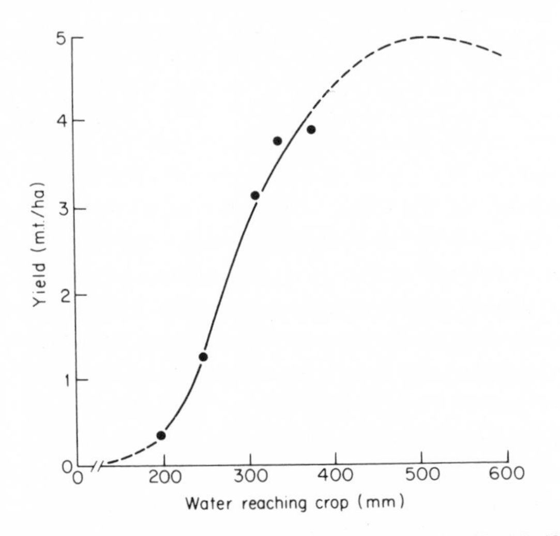

Figure 3.1. The yield of wheat as a function of applied water, Tunisia (Centro Internacional de Majoramiento de Maiz y Trigo, CIMMYT report, 1968-69).

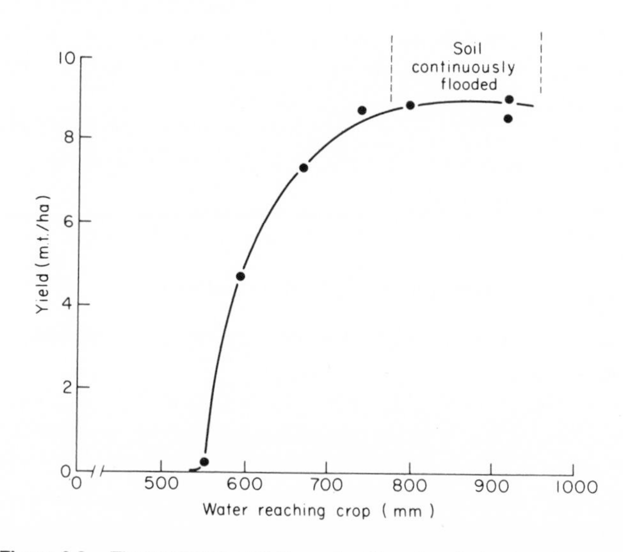

Figure 3.2. The yield of rice (IR 8) as a function of applied water, dry season, 1969, Los Banos, Philippines (R. Reyes, unpublished data, International Rice Research Institute).

centimeters) or sometimes cubic meters (m^3) of PET. The irrigation or net design water requirement is the additional water needed to supplement probable local rainfall or soil water reserves up to the level of PET. (Note that moisture exceeding PET can, with many crops, lower yields.)

In practice, however, the amount of water delivered by an irrigation system to a "turnout unit" (e.g., water gate) on a farm or village has to exceed the net design water requirement because of the losses in water distribution after the water has left the system, i.e., been delivered or sold to the farmer or the local farmers' organization. (The *gross* design water requirement is the net requirement plus allowances for loss by seepage from, and evaporation during, storage or transit along the channels of the systems.)

Our knowledge of the *crop* water requirements for *maximum* yield is good. But our knowledge at *field* rather than *crop* level is less satisfactory because of the variability of local physical conditions, particularly of soil and water table conditions. However, from appropriate on-site measurements, it is possible to make reasonably precise estimates for assumed operating conditions. Adding together the losses before and after turnout (i.e., delivery to farm), the amount of water to be diverted is somewhat less than twice the amount needed by the crop in field (i.e., the overall efficiency is 55–60 percent) in some recent systems. But in some arid areas the efficiency is nearer 30 percent than 60 percent. Examples from Southeast Asia range from 25 percent to over 90 percent (Levine[1]). (See Fig. 3.3.) With a basic water requirement of 650 mm per crop, the Tou Liu system in Taiwan is over 90

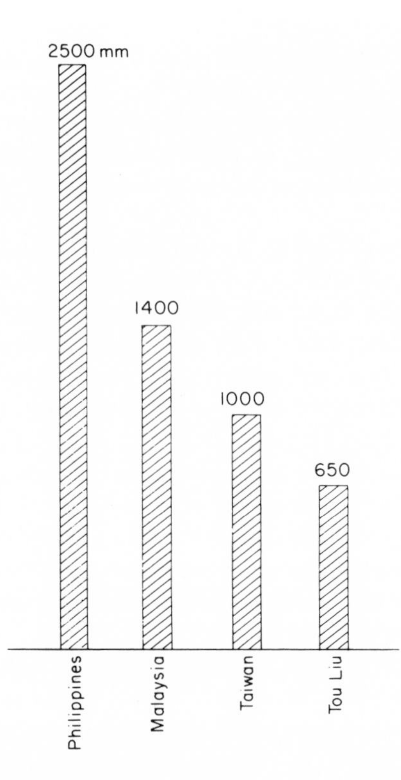

Figure 3.3. Typical irrigation system water requirements (mm/season).

percent efficient, and efficiency is less than 25 percent in the Philippines.

Why does the Tou Liu system in Taiwan have an efficiency of over 90 percent (with an estimated basic water requirement of approximately 650 mm), and systems in the Philippines less than 25 percent? Why are the majority of the systems in Taiwan in excess of 60 percent, while those in Malaysia are closer to 40 percent?

Superficially, an explanation of the differences can be provided. The systems in Taiwan practice rotational irrigation[2] within fifty-hectare units in accordance with a specified plan; the lateral distribution channels frequently are lined with concrete; there are control gates and Parshall flumes at each fifty-hectare turnout; there is an extensive system of farm ditches; on-farm water distribution is handled on a twenty-four-hour basis. The Malaysian systems practice continuous irrigation; control within the primary and secondary conveyance channels is centralized and effective; water policy is specified by ordinance each year; turnouts serve relatively large farming areas; distribution beyond the canal systems is in the hands of the farmers and few farm ditches are used. The Philippine systems are based upon continuous irrigation; there are few effective controls in the conveyance channels and turnouts; channel maintenance is at a low level; there are essentially no measuring devices; system control is exercised eight hours a day, five days a week; farmer cooperation in water distribution is variable and frequently of a low order. Thus, very substantial differences exist among the systems described. These can be visualized in terms of the paradigm shown in Figure 3.4, with the operative differences combined in terms of their impact on the ability to control the water.

Thus, we can relate in very broad terms a combination of physical facilities and set of actions to water-use efficiency within the Southeast Asian context, but we are not in a position to separate them or to establish the reasons for their effective combination, although attempts are being made to do so.[3] The fact that facilities and actions are not combined with equal effectiveness is easily illustrated in many modern systems. An outstanding example of low effectiveness is the Dez Pilot Irrigation Project in Iran. Here the traditional systems, with minimal facilities, had water-use efficiencies approximating 25 percent. The Dez Pilot Irrigation Project is a comprehensive system, with a full range of controls, measuring structures, organizational structure, and all the other accouterments of a large modern system. But the average water-use efficiency in the pilot program area, after six years of operation, was between 11 percent and 15 percent.

By contrast, the Tou Liu system in Taiwan, using the same basic facilities and policies as other systems in Taiwan, achieves substantially higher water-use efficiencies than the typical system. A very high degree of farmer cooper-

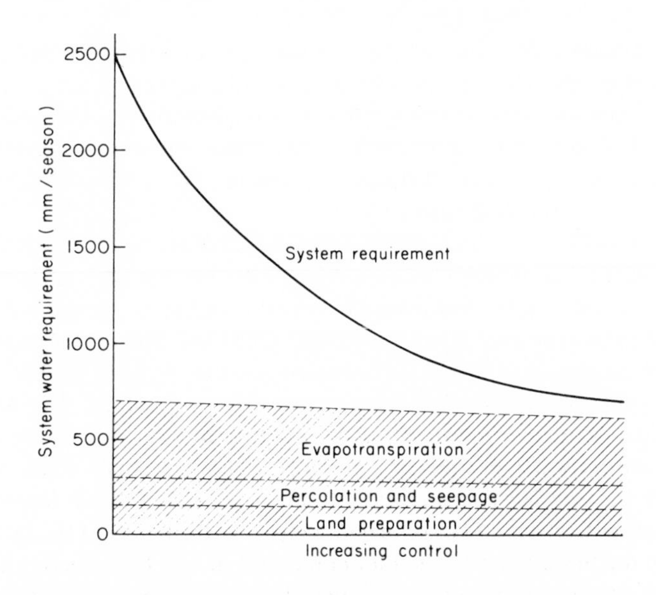

Figure 3.4. Irrigation water requirement for lowland rice as affected by the level of control inputs.

ation, reflected in the joint hiring of common irrigators to whom complete responsibility for water management is delegated, plays an important role in this difference.

I think it is valid to say that we have used the concept of physical or engineering irrigation system efficiency as a substitute for knowledge of the effect of the human element in water management. That some mechanism was—and, at this stage in our understanding, still is—necessary to fill this gap can be accepted. Unfortunately, reliance on the efficiency concept may have seriously inhibited study of the human components of water management.

Water as a Factor Substitute

In addition to the potentially inhibiting effect on the identification of relevant components in water management, reliance on the water-use efficiency concept tends to mask the value of "excess" water as a substitute for other resources in more limited supply. The Southeast Asian examples illustrate this substitution clearly. The Philippines, up to recent years, has had land available for agricultural expansion,[4] rice varieties with relatively low yield potential, easily developed water supplies, and relatively low farm and governmental income. The investment necessary to use the water more efficiently (in terms of net water requirement/amount diverted) has not been considered justified. Within the past few years the availability of new land has decreased,[4] rice varieties with greatly increased yield potential have been developed,[5] and new water sources have increased costs associated with them. There is now evidence that both governmental and farmer attitudes toward the more efficient use of water are changing.

In Taiwan, conditions only recently appearing in the Philippines have existed for almost fifty years. Limited agricultural land, relatively high yielding varieties, and a scarcity of easily developed water sources have put a value on efficient water use that is high in relation to the costs of providing this efficiency. Even with these conditions, it was not until the 1954–55 islandwide drought that widespread measures were taken to increase water-use efficiency to the current high levels.[6]

To accomplish the necessary changes in system and farmer action, subsidies and technical assistance were provided, but in some cases force was required. Thus, political commitment, as well as financial resources, was necessary in order to achieve the efficiency objective. It took very special conditions, however, to raise the value of the water in the Taiwan system to the point where efficiency is essentially 100 percent. Obvious among these conditions is a clearly limited water supply; less obvious are the personal relationships among the farmers and between the farmers and the system. Of significance, however, is the fact that system design, and the associated operational plan, are based not upon an estimated water-use efficiency, but upon specific identification of the water used for the various stages of crop culture. This information has been gained within the context of the local situation (much in the way the early water-use studies were conducted in the United States) and adapted with improved water use in mind.

Operational Equilibria

Implicit in the foregoing discussion is the idea that not only is water used as a substitute for other resources in more limited supply, but that the evolved use is a reasonably optimum use of the combination of available resources. If this argument is tenable, then it must be recognized that the supply of these resources is usually not available to the farmer and to the irrigation systems in equal amounts, at least when the systems are not private. It also must be recognized that the evolved system may not (and probably does not) represent an optimum use of the resources available to each, when considered independently. In addition to this difference in available resources, there are usually differences in objectives. While the public systems are assumed to have goals of serving the farmers in the most effective way, other considerations relating to system performance (apparent command area, cost per unit water delivered, etc.) may actually be the objectives toward which the system strives. Superimposed upon both the farmer and the system may be the government, with a different set of available resources and objectives. The net result of the interaction among these three sets of resources and objectives is the pattern of practices and results that can be seen to be existing.

But with the advent of new high yielding cereal genotypes, fertilizers, pesticides, etc., the situation for those farmers in a position to use, or willing to use, the new technology has dramatically changed their attitude to water and irrigation.

When the range of appropriate crops and the productive potentials of these crops were limited, the farmers were operating with a production function that was relatively flat, and the utility of their resources changed little over a wide range of water service. Thus, system policies that resulted in a relatively low level of service to the individual farmer had a relatively small impact on relations between the farmer and the system. In addition, since the deliveries were, in the case of wheat, usually in the linear range of a single production function (Fig. 3.1), there was relatively little adverse effect on aggregate production from the available supply of water.

However, with the development of these crops with high productive potential, an individual farmer now has the opportunity to shift to entirely new—and steeper—production functions. His investment of labor and other resources can now be of much greater utility with higher levels of irrigation service; thus, his optimal level may be substantially different from that of the system on which he depends for water, if this system has equity or command area objectives. If the farmer's optimum is not recognized (and included in the design and operation of the system), a combination of events can be predicted: aggregate production from the area served will be less than the potential; those farmers with sufficient resources will develop private systems (frequently tube wells where physically possible); those farmers without these resources will put increased pressure on the public system.

This discussion could be extended to consider, for instance, the lowland rice case, which has a different set of physical factors operative (e.g., the rice production function has much more of a "threshold" than a "linear" relationship), or the effect of different land-tenure patterns on the optimum water service for the farm operator. Without doing this, however, I hope the main points have been made: what is optimum from an irrigation system point of view is not necessarily optimum from the farmer's point of view; these optima must be reasonably close for efficient use of the resources available to each; a change in the resources available to either results in a change in the water-use level. Change in the water-use level may in its turn require changes in the organization and administration, i.e., in the control capability of the irrigation system. Efforts to increase this capability usually require significant investment of financial and other resources. In some Taiwan situations, for example, it was necessary to jail farmers for opposing diversion of water serving their areas, even though the amounts remaining were substantially in excess of the physical environment requirements.

System–Farmer Interaction

In most systems the actual amount of water available for delivery is substantially in excess of the basic crop requirement, though there may be considerable variability in the short-term supply of many diversion, or run-of-the-river, systems. In many, if not most, of these systems, even the potentially adequate supply does not serve to encourage farmers to invest in new technology because, as has been suggested before, the "total" system (water source to crop continuum) does not operate with an overall optimum use of resources in mind. In Uttar Pradesh, India, by comparison to the private tube wells, the state tube wells were less effective in meeting the needs of the farmers, and the farmers reacted by neither adopting new practices nor investing financial resources. But the picture is not entirely clear.

The state tube wells provided a basic water supply of approximately 420 mm per hectare per year, with the farmers served from these wells using about 365 mm per ha for the *desi* wheat crop. The private tube wells provided a water source of approximately 500 mm per hectare per year, with the farmers using 485 mm per hectare for the same crop. These ratios of use to supply represent very high levels of water-use efficiency—about 85 percent and 95 percent for the state and private wells, respectively. However, from a net-income point of view, the utility of the different water sources was very different. The question arises to what extent was the difference in net income (approximately 790 rupees per hectare per year) due to the difference in quantity of water available to the individual farmer, and to what extent was it due to differences in other elements of irrigation service?

The data show major differences in the use of new dwarf wheat by the two groups of farmers, as well as differences in other practices reflecting more intensive cropping, with private-well farmers being more intensive in their operations. If we confine our considerations to wheat alone, and assume the production function shown in Figure 3.1 to be reasonably applicable to the Indian environment, we can see that the delivered water supply to private-well farmers is close to the peak of the production function, while state-well farmers would be operating approximately 25 percent lower on the yield scale, if they grew dwarf wheat. This, however, would still be substantially higher than the yield of traditional varieties, and more than sufficient to pay for the additional costs associated with dwarf wheat production. Some of the state-well farmers did grow the new varieties, but "farmers using state tube wells and growing drawf wheat were the more influential farmers with larger farms located near the wells; they were, therefore, able to provide an extra irrigation for their dwarf wheat."[7] It would seem, therefore, that there were problems of equity of distribution in addition to level of supply. In addition, the stress on efficiency of equipment utilization (evidenced by a 4,000-hours-per-year running time, in contrast to 1,500 hours per year for the private wells) resulted in a number of equipment breakdowns and unpredictable delays in repair.

Added to the problems of delivery equity and reliability of service was the complete lack of mechanism to adapt water deliveries from state wells to meet farmer needs. Water deliveries were on a rigid timetable with no provision for coordination with the farmers' abilities to utilize water. This combination of elements made it unattractive for most of the state-well farmers to take the risks accompanying the higher investment for dwarf wheat and for other forms of crop intensification.

By contrast, in Taiwan, when government policy emphasized water-use efficiency, but also stressed productivity, equity, and payment of irrigation costs by the farmer beneficiaries, a series of interacting efforts was conducted to implement the policies[6] and interaction was built into irrigation system operating mechanisms. These latter included regular meetings between system personnel and farmers to explain broad policies, specific policies relative to water delivery and operational procedures, election by the farmers of an honorary group leader to act as the intermediary between small groups of farmers and the irrigation system (on a continuing and operational basis), performance rating of all system personnel, a program of field performance data collection, and a number of others. The impact of these types of interaction can be illustrated with a few examples.

When the engineering personnel of one system planned a channel lining program they anticipated major reductions in the losses of water in transit and recommended a 40 percent reduction in the amount of water diverted into the channels. Management personnel of the system objected to the reduction pending actual measurement of anticipated savings. The field data collection program revealed much smaller water savings, and the diversion reduction actually implemented reflected field information, rather than the projections. There was essentially no adverse effect on the farmers. The lining program, however, did result in improved timing of deliveries and in reduced maintenance costs. The improved maintenance situation was noted by the farmers, who have responsibilities for maintenance of many of the smaller channels, and they embarked on lining programs for the smaller channels, with the assistance of regular system personnel.

In the Tou Liu system, previously mentioned, the normal pattern of farmer–system interaction was insufficient to provide effective service due to a very limited water resource. While the system could provide the limited water equitably to the ten-hectare turnout units, water distribution within the units (farmed by eight to ten farmers) was done with the individual farmer objectives uppermost. Discussions among the farmers, with the honorary group leader and system technical personnel participating, indicated possibilities for improved service provided within-unit water distribution was delegated to common irrigators. These individuals, responsible to and paid by the farmers, were trained by system personnel. The result has been a very effective water-utilization program, with *equity* of water distribution being recognized as differing from *equality* of water distribution. For example, in the gently sloping areas of the system, upper paddies are irrigated first, with specified quantities. As water is rotated to lower lying paddies in subsequent days of the irrigation cycle, the amount of water turned into these paddies is less, in recognition of seepage from higher paddies. Thus, the overall system operation provides a water environment that permits equity of crop growth, rather than providing equal amounts of water. This represents a sophisticated management practice which is not found in many systems in developed countries. All of this is done within the crop constraint of relatively high sensitivity to water stress (by comparison to most upland crops), and evidences both dependable system performance and farmer acknowledgement of that performance.

The interaction and response illustrated are not only to be found in water delivery aspects of system operation. Within a given system the special fee charges are based upon capital costs associated with individual channels to reach the areas of the system. Thus, farmers within one system will have different irrigation fees, depending upon their location. In some cases these charges may be as high as 5,000 $NT ($125) per hectare per annum.[8] Notwithstanding these rates, the percent of farmer repayment has been very high—frequently above 95 percent. However, when it has not been possible to adequately serve an area, and a drop in production results, adjustments are made in the fee schedule.

In much modern irrigation system design the importance of feedback-response interaction is recognized, in principle, and a delivery-on-demand program is assumed to be the way to implement this interaction. Within a variety of constraints (water rights, requirements for multiple requests on the same channel, etc.), water is delivered upon request of the farmers. That this will not always lead to anticipated results is amply illustrated in the Dez Pilot Project. That it is not the only appropriate mechanism is clearly evident from the Taiwan experience.

Probabilities and Possibilities

The problems experienced by farmers served by existing irrigation systems in tropical low-income countries and the problems experienced by the systems themselves are manifold and manifest. As a result, but not solely due to these problems, a philosophy has evolved to the effect that irrigation modernization can take place only with radical departures from traditional practice. These changes usually are viewed from the perspective of engineering efficiency: efficiency in use of water, efficiency in mechanics of operation and maintenance, efficiency in irrigation system costs. Where causes of problems are identified as being in the realm of the physical environment, design usually includes specifics for operation within that environment.

Attempts are made to tailor a physical solution to a specific problem. In doing so, social constraints are either ignored or treated in a general, nonspecific manner.

The pattern of looking for general rather than specific solutions to social constraints, and for emphasizing change rather than adaptation, is evident even when problems are identified as being primarily in the social area. Almost all new irrigation projects recommend some form of public institution similar to that found in developed countries. There is very little attempt to work within existing institutions or to develop adaptations based upon indigenous social relationships.

The argument frequently is made that adapting to existing social constraints reinforces the existing social structure. From a development point of view, or from an equity point of view, this may be undesirable. In principle, there is no disagreement. However, I would argue that moderate adaptation with specific equity or development goals in mind is a more effective and efficient mechanism for achieving these goals than is the introduction of foreign institutions with a high probability of failure.

Compounding the problem is the emphasis on large projects. While there frequently are major economies of scale associated with irrigation projects, there are attendant diseconomies associated with the added and more stringent requirements for data, for accurate projections, and for specialized skills, as well as with the more complex nature of relationships within large projects.

Conclusions

1. There are serious gaps in our understanding of irrigation in the low-income tropics.

2. These gaps exist not only in our knowledge of the physical environment but even more broadly in the social and economic environment.

3. Given these gaps, research to bridge them should be given high priority.

4. Recognizing that irrigation investment will continue to be made before substantive answers are obtained, the role of smaller systems vis-à-vis the larger should be evaluated with the tools at hand.

5. Existing irrigation systems represent research infrastructure. Comparative research using this infrastructure may be the most efficient route to identification of pertinent problems, and toward their solution, in the technical engineering and biological fields as well as in the social field.

References

1. GILBERT LEVINE. 1971. Paper presented to a seminar sponsored by the United States Agency for International Development (USAID).
2. TSAI TSUI-YUAN. 1964. "Rotational Irrigation Practice." Chianan Irrigation Association.
3. SENEN M. MIRANDA. 1975. "The Effects of Physical Water Control Parameters on Philippine Lowland Rice Irrigation Performance." Ph.D. thesis, Cornell University.
4. F. H. GOLAY and M. E. GOODSTEIN. 1967. "Philippine Rice Needs to 1900: Output and Input Requirements." USAID Mission, Manila.
5. R. L. CHANDLER. 1973. "The Basis for the Increased Yield Capacity of Rice and Wheat and Present and Potential Impact on Food Production." In *Food, Population, and Employment: The Impact of the Green Revolution,* ed. T. T. Poleman and D. K. Freebairn. New York: Praeger.
6. T. S. LEE and L. T. CHIN. 1961. "Development of Rotational Irrigation in Taiwan." Far East Regional Irrigation Seminar, Taipei.
7. J. W. MELLOR and T. V. MOORTI. 1971. "Dilemma of State Tube Wells." *Economic and Political Weekly* 6 (March 13): 27.
8. L. T. CHIN, water management specialist, FAO, Rome. Personal communication.

PART II

Community Irrigation Systems

Introduction

Among the irrigation systems of the world are many examples of what I categorize as community irrigation systems. These are systems in which the water users are directly responsible for maintaining and operating the system, which they or their predecessors have usually built.

The fact that the responsibility for the system lies with the local water users and not with some agency of the state does not mean, however, that community irrigation systems are completely outside the orbit of the state. Irrigators in these systems may from time to time call on local administrators or politicians for assistance in repairing canals, replacing a temporary head-gate, or settling a dispute regarding water rights. Indeed, irrigation systems that now are the responsibility of the community may originally have been constructed under the direction of powerful regional elites.

Nonetheless, as an ideal type, community irrigation systems are characterized by independent "in-system" authority and responsibility for operation and management, and their contacts with the state are usually initiated by the community. In this sense, the term "community" refers to the community of water users and not to the residential community or village. Since irrigation systems frequently are hydrologic units that intersect several village boundaries, the irrigation community often draws its members from a number of residential units.

In general, social scientists' knowledge of community irrigation systems is limited. While individual researchers often have considerable information about the particular system they have observed, they tend not to have detailed information on similar systems elsewhere. Indeed, the number of comparative studies of indigenous irrigation systems has been relatively few (important exceptions are Beardsley 1964, Millon 1962, Geertz 1972, and Hunt and Hunt 1976).

The practicing engineers' and agency policy makers' knowledge of community irrigation systems is even more circumscribed. The generally prevailing notion is that community irrigation systems are minuscule, inefficient, and unsophisticated.

The chapters included in Part II are intended to enhance the information base both of social scientists concerned with the organizational aspects of agricultural development and of irrigation practioners faced with serious problems of local organization for water management. In assembling these chapters I have not meant to imply that all community systems operate as smoothly as these cases seem to do or that the arrangements observed in Bali or Sri Lanka are directly transferrable to systems elsewhere. As is discussed in Chapter 10, some general themes and principles of irrigation organization may be derivable from these cases and others. Even so, such derived themes would need to be provisioned with the appropriate accouterments of the sociocultural setting in which the irrigation systems were to be located.

The *subaks,* or irrigation societies, found on the Indonesian island of Bali have become well-known examples of community irrigation systems. In Chapter 4 Clifford Geertz provides a detailed account of one such *subak.* A major feature of the *subak,* as Geertz notes, is the organizational separation of the irrigation community from the residential community. An individual water user is a citizen of a village and a member of an irrigation society (one might, in fact, be a citizen of more than one irrigation society, depending on the location of one's fields).

A second important feature is the complex internal structuring of the *subak* in intermediate administrative units (*tempek*), each with its own group leader, and the basic accounting unit, the *tenah.* The *tenah* can be a unit of water, a unit of land, or a unit of rice seed, and provides the basic measure for calculating both system benefits and responsibilities.

Edmund Leach set out to analyze patterns and consequences of the kinship structure in an agricultural village in Sri Lanka (then Ceylon). In doing so, he has provided us with an admirably detailed account of the organization of irrigated agriculture in a small village in the dry zone. Chapter 5 is based on excerpts from his larger study. Of special interest in the case of Pul Eliyah is the manner in which the irrigated lands have been arranged to facilitate the equitable distribution of water in times of scarcity. By dividing the total service area into several zones and then organizing each irrigator's land so that a portion of his holdings are located in each of these zones, the Pul Eliyah system is able to avoid many of the structural problems associated with systems in which some irrigators have all of their land at the head of the canal while others have all of theirs at the tail of the watercourse.

A less apparent, though critically important, point that derives from

Leach's work is the fundamental impact of the arrangement of land and irrigation structures on the patterns of organization that irrigators devise. The relationship is not highly determinate, but it appears that the ordering of the technical irrigation apparatus does set limits on the range of appropriate social forms.

Chapter 6, based on the work of Richard K. Beardsley, John W. Hall, and Robert E. Ward, discusses contemporary irrigation arrangements and reviews the developmental sequences preceding them in a small settlement in southwestern Honshu, the main island of Japan. Niiike has multiple sources of water for its fields, including several ponds (tanks) and access to a large canal system that diverts river water. In this setting, too, residential units and irrigation units overlap only partially. As the authors note, households in Niiike participate in several water-use communities that vary in size, though all are larger than the settlement itself.

The largest of these water-use communities is the Twelve-Go Irrigation Association, which provides water to users in thirteen villages and towns. It has a complex organizational structure, with formal and informal subgroups associated with specific service zones in the command area. It also is linked with appropriate units of local government, including various mayors. The canal system is large, technologically complex, and capital-intensive in its construction, and thus is operated and maintained through the shared responsibility of local units and higher levels of government.

In the Ilocos region of the Philippines are found the stable, multivillage irrigation communities locally called *zangjeras*. A contemporary discussion of the *zangjeras* has been provided by Henry T. Lewis in Chapter 7. In the larger study from which this chapter is excerpted, Lewis explores the continuity of fundamental Ilocano institutions and organizations as Ilocanos move from their original locations in the northwest corner of Luzon to the more expansive and less densely populated areas in the region of the Cagayan valley in northeast Luzon. Thus, his chapter discusses irrigation societies in Buyon (a location in the original homeland of Ilocos Norte) and in Mambabanga (a location in the newly settled region).

The internal organization of the *zangjeras* is variable across systems, displaying numerous seemingly idiosyncratic patterns. Two general characteristics are observable, however. First, *zangjeras* are nearly always composed of members from more than one village. They are discrete social entities, often interlocked with village entities (for example, through overlapping leaders), but separate from them. *Zangjeras* are named units occupying well-defined territories and coordinated by various norms enforced through specified authority roles.

Second, the *zangjeras,* because of their frequent proximity to one another and their dependence upon common natural water supplies, have established intricate inter-*zangjera* ties. These linkages recognize the basic interdependencies of the separate irrigation communities and regularize their ability to manage conflicts as well as mobilize scarce resources such as labor. Lewis's discussion of the *zangjera* is an important statement of the ability of local irrigation communities not only to operate and manage internal system processes but also to manage more comprehensive river systems through such organizational devices as federations and other intergroup compacts.

Highland irrigation is practiced in the Philippines in the well-known rice terraces of northern Luzon. Chapter 8 is Albert S. Bacdayan's report of irrigation activities in this general region. Several features observed in the previous readings are repeated in this Bontoc system. There are elaborate internal structures and patterns for operating the irrigation system. The use of the traditional *dap-ay* groups (organized subunits of the village) for the mobilization of labor and materials and for the discussion and dissemination of irrigation management decisions is significant. The actual allocation of the system's water is handled by a number of water distributors selected by the water users.

The focus of Bacdayan's report, however, is on the intricate exchanges that occurred between Tanowong and its neighboring villages as the people of Tanowong attempted to improve their water supply by extracting water from a new source. Ultimately, they dealt with the competing claims for this water supply by artful negotiations with their neighbor villages and opportune involvement of the central government for the confirmation of their water rights. The Tanowong case illustrates the point made earlier—that community systems are not unarticulated with the central government; the initiation and extent of the linkage, however, is often determined by the community.

While one can marvel at the intricate arrangements that local groups have devised both to organize their internal irrigation activities and to interact with competing systems, one also might wonder about the relevance of this local organizational ingenuity to contemporary problems of irrigation development. The last two chapters in this section are comments on this point.

In the latter part of the nineteenth century, as Michael Roberts explains in Chapter 9, British policy makers in Ceylon were concerned with the rehabilitation of the island's irrigation systems. They recognized that rehabilitation would require both improvements of the physical infrastructures and the creation of community-level institutions for their use and persistence. To deal with the communities, they instituted the Paddy Lands Irrigation Ordinance (1856), which provided for the operation of minor irrigation systems based on "ancient customs" and supervision by the village councils (*gansabhawas*). Furthermore, it was recognized that the "ancient customs" would vary from place to place, and in later versions of the ordinance, options for local organization were provided. These options included supervision of the local systems by an irrigation headman (*vel vidane*), by the village councils, or both.

As Roberts remarks, the success of the irrigation ordinance was judged by many as "patchy." Evaluations often were mixed because of the divergent yardsticks used. Particularly important were the contrasting judgments of those who looked for success in the improved physical state of the systems and those who sought success in the improvement of conflict resolution; the latter measure provided the more positive assessments.

Chapter 10 is an attempt to extract several organizational themes that tend to recur in community irrigation systems. The three themes discussed are the establishment of accountable leaders, the creation of functional subunits within the irrigation system, and the arrangement of irrigation organizations that conform with the physical design of the system.

This chapter also discusses situations in which knowledge of community irrigation systems can be useful to irrigation policy makers. One obvious situation is that which occurs when a nation's irrigation policy includes assistance to existing community systems. Without an understanding of the manner in which such systems are organized and operate, well-intended assistance may go awry.

A second instance is national policy that includes the development of new small-scale irrigation systems. Here an understanding of the existing irrigation organizations in the nation's sociocultural context will be of direct use in fostering organization in the new systems.

Third, as large-scale irrigation systems increasingly are being designed and constructed with extensive terminal facilities, relatively small service units are being created. In the organization of these small service units, consideration can be given to the use of a pattern of organization similar to that found in community systems that exist within a similar sociocultural context.

References

BEARDSLEY, RICHARD K. 1964. "Ecological and Social Parallels between Rice-Growing Communities of Japan and Spain." In *Symposium on Community Studies in Anthropology,* ed. Viola E. Garfield, pp. 51–53. Seattle: American Ethnological Society.

GEERTZ, CLIFFORD. 1972. "The Wet and the Dry: Traditional Irrigation in Bali and Morocco." *Human Ecology* 1:23–29.

HUNT, ROBERT C., and EVA HUNT. 1976. "Canal Irrigation and Local Social Organization." *Current Anthropology* 17:389–411.

MILLON, RENÉ. 1962. "Variation in Social Response to the Practice of Irrigation Agriculture." In *Civilizations in Desert Lands,* ed. R. B. Woodbury. Salt Lake City: University of Utah Anthropological Papers, no. 62.

4

Organization of the Balinese *Subak*

Clifford Geertz
INSTITUTE FOR ADVANCED STUDY

Bali

Bali is Indonesia's most famous island. As possessor of the only explicitly Hindu culture remaining in the archipelago, its temples have been photographed, its rituals described, its art analyzed, its psychological depths probed, and the beauty of its women praised by a long series of gifted ethnographers until today the anthropological literature on it is more developed than on any other part of the country.[1] But, for all this, aside from Korn's monograph on Tnganan (1933), a community as atypical as it is interesting, and a few scattered papers on so-called Bali Aga mountain settlements, detailed and systematic descriptions of particular villages from a sociological point of view are still lacking. It is toward the filling of this gap that the following brief description of a South Bali gong-making village is intended to contribute. Though something will be said about religious and artistic life, as breathtaking here as elsewhere on the island, the main emphasis will be on political, social, and economic structure.

Environment, House Land, and Field Land

In 1957–58, the year in which my wife and I studied Bali, the island's population was about 1.7 million, of which some 80 percent or so was concentrated in the southern heartland of Tabanan, Badung, Gianjar, Bangli,

Reprinted from Koentjaraningrat, ed., *Villages in Indonesia*. Maps, tables, and footnotes have been renumbered.

1. For references, see Kennedy's bibliography (1955, II, 544–67).

Klungkung, and Karengasem.[2] Geographically, this heartland consists of a sloping, volcanic hill mass running down from a complex of huge fifteen-hundred- to three-thousand-meter cones in the north center of the island and leveling out to a flat piedmont only within eight to sixteen kilometers of the southern coast. The whole area (some 3,450 square kilometers) is striated lengthwise by dozens of narrow, very deep-cut river gorges a few hundred meters apart, spreading out from one another like the ribs of an opened fan as they run downward to the sea. Settlements are strung one below the other down the slope, along the spurs between these gorges, separated by wet rice fields and coconut gardens. North–south—or up-down—relationships between settlements have consequently always been much closer than east–west ones, and each spur, drawing water from a common source, has tended to form a natural ecologic unit. The main traditional (i.e., pre-1906) court centers tended, in turn, to be located toward the lower ends of the more important of such spurs, but still well back from the coast at about the place where the hills begin to level out into plains, from which vantage points they vied ceaselessly with one another for control of peasant communities both above and below them. Some seven kilometers mountainward from Klungkung, one of the most illustrious of these ancient courts but today a sleepy small town and regional (*swapradja*) seat, lies Tihingan: population 720.

You cannot see Tihingan until, coming upward along the road from another settlement a few hundred meters across the rice terraces below it, you are already in its midst, for it is wrapped in a blanket of coconut trees. Even then you will not at first see very much—the people live concealed behind high brick and mud walls, which, lining the narrow, crisscrossed streets and pathways, give the forbidding effect of a repeating labyrinth. Halfway up the north–south road the monotony is relieved momentarily by an open-sided, vaguely Polynesian-looking meeting house (*balé bandjar*), a few temples (*pura*) also largely concealed by walls, a towering waringin tree, a small cleared space for petty trading, and a ramshackle bamboo-shed coffee stand. And then the unbroken walls—unbroken save for narrowed, screened-off doorways—begin again. A further climb of a few hundred meters and you are out of Tihingan as suddenly as you came into it; another few hundred and you are in another settlement, physically (but not socially) its twin. Except for a few people moving self-absorbing along the roadway, grooming fighting cocks in the shade of the meeting house, or chatting indolently at the coffee stand (and if it is midday, even these may not be present), you would hardly know, despite the frenzied yapping of the emaciated dogs, that you had been in a village.

Actually, whether or not Tihingan *is* a village depends upon how you

2. These estimates are based on 1954 election statistics data as given in I Gusti Gde Raka's monograph (1955, 10). Our study was supported by a grant from the social sciences section of the Rockefeller Foundation, administered by the Center for International Studies, Massachusetts Institute of Technology.

choose to define the term, for, as I have tried to show in more general terms elsewhere, the notion of a single, uniform social referent for this word (*desa*) cannot be defended when actual social relationships and not merely ideal cultural conceptions are considered (C. Geertz, 1959, 991–1012; 1961, 498–502). Tihingan has an origin temple and a death temple of its own (Pura Puseh, Pura Dalem), but it shares its third main temple (Pura Balé Agung) with the settlement below it; it is the headquarters of a man the government (and, on occasion, the population) refers to as a ''village head,'' but his bailiwick includes three other settlements besides Tihingan; the rice fields of its inhabitants are scattered around among several nearby irrigation societies (*subak*), mixed in with those of people from dozens of other communities; and so on. But the people of Tihingan call it a village, as do their immediate neighbors, and for practical purposes this will perhaps suffice. It must not be assumed, however, that the social entity referred to as a village in other Balinese communities will necessarily be the same sort of entity Tihingan represents. For all the visual monotony of its settlements, Bali's social structure is a complex of contrasts, a tissue of irregularities.

The definition of the ''village'' of Tihingan (or of any Balinese ''village'') is further complicated by an odd, peculiarly decisive structural fact: unlike peasant communities in most parts of the world, there is not in Bali any simple enfoldment of both house land and field land into a single corporate or semicorporate unit. On the contrary, a sharp distinction is drawn between the settlement unit, the *bandjar,* or hamlet, and the agricultural unit, the *subak,* or irrigation society. The organization and regulation of daily social life is severed from the organization and regulation of cultivation; though the hamlet and irrigation society are built on similar principles, they are built separately and function autonomously. We have two sorts of customs, the Balinese say: dry ones for the *bandjar* and wet ones for the *subak*.

This division can be seen in ecological terms if one looks to Map 4.1, which depicts a segment of the Tihingan spur running from about 75 meters above sea level to around 300. The various *subak* line either of the bounding river gorges, growing generally larger as one moves down the slope—a simple function of topography. The various *bandjar* are strung up the center of the spur along a dirt, but usually passable, road branching off from the main east–west Dutch-built highway which now connects the main court-center towns. Though there is a natural tendency (but only a tendency) for the land of people living in any particular *bandjar* to be located in the more near-at-hand *subak,* there is no unique or well-defined tie between any given hamlet and any given *subak*. From the hamlet side, land is spread out through a number of different irrigation societies; from the *subak* side, land is owned by members of a number of different hamlets. To get a perspective image of Balinese ''village'' life one must take, therefore, a stereoscopic view, looking at it on the one hand through the lens of the *subak* and on the other through that of the *bandjar*.

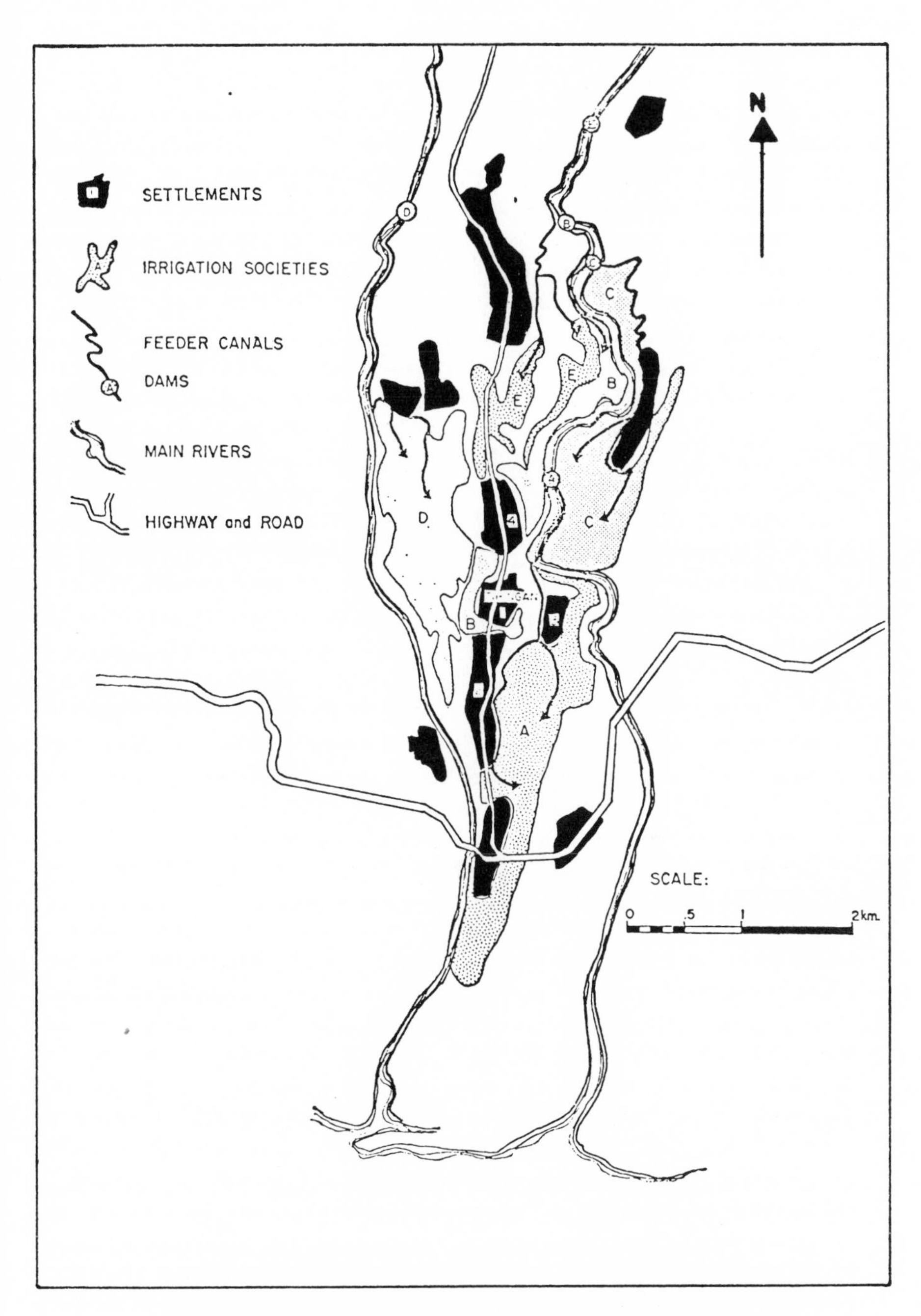

Map 4.1. *Bandjar* and *subak* in the Tihingan area (simplified). From Peta Pengairan, Swapradja Klungkung, Djawatan Pengairan, Den Pasar, Bali.

The Settlement Unit

In consisting of only one *bandjar,* Tihingan (see Map 4.2) is somewhat atypical for a Balinese settlement cluster. The cluster to the immediate south consists, for example, of three; that to the immediate north, of four. As a public corporation, the *bandjar* is centered in the *balé bandjar* meeting house, where once in every 35-day Balinese month a meeting of the male heads of all the independent families of the hamlet—some 138 people—is held. All such family heads must attend (or, if ill, send a substitute) on pain of fine, and all official policy decisions for the *bandjar* are taken at such meetings, entirely by unanimous decision. The meeting—and the *bandjar*—is led by the *klian bandjar* (literally, hamlet elders), of whom there are here five, chosen also on a consensual basis by the meeting for five-year terms, at the end of which they may not be reelected but must nominate their successors in order to avoid electioneering. The rights, responsibilities, and limitations of the scope of the meeting's authority are strictly defined in a written "constitution" (*awig-awig bandjar*) inscribed at some forgotten time on palm leaves (*lontar*).

As a social institution, the *bandjar,* or the *bandjar* government, is but one corporate unit among many others in the village, and its powers, though great, are neither unbounded nor all-pervasive. Its major powers focus around collective worship, public works, and civil order. The *bandjar* as a whole is responsible for maintaining and repairing the origin and death temples and for carrying out the elaborate ceremonies which are performed in them every 210 days. The *balé bandjar* must also be kept up by the hamlet, as must various other public facilities: the house in which the statues of the gods are kept, the sheds in the small market place, the graveyard, the roads and paths, and so on. In 1957–58 a school building—located in the next settlement southward—was built by hamlet labor and money in cooperation with several neighboring *bandjar*. The hamlet membership is responsible for night guard and must respond to alarms for fire and theft. If any hamlet member is holding a cremation (an increasingly rare, but still occurring, event), each family unit must give ten days' work toward its preparation to the family involved, and at death itself the hamlet as a body is responsible for making the death litter, digging the grave, and sitting all night with the bereaved.

All land in the *bandjar* belongs to it as a whole, and when house sites or coconut groves fall vacant they are allotted by the meeting to certain families, the rules and regulations concerning such matters being very precise and very complex. The *bandjar* owns a *gamelan* orchestra, as well as some dance costumes, which members of the *bandjar* may use, contributing a fixed share of any income gained from performances to the hamlet treasury. It also has the right to plant and/or harvest, as a group, the fields of any member of the village, the one-twelfth of whatever crop-share payment for this work becom-

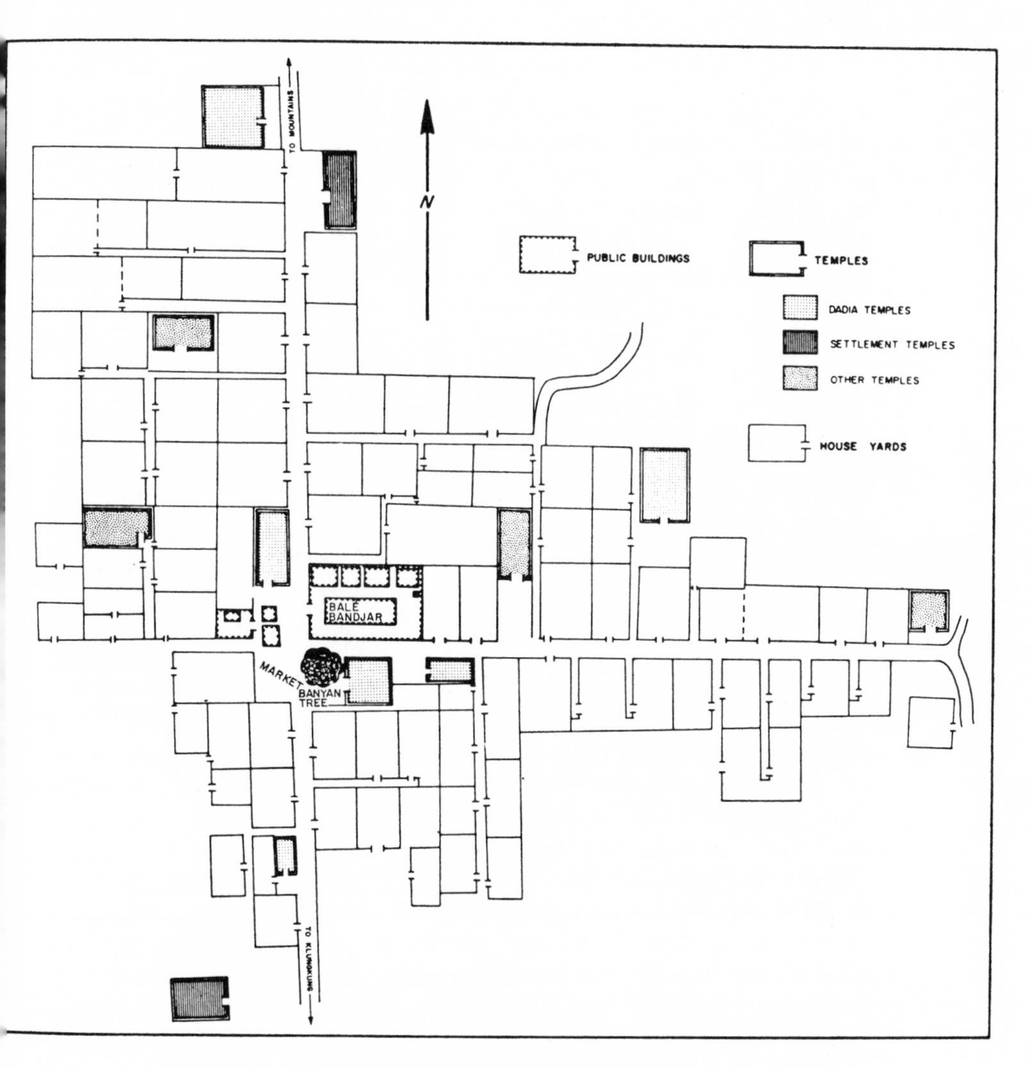

Map 4.2. Settlement pattern of Tihingan.

ing part of the public fund also. It even owns a one-third interest in a hand cart (the other two-thirds are owned by a private coconut-picking group in the village and by a single rich man), which is rented out to whoever may need it for transporting goods.

All important legal matters are made official by being announced to the assembled household heads: marriage, divorce, inheritance, transfer of property (except rice land), and so forth; and oaths—a very serious matter—are sworn before the same body. It has important sanctions with respect to crimes, ranging from fines all the way up to total exclusion from the hamlet. (A few years ago, in a hamlet a few kilometers from Tihingan, a man was even stoned to death at a *bandjar* meeting for certain improprieties.) And finally, the *bandjar* sponsors dance and drama productions, hiring outside groups from the public treasury, holds cockfights, and conducts special purification rituals in the case of epidemics, social upheavals, crop failures, or other disasters. All in all, the Tihingan *bandjar,* like most of those of Bali, seems hyperactive. It is always planning something, building something, debating something, performing something.

Actually, this heavy burden of public duty is somewhat lightened so far as the individual family is concerned by dividing the *bandjar* on the one hand into two halves and on the other into three thirds on a simply territorial basis (see Map 4.3). For very large jobs—for example, repairing the road or building a new wall around a temple—the whole hamlet will be used as a unit in a mass attack on the task. For others—such as preparing offerings for the origin temple—only one "half" will be used, the other "half" being used the next time around. And for smaller, more routine tasks—grave digging, for example—only one of the thirds will be activated, in a similar rotation system. The result is a very complex and shifting pattern of almost continual public activities regulated by the several *klian* under the general policy direction of the meeting. And as these territorial divisions coincide, as we shall see, with no other social grouping in the hamlet, but have been in fact deliberately designed to crosscut them, they act, as does the *bandjar* government in general, as an important integrative mechanism in this otherwise faction-prone village.

Citizenship in the hamlet (or, as the Balinese put it, in the *adat*—roughly, the legal community) always involves an adult man and woman, bound together as an indivisible unit, because there are two kinds of work to be performed: men's and women's. Usually when a young man has been married for about a year he and his wife "go down into the *adat*" (*tedun ke adat*) per decision of the hamlet meeting. At divorce or at the death of one of the pair, those involved will usually leave the *adat* because, now only a half person, so to speak, they are no longer competent to carry out the duties involved. However, several other arrangements are possible for adults who, for one reason or another, lack spouses. Unmarried adult brothers and sisters, a widow and her son (if he is at least adolescent), even two unrelated people living, for some reason, in the same compound, may pair up to enter the *adat.* When a couple's youngest son is married and enters the *adat,* they may, if they wish, leave it—a process called *ngarepan panak,* "pushing one's child forward." But even though this means an escape from a good deal of labor and other obligations, most people are reluctant to take advantage of it; "to leave the *adat* means to lie down and die," and there is therefore a tendency to stay in as long as physically possible, even though the work is explicitly not

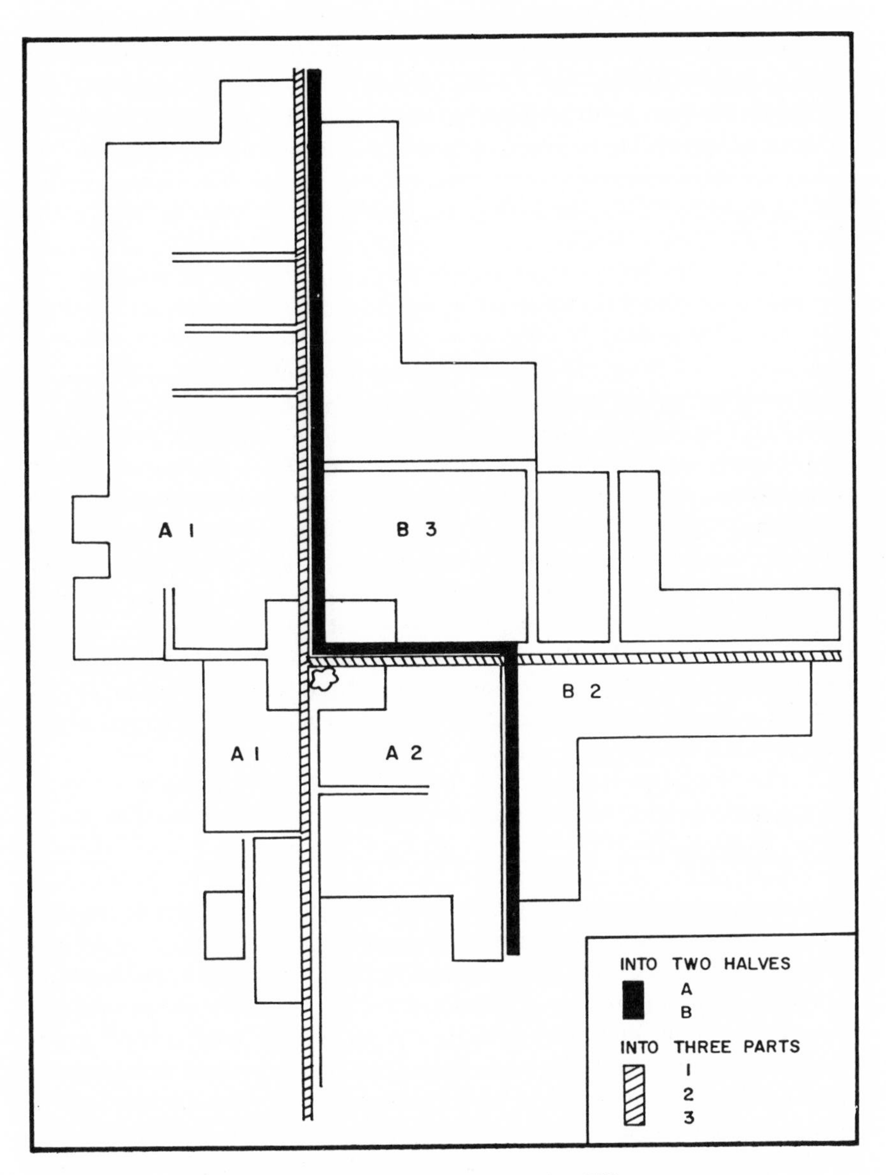

Map 4.3. Divisions of the *adat bandjar*, Tihingan.

lightened in deference to age or infirmity—a social policy toward senescence upon which a good many of the aged seem, indeed, to thrive. In any case, the parents' remaining in the *adat* does not prevent their married youngest son from entering himself.

Beside the five *klian,* there are a number of other officials chosen by the hamlet (secretaries, messengers, etc.), but the administrative apparatus is in general quite rudimentary and the *klian* follow the public will rather more than, at least as *klian,* they lead it. The actual political process, here a very lively one, revolves more around kin groupings, "caste," class, and, latterly, education and national parties than it does around the formal governmental structure. The hamlet is a legally very carefully defined arena (the "constitution," a sacred as well as a secular object, runs to more than 6,000 precise words), within which the political factions of the hamlet compete rather fiercely for influence, wealth, and, especially, prestige.

Finally, superimposed over the top of this whole, very localized, political system (Tihingan is 70 percent endogamous, the three neighboring settlements 77, 80, and 95 percent) is a central government-appointed "village" bureaucracy headed by a salaried official called *perbekel.*[3] His domain (called a *perbekelan)* includes not only Tihingan but the immediately surrounding settlements as well (numbers 2, 3, and 4 on Map 4.1). In each settlement there is a *pengliman* appointed to assist the *perbekel.* (In Tihingan this is at the moment one of the present *klian,* but such a coincidence is not inevitably, not even usually, the case.) The main function of this miniature bureaucracy is to relay information, regulations, exhortations, and commands from the central bureaucracy to the populace—that is, to talk rather more than to listen—and its role in village life remains, as a consequence, rather limited.

The Agricultural Unit

Each major drainage of Bali—running again lengthwise from the mountains toward the sea—is considered by the population to be a single self-contained ecological unit. This unit, a long, narrow strip centered around a larger river, is called a *sedahan,* after a traditional tax collection official now transformed into a government functionary, and over all such units in the general region where once a local court held sway and today a regency (*swapradja*) government does, there is a *sedahan agung* (great *sedahan*), a chief tax collector. Each *sedahan* is broken down into a large number of *subak,* and these are the fundamental elements of the whole system. The term *subak* is commonly translated as "irrigation society," because of the central role this institution

3. In making comparisons I will, for the sake of simplicity, compare Tihingan, in which the *bandjar* and the settlement happen to be identical, with settlements (i.e., nucleated residential units) rather than neighboring *bandjar,* though the latter sort of comparison would be more proper technically.

plays in the regulation of water supply. But the *subak* is in fact very much more: an agricultural planning unit, an autonomous legal corporation, and a religious community. Aside from house gardening, virtually everything having to do with cultivation lies within its purview. Effective power with respect to agricultural matters lies and seems always to have lain in the *subak;* it is thus not a mere appendage of the larger order units, which as political entities are hardly more than tax districts. Theories of "hydraulic despotism" to the contrary notwithstanding, water control in Bali is an overwhelmingly local and intensely democratic matter.

A *subak* is defined as all the rice terraces irrigated from a single dam (*empelan*) and major canal (*telabah gde*). All individuals owning such land (with some minor exceptions, it is all in freehold tenure) are citizens of the *subak,* or *krama subak,* just as all those living on the land of a *bandjar* are its citizens, or *krama bandjar*. As there are *klian bandjar,* so there are *klian subak;* as there are bandjar meetings, so there are *subak* meetings; and as there is a *bandjar* legal code, or constitution, so also there is a *subak* one. Public obligations enforceable by fines, regulations concerning land use, legal transactions having to do with land transfer, collective ritual for group ends (such as fertility)—all these fall within the domain of the *subak* government as their counterparts fall within the domain of the *bandjar* government.

The ecological and engineering details of the *subak* irrigation system are far too complex to be described in any fullness here. The main feature is the one-dam-one-*subak* relationship. As can be seen from Map 4.1 the dams are arranged one below the other down the river canyons, a single canal, usually of some length, carrying the diverted water to the *subak,* often with the aid of overhead aquaducts or long tunnels. After the water arrives at the *subak* it is successively partitioned by an extended series of carefully graduated bamboo water dividers distributed over the whole area in such a way that what was a single broad incoming artery veins out into dozens—in larger *subak,* even hundreds—of small rivulets directly feeding one terrace or a small group of them. This final unit of water, and the amount of land it irrigates and the amount of rice seedlings needed to plant that land, *and* the amount of (unthreshed) paddy harvested from it are known as one *tenah*. Thus, the sum total of *tenah* in a *subak* adds up to its total water supply, to its total area, to its total rice-seed demands, and to its total rice production—depending upon whether you interpret the *tenah* in its water, areal, seedling, or rice-harvest meaning. For any one *subak* the number of *tenah* is fixed by the concrete pattern of successive water division whose form is determined by the *subak* as a corporate group (for the most part relatively few changes in the established pattern are introduced, however), but between various *subak* it will of course, differ. *Subak* B in map 4.1 contains 160 *tenah,* making a single one about 0.45 hectare in areal terms; *subak* A has 530, making a single one about 0.30 hectare areally. Also, as the total water supply varies from *subak* to *subak* according to ecological factors (in general, higher ones tend to be better off), the size of a *tenah* in water also varies. The number of cubic feet per year in a *tenah* may—and commonly will—differ between two *subak* even if their number of *tenah* is the same.[4] The *tenah* is fundamentally a water unit, sliced out of the entire water supply available from the *subak's* main dam and canal by a fixed pattern of successive divisions of flow, and thus varies as that supply and that pattern vary. It is a maddeningly irregular system for one who wants to make any sort of analysis, but a remarkably ingenious, just, and effective one so far as the peasants are concerned.

The *tenah* becomes, then, the basic unit for *subak* taxation (labor or monetary), agricultural planning, and land transfer. Though quite aware that metric units are more comparable between *subak* than *tenah* ones, the peasant nonetheless thinks of his landholdings entirely in *tenah* terms. Inheritance divisions and tenancy grants are expressed in *tenah,* agricultural wealth is calculated in terms of *tenah* (though, naturally, with rule-of-thumb adjustments with respect to their location, etc.), one's rights and obligations within the *subak* are expressed in *tenah.* The one exception is in voting. In determination of *subak* policy and election of *subak* leaders, each owner is legally entitled to a single vote no matter how many *tenah* he holds. In fact, of course, large holders tend to carry rather more weight for the usual economic and social class reasons. In the majority of cases, however, what is remarkable is not how much wealth differences account for but how little. If anything, the *subak* organization as such tends to act against the interests of the well-off rather than in their favor; to counter the political forces, real enough in themselves, stemming from economic inequality. Like the *bandjar,* the *subak* is deliberately blind to any other basis of social status but membership in itself, and like the *bandjar* it is ultimately sovereign.

As mentioned, the *subak* head, elected by all the members of the *subak,* is called, at least in Tihingan, the *klian subak.* Under him are a series of (also elected) *klian tempek.* A *tempek* is a territorial (that is, water) subdivision of the *subak* but is in no way independent of the whole; its role is merely administrative. In *subak* A there are four such *tempek;* in *subak* B, seven. Below the *klian tempek,* and under their direction, are a large number of men called *pekaseh,* a term that in some parts of Bali is synonymous with *klian subak* but in Klungkung denotes a member of the "water group," the *seka jeh.* In *subak* A, 160 of the 455 members are *pekaseh;* in *subak* B, 60 of 222. It is this group that, as its name implies, performs the actual tasks of water regulation and irrigation system upkeep.

4. Owing to intra*subak* ecological differences, custom based on assumed precedence in clearing, and other factors, the areal extent of a *tenah* may vary slightly within a single *subak* also. And as field fertility naturally varies somewhat, a *tenah* as a volume of rice also, of course, varies—but generally not too widely.

These tasks range from the constant clearing of the small field canals and minor repairs on the water dividers, and the daily, even hourly, shifting of channels of flow that is necessary to make the whole system work, to the guarding of the fields at harvest to prevent theft. The members of the *subak* as a whole are taxed, according to number of *tĕnah* owned, for these services, the rate differing from *subak* to *subak* but being, in any case, far from nominal. (In *subak* A, for example, it is 30 *rupiah* a season, in *subak* B, 20 *rupiah*.) The proceeds from this "water tax" (*pengot*) are first applied to general *subak* needs, and then any remainder is divided among the *pekaseh*, who are, in any case, excused from one *tenah's* worth of tax. For performance of the unending everyday tasks, the *pekaseh* usually serve a daily "watch" in pairs. Thus, in a *tempek* in *subak* A, for example, where there are twenty-four *pekaseh*, each *pekaseh* is "on duty" with a fixed partner one day in every twelve. Once every two weeks or so the entire twenty-four are mobilized for some piece of heavier work, and for even larger tasks—such as the clearing of the main canal or repairs on the main dam—all the *pekaseh* of the whole *subak* may be mobilized. Finally, if the task involved is very heavy, the whole *subak* membership may be called upon, but this is quite rare. The role of the *pekaseh* is so great that people often refer to them as the "*subak* members" without qualification, although everyone is aware that legally this is not the case and in important policy decisions (e.g., adding new terraces to the *subak*) all owners, *pekaseh* or not, have a right to participate and usually do so.

There are a number of other important aspects of *subak* organization, but they can only be mentioned here in passing. On the one hand, there is the obvious problem of inter*subak* coordination. By and large this is accomplished through a process of collective bargaining among the various *subak* themselves, under the governance of an explicit code of customary law (*adat*) to which all in one drainage adhere, and not by any administrative dictation from above. Rules for the amount of water that must be allowed at all times to bypass any dam, patterns for the borrowing and lending of water rights, and ritual systems regulating the order of planting (higher *subak* plant earlier than lower ones so as to stagger peak water requirements over time) are all firmly established and strictly observed. On the other hand, there is a complex intra*subak* ritual pattern centering around two sorts of temples, one (*pura mastjeti*) set in the midst of the field and dedicated to the goddess of fertility and one (*pura ulun suwi*) set near the main dam and dedicated to the god of water. In addition there are the usual Malaysian rice-mother rites carried out by individual farmers, seasonal-round ceremonies for each stage of (rice) cultivation usually carried out by all the *pekaseh* as a group for the whole *subak*, and occasional mass processionals down to the sea by the entire membership to cure a crop epidemic or relieve a mice plague, for example. This developed ritual pattern, matching with fine precision the actual flow of agricultural activity, is one of the major regulating mechanisms in the whole, marvelously intricate ecological system the *subak* represents.

The extent of the intricacy is perhaps most clearly dramatized by the so-called *masa* (roughly, month, time) system of planting in effect in the various *subak* around Tihingan. There are four such *masa*—the "eighth," "sixth," "fourth," and "second," named after the Balinese months (how *those* are determined we had perhaps best ignore here)—in which rice is planted in each system. Table 4.1 compares the four systems (in a very oversimplified form). Now, within any one *subak* some *tempek* will, per *subak* decision, follow the "eighth" system, some the "sixth," some the "fourth," and the others the "second," though all of these may not be in force at once. The point of this arrangement is, again, to stagger the peak water demands, this time within one *subak* rather than among a set of them. In *subak* A, for example, there are seven *tempek,* one of them on the "second" system, two on the "eighth," and three on the "sixth"; the seventh *tempek* alternates from year to year, as a sort of wild card, among the "sixth," "second," and "eighth" systems, depending on what the water situation happens that year to be. *Subak* B, on the other hand, is divided right in half, two of its four *tempek* being on the "fourth" system and the other two on the "eighth." The details of all this—how it is determined who will follow which system, the means of enforcement, comparative advantages of the various systems from the ecological point of view, and the ways in which compensations are made—need not be pursued here. But it should be clear that the control of cultivation around Tihingan is fairly exact. In fact, as—somewhat in contrast to most parts of Bali—the *subak* as a whole often decides even which dry crops are to be planted at which times, the control is exact indeed. To call the *subak* merely an irrigation society is thus to underestimate its scope rather seriously.

Turning to the social composition of the various *subak* around Tihingan, the first point that has to be reemphasized is the lack of any straightforward correlation between the place of a person's residence and the location of his rice fields. Of course, most of his fields will be somewhere reasonably nearby, but they will not as a rule be confined to one, or often even two, *subak.* Thus residents of the *bandjar* of Tihingan own land in significant amounts in five different *subak* lying both to the north and the south of them, as well as having scattered holdings in several others more distant. "Large" landowners (i.e., those with between 1.5 and 3.5 hectares) all have land in more than one *subak,* and most have some in three or four. *Subak* A (see Map 4.1), in which more Tihingan land is located than any other, only contains 47 percent of the total land owned by residents of the *bandjar.* And, to put the matter the other way around, this land accounts for only 5 percent of the total land in that *subak.* Altogether, no fewer than thirty-eight *bandjar* are repre-

Table 4.1.

The *masa* system

Masa system	Month											
	1	2	3	4	5	6	7	8	9	10	11	12
"Second"		plant rice					harvest rice	plant dry crops				harvest dry crops
"Fourth"		harvest dry crops		plant rice					harvest rice	plant dry crops		
"Sixth"				harvest dry crops		plant rice					harvest rice	plant dry crops
"Eighth"	harvest rice	plant dry crops				harvest dry crops		plant rice				

Table 4.2.

Location of land owned by members of Tihingan *bandjar*

Subak	Percent of land
A	47
B	21
C	7
D	13
E	10
other	2

sented in the membership of this *subak,* and the *bandjar* with the largest amount of land in it owns only about a quarter of its total. This dispersive pattern is summed up in Tables 4.2 and 4.3, in which the scatter of holdings is depicted from both the *bandjar* and *subak* perspectives.[5]

As Tihingan is basically an artisan *bandjar,* its relations with the realm of the *subak* are actually somewhat anomalous. In the first place, the proportion of its members holding wet rice land is comparatively low: of the total 138 family heads, only 46 or 30 percent own rice land, as compared to 47 percent, 67 percent, and 72 percent in settlements 2, 3, and 4 respectively. In per capita terms, this works out to 0.19 hectare in Tihingan, against 0.24, 0.32, and 0.53 hectare for the neighboring settlements; so that, in gross terms, Tihingan is not in any sense a land-rich village. So far as rice land *per owner* is concerned, however, Tihingan is around, or even slightly above, the local average: 0.61 hectare against 0.50, 0.59, and 0.73—a fact that indicates, of course, that in Tihingan one tends either to have a significant amount of land (though the term ''significant'' is a most relative one: even the largest holder has only 3.3 hectares) or to have none whatsoever, in contrast to the more graded distributional patterns characteristic of the other settlements. And finally, those Tihingan people who do own land—mostly the more accomplished smiths—tend not to work it themselves but to give it out in various sorts of tenancy arrangements, mostly to members of surrounding *bandjar*. In Table 4.4, which compares the percent of *bandjar*-owned land worked by members of the *bandjar* as against the percent worked by nonmembers, Tihingan's role as a ''landlord village'' is quite apparent.

The tenancy patterns through which the relations between landholder and tenant (*sakap*) are actualized (wage labor in agriculture is very rare around Tihingan) are, again, complex in the extreme. There are four main sharecrop systems found in Bali: *nandu,* under which the tenant receives half the crop; *nelon,* under which he received two-fifths; *ngapit,* under which he receives one-third; and *merapat,* under which he receives one-quarter. In the Klungkung

5. Were the percentages to be adjusted to *bandjar* population on the one hand and *subak* areas on the other, the picture would be even more striking.

Table 4.3.
Distribution of land ownership in *subak* A

Bandjar	Percent of *subak* A land
Most-represented	24
Tihingan	5
36 other	71

region the *ngapit,* or one-third, system is standard; but, as Balinese only maintain standards so as to be able to vary them, the actual arrangements differ almost from case to case. Social relations between the contracting parties, location and quality of the land, type of crop involved, current economic conditions, source of such capital as seeds and cattle, and sheer love of complication combine to produce a range of tenancy institutions whose adaptability to particular circumstances is endless.

Aside from the basic two-to-one crop-sharing pattern, three other institutions regulating access to land ought to be briefly mentioned. The first is the pawning or *gadé* system. Under this practice the owner of a piece of land surrenders it to another man for a cash sum, but the title is not transferred. Instead, the pawn taker works the land until the owner is able to repay the loan, at which time the land returns to the control of its proper owner. The amount of the repayment is the same as that of the original loan; the lender's interest is considered to consist of the returns from the land while he has it in his possession. In many cases (about 80 percent of those in Tihingan and the three surrounding settlements), the owner will only consent to pawn his land with the express understanding that he will be employed as a *sakap* sharecrop tenant on that land for the duration of the pawn, in which case the amount of cash involved is much smaller. Again, as a relatively well-to-do *bandjar*—or rather, as a *bandjar* with a number of well-to-do individuals in it—Tihingan is primarily a giver of money and a receiver of land in *gadé* transactions. Over all, it shows a net gain of seventeen *tenah* of land (or about 12 percent of the total amount of land owned by its residents as such) by means of the *gadé*

Table 4.4.
Owner-tenant percentages

		Percent of tenant-worked land	
	Percent of owner-worked land	Inside *bandjar*	Outside *bandjar*
Tihingan	11	36	53
Settlement 2	73	26	1
Settlement 3	57	36	7
Settlement 4	63	27	10

process, as against net losses of fifteen, seventeen, and nineteen *tenah* for settlements 2, 3, and 4.

Second, there is *plais,* a kind of key-money system with respect to tenancy. The man who wishes to become a tenant must "lend" a fixed sum of money to the owner at the outset of the relationship. Usually this sum is returned in full to the tenant when and if the tenancy is ended; less often the principle slowly depreciates over time until after a number of years the owner's debt is liquidated entirely. Owners claim that the *plais* system—particularly the nondepreciating sort—protects the tenant's rights, for the owner will not usually wish, or often even be able, to repay the loan, as he must do if he displaces the tenant. Tenants claim it is mere extortion by landlords. As land is short, the landowner may accept a higher *plais* from another bidder, expel the tenant already on the land, and pocket the difference between the old *plais* and the new one. This can be a quite vicious system once it begins to roll; but such auctioning happens less often than one might expect, in part because it is so universally disapproved of and in part because the reaction of the tenant may well be violent. In any case, not all tenancies involve *plais* (from the government's point of view, it is illegal), for one does not demand such payments from relatives or close friends; and the degree to which it is protective or oppressive so far as the tenant is concerned varies a good deal from case to case.[6]

Finally, there is a very popular system in the Tihingan area—and particularly in Tihingan itself—called *melanjain,* in which the tenant is responsible for growing dry crops but not rice. He receives one-half of the dry-crop product, but after harvesting it he must prepare the field—plot it, etc.—for rice. The owner then takes over, and the tenant receives no share of the rice crop at all. The *melanjain* system is considered to be more a "worker" than a "tenant" system, for what you are really doing is hiring a man's labor and, especially, his cattle for the preparation of one's rice field, giving him half the dry-season crop as payment. Only poor men will accept *melanjain* contracts, and the system is generally regarded as a means by which people who are not themselves farmers can nevertheless avoid sharing their rice returns with a tenant. In fact, as planting, weeding, and harvesting are all commonly done by *seka* of one sort or another—*bandjar,* kin group, voluntary—one can, by using the *melanjain* system, "cultivate" a rice field without ever actually doing any serious work in it oneself. For Tihingan smiths, busy in their forges, such an arrangement is very attractive; though many Tihingan people do themselves farm and some of the poorer ones even serve as tenants or *melanjain* workers.

6. The very existence of *plais* might be held as evidence for the fact that share rents are, in purely economic terms, "too low." And, to maintain perspective, it must be remembered that a landlord here will be a man holding one or two hectares of land and may well be himself a tenant on someone else's hand.

All these patterns, and a large number of others, are combined in various ways to yield a dense network of land rights and labor obligations that is, of course, at the same time a social network. Sharecrop arrangement, *plais* payment, pawn contract, *melanjain* relationship—all have different meanings in terms of what they imply with respect to economic class, mutual obligation, and social alliance. In the context of "village" life generally, the ties between economic patron and economic dependent growing out of the structure of productive organization in agriculture entwine with those growing out of residence, age, title, kinship, and personal friendship to join *subak* and *bandjar* together, not into a single, bounded, self-contained unit but as strands in a web of social interconnections that spreads out in countless crisscrossing lines and in all directions over the entire countryside.

The Temple System and Social Integration

The begetter of order in this otherwise rather particulate social field is the temple system. Arguments from functional potency to functional necessity are both empirically dangerous and logically suspect, but it is nonetheless difficult to see how a social system of the Balinese sort could possibly operate without something very much like the temple system to give it form and outline. A society consisting of a multiplicity of overlapping groups, each directed to a distinct and fairly specific end—a pattern I have called elsewhere "pluralistic collectivism"—would seem to need some ritual expression of the elemental components of its structure in order to maintain a level of conceptual precision sufficient to permit its participants to find their way around in it. The temple system provides both a simplified model of Balinese social structure and a schoolroom in which the kinds of attitudes and values necessary to sustain it are inculcated and celebrated. Without it, it seems certain that Balinese society would be a good deal less intricate and a good deal less interesting.

On the other hand, all Balinese temples are pretty much alike. The architecture—the split gate, the walled courtyard, the small altars to various gods and godlings—is about the same from one temple to the next; only the scale and, to some extent, the particular altar vary. The ceremonies—the invocation of the appropriate gods of the temple's "birthday," the presentation of offerings, the performance of obeisance rituals by the (usually hereditary) temple priest (*pemangku*)—differ mainly in their degree of elaborateness and in the number of people caught up in them. But yet, for all this outward similarity, the temples differ sharply with respect to the purposes to which they are dedicated and the social composition of the congregations they serve. There are house-yard temples to honor direct ancestors; descent-group temples to proclaim the importance of the patriline; *subak* temples to ensure fertility and an adequate water supply; "state" temples to legitimize traditional patterns of political loyalty; "title" (or "caste") temples to uphold the privileges of rank; and special "voluntary" temples to mark a holy place, commemorate a historical event, or fulfill a personal vow. Each of these temples marks out the boundaries and stresses the significance of one or another sort of tie, whether of kinship, ecology, status, or whatever, and taken together they sketch out a general map of the social terrain through which, in his secular life, the Balinese moves. In these terms, perhaps the most important landmarks of all are the so-called "Three Great Temples" (Kahyangan-Tiga)—the origin temple (Pura Puseh), the death temple (Pura Dalem), and what can be translated, a little awkwardly, as the great council temple (Pura Balé Agung).

In Tihingan, the origin temple is located at the extreme northern edge of the village, the death temple at the extreme southern edge (see Map 4.2), a pattern that is, however, in no way standard. In the every-210-day ceremonies (called *odalan* and falling at different times for the two temples), the entire *bandjar* participates, for the origin temple commemorates the first settlement in the area and is dedicated to the general welfare of the inhabitants, and the death temple is concerned with honoring the dead and appeasing the divinities of evil and destruction. It is not always, however—not even usually—the case that the congregations of an origin or a death temple coincide with the citizenry of a single *bandjar,* or even with one another. In most cases, several *bandjar* are included in one congregation, and sometimes congregation lines actually cut across *bandjar* lines, dividing the citizenry with respect to temple allegiance. The ceremonial pattern carried out in such temples has been described several times in the literature, and the Tihingan pattern shows no significant deviations from the norm—there is the same joyous reception of the visiting gods, descendent from the holy mountain, Agung, in the center of the island; the same Hindu-style ritual obeisance under the leadership of the *pemangku;* the same sad, muffled-drum departure of the gods after three days; and so on. The famous ritual combat between the witch Rangda and the dragon Barong which is associated with death-temple *odalan* in many places in Bali is absent here, because, the villagers say, the king at Klungkung took their Rangda and Barong paraphernalia away at some distant time in the past for rebellious activity; but in settlement 2 there is such a combat, known over much of Bali for its fervency, which virtually all Tihingan people attend.

As for the third of the great temples, the great council temple, it is, as noted in section 2, shared with settlement 2 and is, in fact, located midway between them. This temple has an *odalan* not every 210 days but only once in a lunar year and is, rather vaguely, dedicated both to the fertility of the fields and the vigor of the *bandjar*. (Not only members of the two *bandjar* which "carry" it participate in its *odalan* but, secondarily, representatives from the various surrounding *subak* as well.) It thus relates, in symbolic terms, the, in structural terms, disjunct entities of hamlet and irrigation society and expresses the general unity, or at least the continuity, of what I referred to as the extended field of village life.

From the social point of view, what is perhaps most important to emphasize is the extraordinary preparatory effort that goes into the making of an *odalan*. The Pueblo Indians of the southwestern United States have often been remarked as considering the really crucial part of a ritual the extended and meticulous preparations leading up to it; and, though the Balinese do draw great rewards from participation in the ceremony itself, the same sort of thing is true of them. For literally weeks ahead of time, the members of the congregation are preparing offerings, repairing and decorating the temple, discussing who will do what, and practicing *gamelon* pieces, so that by the time the ritual itself comes it is something of an anticlimax, or at least hardly more than the final phases of preparation. During the period prior to the *odalan* proper, the intense cooperation among female members of the congregation (who make most of the offerings) and male ones (who do most everything else) clearly acts to bind them together into a whole. In fact, given the great number of temples in the area (all of whose *odalan* usually fall at different times), Tihingan people seem to be always either preparing for a temple ceremony or cleaning up after one, and the integrative force of this continual collective effort, as it moves from one social context to another, is the linchpin of the entire system.

The Balinese Village

Tihingan shares with the other "villages" of Bali not so much a specific concrete form as a set of characteristics, general principles of social organization, which, though they work out variously in different places—even, as we have seen, in immediately adjoining settlements—are nonetheless common over the whole southern heartland area. Among the more important of these are the following:

1. There is a marked tendency to perform particular social activities in groups specifically designed for those activities and for them alone. The view of a peasant village as a functionally diffuse, all-purpose social form does not, whatever its value may or may not be elsewhere, apply at all in Bali.

2. There is a related and equally marked tendency to deny all other bases of social demarcation save those immediately relevant to the situation or task at hand. In a *gamelan seka* it doesn't matter what your title is; at a *bandjar* meeting it doesn't, at least legally, matter what your kin group is; in a *subak* it doesn't, or at least shouldn't, matter where you reside.

3. As a corollary, social activities tend to be relatively "pure" in nature: status (or, better, rank) relationships are a matter of pure prestige; economic

relationships of pure wealth; political relationship of pure power; religious ones of pure ritual. The various systems interact and influence one another significantly, but they never (or almost never) fuse.

4. There is a very sharp line between the worlds of private life, conducted behind the house-yard walls (and which I have not been able to describe in any detail here), and that of public life, which goes on outside of them. Public obligations are heavy, unevadable, but nonetheless strictly limited, and there are specific mechanisms (also not described here) for adjusting between them.

5. Activity and elaborateness are valued for their own sake. A Balinese "village" is a very busy place, and the complexity of the ways in which people are, even in formal terms, related to one another is staggering. If one were to apply stylistic categories to social structures—probably not too good an idea—the Balinese would surely be classed as rococo.

6. The crisscrossed maze of specific, strictly defined, highly autonomous, and largely collective social ties that all this subtlety and industry produce is held within a general form by the temple system, in which the elements that make it up are both expressed and strengthened.

There is much more to Balinese village structure than this, to say nothing of the actual processes in terms of which it functions. But in a microscopic examination of what is really only one, in no sense independent, part of a much larger whole, perhaps some starting points for a more thorough analysis can be located. And knowing where, in the whole mass of overlapping alliances, to begin is half the battle in any effort to lay bare the underlying form of Balinese social organization. In Bali, the term "village" is perhaps best reserved (as the Balinese themselves often reserve it) not for a particular social body but for the entire plexus of discrete interpersonal and intergroup relationships that honeycombs the whole island from Djembrana on the west to Karangasem on the east. Properly conceived, the Balinese village is not a circumscribed community but an extended field.

References

GEERTZ, CLIFFORD. 1959. "Form and Variation in Balinese Village Structure." *American Anthropologist* 61: 991–1012.

———. 1961. Book review: "Bali: Studies in Life, Thought, and Ritual." *Bijdragen tot de Taal-, Land- en Volkenkunde van Nederlandsch-Indie* 117:498–502.

KENNEDY, RAYMOND. 1955. *Bibliography of Indonesian Peoples and Cultures*. Rev. ed. Ed. T. W. Maretzki and H. T. Fischer. 2 vols. New Haven: Human Relations Area Files.

KORN, V. E. 1933. *De Dorpsrepubliek Tnganan Pagringsingan*. Santpoort: C. A. Mees.

RAKA, I GUSTI GDE. 1955. *Monografi Pulau Bali*. Djakarta: Pusat Djawatan Pertanian Rakjat.

5

Village Irrigation in the Dry Zone of Sri Lanka

Edmund R. Leach
CAMBRIDGE UNIVERSITY

Pul Eliya: The General Background

Topography and Historical Continuity

Pul Eliya is a village in the Kende Kōralē of the Nuvarakalāviya District of the North Central Province of Ceylon. It lies about 12 miles to the north of Apurādhapura. It is a Sinhalese-speaking village inhabited by members of the Goyigama caste. Most other villages in the immediate vicinity are similarly Sinhalese-speaking, but to the southwest there are some Tamil-speaking villages of which the inhabitants are Moslem (*marakkal*); Hindu Tamil villages are to be found a few miles to the northwest.

Because the ancient Sinhalese kingdom was for many centuries centred on Anurādhapura and Mihintalē, the region is one which is of very special interest to Ceylon historians. Contemporary documents of the early Sinhalese kingdom scarcely exist. Far more than in Europe the medieval historian must rely on archaeological evidence and the decipherment of crytic, fragmentary stone inscriptions. In such circumstances it is not surprising that scholars, in endeavoring to find meaning in these defective records, have looked for clues among the modern practices of the Sinhalese villagers living nearby.

In using twentieth-century materials as evidence for ninth-century custom it has come to be assumed that rural Ceylon continues from century to century almost unaffected by the passage of history. Documents from the tenth, twelfth, sixteenth, eighteenth, nineteenth and twentieth centuries are repeatedly cited side by side as if they all referred to the same thing. This is particularly true with regard to the region covered by the modern North Central Province.

I must emphasize very strongly that no such assumptions are implicit in the present volume. This chapter is concerned with the state of affairs in Pul Eliya in 1954. There is archaelogical evidence that a village also existed on the same site in ancient times, and there are grounds for thinking that in certain respects this ancient village must have been rather similar to the present one. But that is all. I do not claim that Pul Eliya is a typical Nuvarakalaviya village, still less do I suppose that the social system which I describe was once universal throughout the Kandyan kingdom or any of its predecessors.

The historians' assumption of long-term structural stability in the society of the Nuvarakalāviya District can be justified by certain ecological arguments. Geographically considered, the region is a very special one. It forms part of the Dry Zone of Ceylon. Rainfall is around 50–75 inches a year, which, under tropical conditions, is meager. Most of this rain is concentrated into two periods, October–December and April–May. The terrain is nearly flat. The soil cover is thin and markedly lacking in humus. Rain, when it comes, is usually heavy; there is then a rapid runoff, the rivers become torrents for a few hours and then relapse. In a few days everything is just as dry as before. The natural cover is forest, but it is very dry forest, which, once it has been felled, grows up again only very slowly. To anyone accustomed to the lush pastures of Europe, such conditions offer an uninviting prospect. Robert Knox, writing in the seventeenth century, treated the whole area as an uninhabitable wilderness. Yet, paradoxically, this was the center of the first Sinhalese civilization. The whole region is thick with the archaeological remnants of its romantic past.

The ancient civilization endured from about the second century B.C. to the middle of the thirteenth century A.D. Political decay and depopulation then followed and continued for several centuries. Only in quite recent times has the land once again been required to support a large and growing population. Chronologically, there is thus a wide discontinuity between the prosperous present and the glories of the historical past. Nevertheless, the newly emergent society is, in a curious way, conditioned and restricted by the civilization that flourished a thousand years ago. If it be true that the present-day society is in any sense similar to the ancient one—and that is very difficult to demonstrate—it is because the new society has had to adjust itself to a rather special and difficult set of ecological conditions which have remained constant through the centuries. There has been no substantial historical continuity in the society itself.

The classical Sinhalese kingdom, with its capital at Anuradhapura, was a striking and characteristic example of what Wittfogel has called "hydraulic civilization" (Wittfogel, 1957; but cf. Leach, 1959). A region of poor natural fertility was made to support a large and flourishing population by resort to irrigation engineering. Statements about demographic conditions in the ancient kingdom tend to be fabulous rather than exact, but, around the tenth century A.D., the population of the Nuvarakalāviya District can scarcely have been less than it is today.

The Chinese traveler Fa Hsien described the Anurādhapura of the late fourth century A.D. as a large and prosperous city. He had been told that there were then over sixty thousand monks in the Sinhalese kingdom and he accepts the possibility that in Anuradhapura alone there might have been five or six thousand monks directly dependent on the king's bounty. Even allowing for some exaggeration, it is evident that the Sinhalese economy of this period was such that the state could afford to maintain a large number of unproductive individuals.

The high productivity per unit of labor which Fa Hsien's account implies had gradually become possible through the accumulated capital investment represented by ever extending irrigation works. Although in later centuries most of these works were abandoned, it is essentially the same system of irrigation which has now been restored and refurbished. It is thus probable that the distribution of population on the ground is today rather similar to what it was in Fa Hsien's time. (See Brohier, 1934/35.)

The present-day villagers go much further than that; they claim that their whole system of organization has been handed down intact from the most ancient times. That they should hold this opinion and thus put a value upon traditional ways and social stability is a fact of sociological significance, but it is not a fact of history. We know nothing at all about the organization of village life in the ancient Sinhalese kingdom.

The modern irrigation works of Nuvarakalāviya, like their ancient predecessors, fall into two distinct categories. There are the small reservoirs (tanks) associated with individual villages and the very much larger central reservoirs and feeder canals which now, as formerly, are under the control of the central government. The latter class of works do not immediately concern us. Pul Eliya is not today connected with any central irrigation system and, so far as can be judged, it never has been so connected in the past.

Of the smaller reservoirs—the village tanks—there are today several thousand in actual use. Almost all are of ancient origin, but only a few have been in continuous use over the centuries. The great majority have been abandoned at various times and then restored again.

In this economy the basic valuable is scarce water rather than scarce land; it is the total water supply available to a community which sets a limit to the area of land that may be cultivated and hence to the size of the population which may survive through subsistence agriculture. The same is true of the district as a whole. In 1954 additional sources of irrigation water were becoming very scarce and expensive to control, and this suggests that by 1954 the population of this district was rapidly approaching the maximum that could be supported from local resources alone. A similar situation seems to have arisen around the eleventh century A.D.

A village tank is created by damming up a natural stream and building a long earthwork wall to hold the water up behind it. The resulting reservoir (when full) is usually about seven feet deep immediately behind the earthwork (bund). Very roughly, the full tank covers much the same area of ground as the land below it which it is capable of irrigating. Clearly, the location of tanks must conform to the natural lay of the ground. Although, in a generally flat terrain, the villager has some choice about how he constructs minor works, the site of the larger tanks is predetermined from the start.

The village of Pul Eliya today has a main tank of about 140 acres. Archaeological evidence in the shape of an ancient spillway and two *bisōkotuva*-type sluice works demonstrates conclusively that the present bund occupies exactly the same position as the bund of an ancient tank of the eleventh century or earlier.

This particular village has been occupied continuously since the British records begin around 1838. For earlier periods we have no evidence. It is a tank that is exceptionally well situated, the length of the bund in relation to the area of water being relatively short. This means that the tank can be fairly easily kept in good order. Moreover, it is a tank which very seldom dries up completely even in years of drought. A reliable tank such as this would seldom have been abandoned for very long even in periods when the population was at a minimum, and it could be the case that Pul Eliya has been continuously occupied from the beginning. This we cannot know. On the other hand, the remnants of various subsidiary irrigation works show that the overall system of water control has not always been quite what it is today.

The present villagers have a romantic tradition that their village was founded in the time of Dutthagāmani, which would imply a date round the second century B.C. They also maintain that they formerly possessed a palm-leaf "book" which was kept in the temple and which recorded that the original grant of their land dated from remote antiquity. This book is said to have been "stolen" by a certain government official some years ago.

Administration

In most villages governmental authority is exercised through the Vel Vidāne (Irrigation Headman), who is, in theory, elected by the villagers themselves as their spokesman and leader. Once a man has been appointed to this office, he is likely to hold it for many years, until he either resigns or is dismissed for malpractice. It is an office which entails a large amount of tedious clerical work, for which the direct rewards are small, but in a prosperous village the indirect advantages which accrue to the Vel Vidāne through his position of influence can be very great. No Vel Vidāne of Pul Eliya has ever vacated his office except on grounds of ill health or old age.

Although Coomaraswamy [1950] writes of this office as if it were ancient, I cannot trace any reference to it prior to 1867 and I am inclined to think that it is a British administrative invention instituted for the express purpose of creating a single responsible village head. The Vel Vidāne's authority is mainly economic. His first responsibility is to see that government regulations regarding the fair distribution of tank water are fully adhered to. In such matters the Vel Vidāne's immediate superior is a government official, the V.C.O. (Village Cultivation Officer). Since one V.C.O. may be responsible for fifty or more villages, the Vel Vidāne is in general left to his own devices. He can, if he so chooses, exercise wide and autocratic powers. The office of V.C.O. is a quite recent creation and in the days of Ratēmahatmayā before 1938 the Vel Vidāne's local authority must have been complete.

In legal as opposed to agricultural matters, the Vel Vidāne is only an informer. He is expected to report to the *tulāna* Headman. The *tulāna* Headman (misleadingly entitled "Village Headman") is responsible for a dozen or more villages. His duties are mainly clerical.[1] Almost every matter which calls for administrative decision, even of the most trivial kind, has to be referred to still higher authority. This higher authority is the D.R.O. (Divisional Revenue Officer), of whom there were in 1954 five for the whole of the North Central Province. The ordinary villager seldom has any direct contact with any higher officer of the adiministrative hierarchy. Above the D.R.O. stands the Government Agent, but to the ordinary villager this high official seems remote. For most practical purposes the D.R.O. personifies the government. In this respect he has taken over the role of the former Ratēmahatmayā.

Today the D.R.O. is a civil servant in permanent government employ who is liable to be transferred from one district to another, but in the Pul Eliya area, it was only in 1938 that the Ratēmahatmayā relinquished his D.R.O. functions. In 1954 this same Ratēmahatmayā was serving as Minister of Lands in the Central Government. At this date the local D.R.O. was an extremely efficient regular civil servant but he also, as it happened, was a member of the local aristocracy affinally related to the former rulers. His sympathetic understanding of traditional custom was greatly appreciated by the villagers.

In 1954 the local *tulāna* Headman was K. V. Appuhamy, himself a resident of Pul Eliya, while the office of Pul Eliya Vel Vidāne was held by V. Menikrala.

1. Unlike the Vel Vidāne, who was (until 1958) rewarded with a commission on the crop, the Village Headman is a salaried government official.

The two men were cross-cousins, the Vel Vidāne being the older man. So far as village affairs were concerned, it was the *latter* office which mattered most.

Over the years the Ceylon Government has endeavored to soften the rigidities of direct colonial administration by creating a supplementary system of local self-government. In the North Central Province these legislative measures have led to a great deal of administrative duplication, but have seldom brought any obvious benefits to the individual villager. In 1954 the term Village Committee represented an institution roughly comparable to an English Rural District Council with greatly reduced powers. Members of the Village Committee were elected on a constituency basis, each member representing a group of villages. This Village Committee had control of certain funds intended to be used for village improvements such as roadmaking, well digging, and so on. The Village Committee which was supposed to look after the welfare of Pul Eliya village had its headquarters some seventeen miles away and its incursions into Pul Eliya affairs were very slight. It was said to be an institution of great profit to its chairman. In 1954 there was no member of the Village Committee actually resident in Pul Eliya itself. The local constituency member was from Tulawelliya village, two miles to the south.

The Pul Eliya Land Map

My purpose in this section is to give a detailed picture of the ecological situation in Pul Eliya village as it affects the pattern of land use and residence. My account is primarily an elaboration of the two village maps. Map 5.1 shows the whole of the village land, while Map 5.2, on a larger scale, shows the detailed position of the various buildings in the village as they were in 1954. These buildings are mostly clustered together on the damp land immediately below the main tank. This is the usual pattern of all villages in this part of Ceylon.

Tanks

There are two tanks. The larger one is deemed to be on Crown land; the smaller one belongs to the temple. Here again we are dealing with a typical situation. It was the general policy of the British administration to presume that, in the absence of very specific evidence to the contrary, all village tanks belong to the Crown. It is by virtue of this Crown ownership that the customary *corvée* work known as *rājakāriya,* which is used to maintain the tank in good condition, is given a legal enforcement.

A minority of tanks are *not* the property of the Crown. These fall into three classes. First there are a few which have been from ancient times the property of feudal landlords. Mostly the villages associated with these tanks are those of lower castes such as Palanquin Bearers, Drummers, and the like.

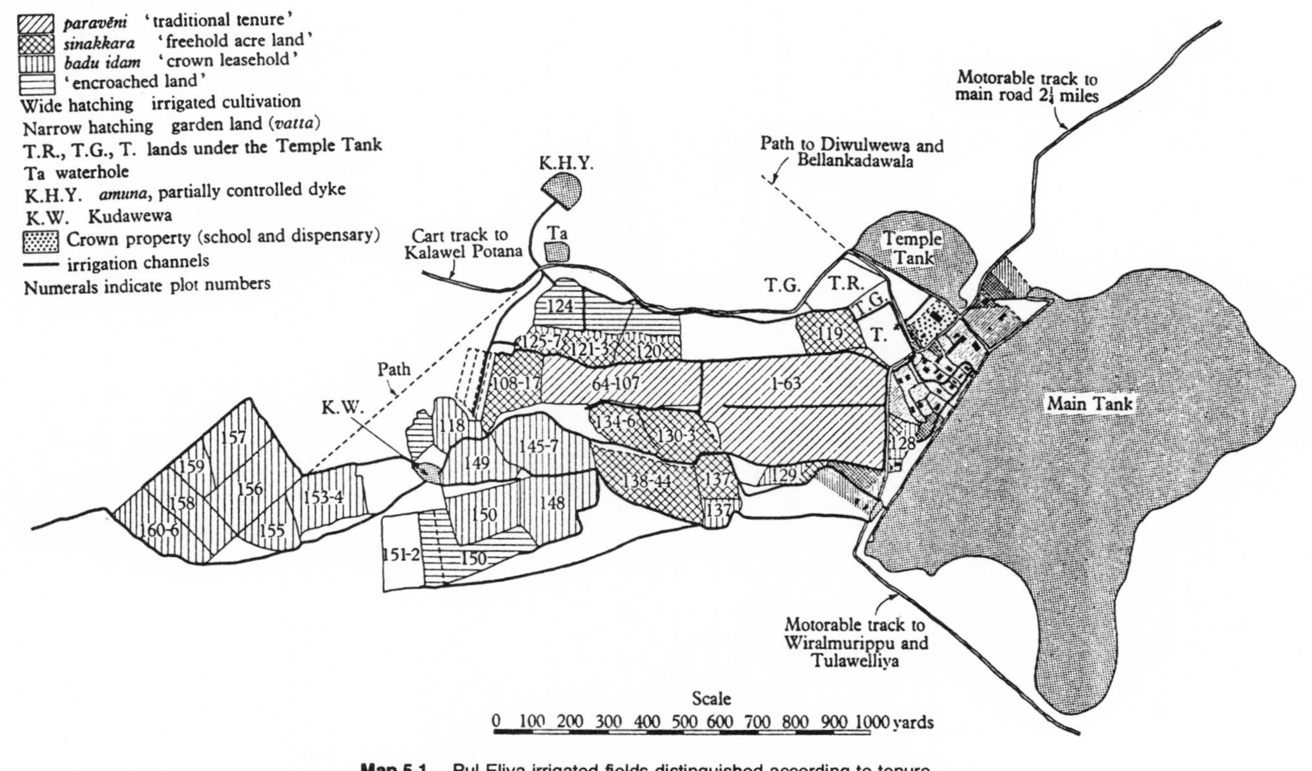

Map 5.1. Pul Eliya irrigated fields distinguished according to tenure.

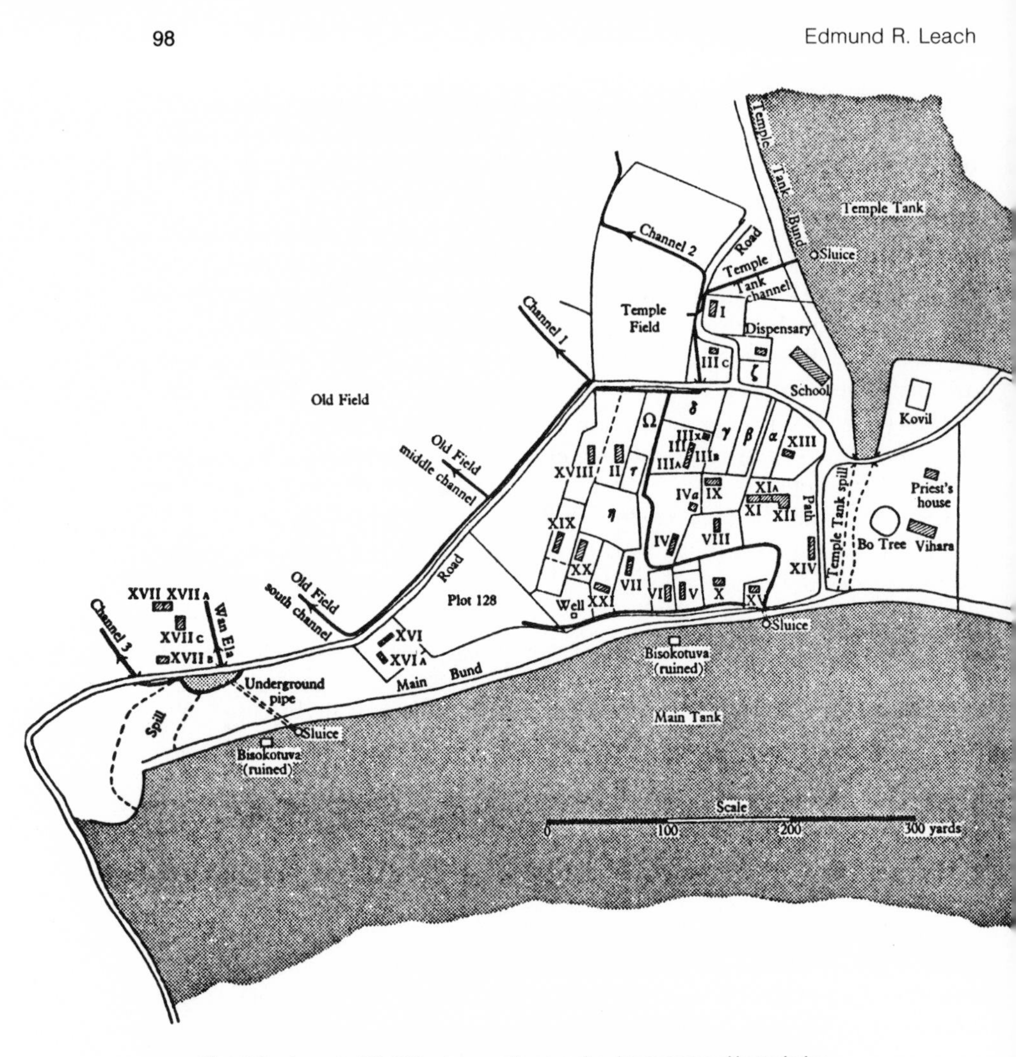

Map 5.2. Layout of Pul Eliya house-site area showing compound boundaries. All permanent buildings are shown and distinguished by Roman numerals (see Table 5.1). The Greek letters indicate plots of garden land (*vatu*) lying within the *gamgoda* area but not containing any residential building in 1954.

In former days a considerable proportion of all village tanks fell into this category, but in the course of the nineteenth century the British Colonial administration managed to whittle down their number to a small minority. None of the village tanks in the immediate vicinity of Pul Eliya falls into this category.

Secondly there are tanks which belong to temple authorities. Acting on the general principle that interference with religious affairs should be avoided whenever possible, the British authorities were always extremely cautious about questioning the title of any lands or tanks claimed by temple authorities.

Consequently the number of such tanks is considerable. Temple tanks receive no financial assistance from the Government Irrigation Department and most of them are now in a very bad state of repair. The associated villages are commonly both poor and small.

Besides owning whole villages in this way, the temple authorities sometimes own a subsidiary tank adjacent to a large one. This is the case in Pul Eliya, where the small size tank is temple property, as are the four acres or so of land immediately below it. It is probable that down to about 1900 there was no substantial difference between the state of upkeep of the main village tank and that of the temple tank. But today the former is in excellent repair while the latter is nearly derelict. In the case of the main tank, major repairs are now supported by government subsidy and every year the annual *rājakāriya* repair work is carried out with official backing. In contrast, for many years past, the upkeep of the small temple tank has been almost entirely neglected.

The third category of tank which is not regarded as Crown land is that known as *olagama*. There still exist in the jungle a very large number of ancient abandoned tanks. It has long been official policy that anyone who goes to the trouble and expense of repairing such a tank shall be entitled not only to the land of the tank itself but also to any area of land below it which can thereby be irrigated. Kudawewa and Ulpathgama, just to the east of Pul Eliya, are privately owned *olagama* of this type; Bellankadawala, to the northwest, was also an *olagama* in origin, but at the present time it is treated as a Crown tank of normal type.

Land Categories

The present-day categories of land within the Pul Eliya village boundaries derive from a cadastral survey made about 1900. At this date, all villages in this area were officially surveyed by the government and the principle was adopted that only land in active use should be recognized as private property, everything else being treated as Crown land. By this rule, house sites and gardens which were in good condition were recognized as being *paraveni* property, that is, "private property from ancestral times." Likewise, land that was already "asweddumized," that is, laid out in rectangular flat terraces for rice cultivation, was also mostly recognized as private property. Where such private tenure was admitted, the government made no inquiry as to how these lands were divided up among different individual holdings; nor has there been any later inquiry as to how they have been inherited or otherwise disposed of. Production statistics, collected on behalf of the Agricultural Department, record indirectly the ownership of such holdings, but such returns have no legal standing. In theory, the transmission of *paravēni* land is still governed by traditional Kandyan customary law.

Disputes as to title are, in practice, settled by the Vel Vidāne acting under the aegis of the Village Cultivation Officer (V.C.O.). The Vel Vidāne submits annually to the V.C.O. at a public village meeting two elaborate returns. The first, the "Pangu List," specifies the total irrigated area held by each shareholder in the village lands and provides the basis of assessment for *rājakāriya* repair work on the tank bund. The second is the "Paddy Census," which purports to record the acreage, seed sown, yield, ownership, and leasehold history of every individual plot in the entire village. Both returns are produced by copying down the record from the previous year and then taking note of changes. The Vel Vidāne is required to justify any changes to the V.C.O. in the presence of his fellow villagers. The village meeting which I attended was a distinctly lively affair and I should judge that the sanctions against "fraudulent conversion" on the part of the Vel Vidāne are reasonably effective. Since errors are self-perpetuating, the Paddy Census figures for acreages, yields, etc. are wildly inaccurate, but the facts regarding title over individual plots appear to be correctly recorded.

Old Field. The original "Old Field" of Pul Eliya, which was held or thought to be held, on *paravēni* tenure in 1900, is shown on Map 5.1 with diagonal hatching and plot nos. 1-63, 64–107.

The British administrators of the latter part of the nineteenth century, influenced by current theories concerning the origins of Aryan society, believed that Ceylon villagers held their land on some kind of communal basis. They supposed that the basic principles of Sinhalese law were essentially those of the patrilineal Hindu joint family as it appears in various parts of India. The following extract from Codrington (1938, p. 3) is in line with this general trend of thought:

> The cultivated land in the village was divided into *pangu* or shares, each *panguva* usually consisting of paddy land, of gardens, and, subject to the reservation made below, of chena. For purposes of service the *panguva,* whatever the number of the co-heirs may be, is indivisible and the co-heirs jointly and severally are liable for the service. It seems to be a survival of the Hindu joint-family estate. Of the joint family all the male members own and have a right to the family property; no coparcener is entitled to any special interest in the property nor to exclusive possession of any part of it. Private property belonging to a single person is unusual except in the case of self-earned property or gifts. The joint-family property is generally managed by the eldest member. Dissolution takes place by mutual consent or by application to a court of law, but this division of the property generally is made only when the relations amongst the coparceners grow remote.

This is represented as a description of the state of affairs in Ceylon in the tenth century A.D. Although Codrington recognizes that present-day Sinhalese tenure is of a more private kind, he seems to imply that a thinly disguised primitive communism is still latent in the system.

The facts are quite otherwise. In Pul Eliya every adult individual, male or female, is treated as a separate economic unit, separately entitled to own property and separately entitled to derive benefit therefrom. Indeed, where livestock are concerned, even small children are independent owners. Even when close relatives team up together they consider themselves to be working a "sharecropping partnership" (*andē havula*) rather than to be coparceners.

Again, as we shall see, there are several plots of inherited land in Pul Eliya which are "undivided." But the owners of such land, "the heirs of X the previous owner," are not coparceners; they do not pool the proceeds of the land and share it out in common—they are rather shareholders or partners, each of whom has a separate title to a particular mathematical proportion of the total. The system has its analogue in English law. When a man leaves his estate in trust to his children with an annuity to his widow with the proviso that the trust terminates on the death of his widow, he does not thereby establish a joint family estate. On the contrary, each of the individual heirs is jealously conscious of his private individual rights. So too with "undivided" property in Pul Eliya.

Although there is no evidence for the existence of anything remotely resembling communal tenure anywhere in Ceylon during any part of the British colonial period, the *belief* that such communal elements were present, plus the general nineteenth-century prejudice against egalitarian ideas, provided the basis for official policy. The supposed village communism was condemned on principle, and it became the publicly declared objective of government to replace this "primitive" form of organization by a system of peasant proprietorship. Official policy was explicitly designed to favor the relatively wealthy peasant at the expense of his poorer neighbor, the theory being that the richer man must be, *ipso facto,* the more enterprising.

Freehold "Acre Land" (sinakkara). Accordingly, rules were introduced whereby individuals could purchase Crown land outright and thereby acquire freehold areas over and above whatever they might hold in the supposedly communal village fields. It was quite consistent with this policy that these Crown lands were sold only in relatively large plots so that the poor peasant was excluded from the market. In theory, no plot of Crown land sold for use as rice land could be less than a five-acre block. The obvious intention was that each village would ultimately consist of a large number of separate smallholding farms.

Such a notion conflicts not only with the traditional theory of landholding but also with technological common sense. It is therefore hardly surprising that, from the start, the villagers resorted to a variety of devices to get around the law, both as regards the purchase of this freehold Crown land and its inheritance.

Land purchased freehold from the Crown in this way was always officially surveyed; formal title deeds were then registered at the government land office in Anurādhapura. These title deeds show land areas in English acres, which explains why this category, as *sinakkara,* which may be translated "freehold acre land." House sites and gardens were also sometimes purchased from the Crown on freehold *sinakkara* title. In this case the plots were much smaller.

The government policy of selling "acre" land outright continued until 1935. By that date, most of the ground in the immediate vicinity of the original Pul Eliya *paravēni* field had passed into private hands. This acre land is shown on Map 5.1 by diamond hatching. The principle of marginal utility applies here. Naturally the Old Field—that is, the land already under cultivation in 1900 (Map 5.1, diagonal hatching)—represents the land which can be most easily irrigated. When Crown lands began to be sold, the plots which were first taken up were those which could most easily be provided with irrigation water simply by extending the original irrigation channels of the Old Field itself. Thus in general the later the date of sale, the more disadvantageous the position of the land, though occasionally the buyers made mistakes; one plot of acre land sold in 1919 proved to be incapable of irrigation and has never been "asweddumized" at all.

*Crown Leasehold (*Badu Idam). As the ideology of self-government began to supersede that of imperialist colonialism, a new socialistic mood came to prevail in government circles, and the concept of the "peasant proprietor" underwent a sea change. The outright sale of freehold Crown lands had been intended to encourage the small-holder capitalist; the new emphasis was on preserving the peasantry from exploitation.

Land regulations introduced under the Land Development Ordinance of 1935 were designed to assist a supposed category of "landless peasants." It was definitely intended that these new regulations should operate to the disadvantage of the owner of freehold land, who now began to be thought of as a wealthy parasitic absentee landlord. In terms of the oversimplified categories of popular politics, all owners of freehold land became tyrannous rent collectors, while all landless peasants were *ipso facto* virtuous and exploited serfs. (For details see Farmer, 1957, pp. 123–28.)

Under the 1935 regulations no Crown land at all could be sold outright. Instead, land was granted in two-acre plots on permanent lease. The individual villager acquired a documentary title to his land and, provided he behaved himself and continued to pay quite a nominal rental to the government, the land was inalienable.

Certain defects in this legislation may well be mentioned from the start. The leasehold land, which in Sinhalese is known as *badu idam,* was to be allocated only to poor peasants. But the regulations failed to provide any adequate definition of a poor peasant. In Anurādhapura District, village headmen were required to produce lists of all adult male individuals owning less than one acre of freehold rice land, and these were deemed to be automatically worthy of *badu* allocations. In practice this meant that the newly married son-in-law of the richest man in the village might be granted a *badu* lease, but his neighbor, who happened to own just over one acre in the Old Field, could be excluded.

Secondly, the rules included the proviso that, while *badu* plots could be inherited, they could only be transmitted to a single heir specified in the lease. On the death of the holder of a *badu* plot, the title would be transferred to this individual without fragmentation.

Finally, the leasehold character of *badu* tenure, whereby the government retained the right to eject the holder in certain special circumstances, was one which the Sinhalese villagers themselves bitterly resented. In 1954 very few *badu* tenants had in fact ever been evicted, but the villagers expressed an exaggerated anxiety lest this might happen. The real source of their hostility to the leasehold element in the system was that the ultimate insecurity of tenure precluded a tenant from using such land as security for a mortgage. *Badu* land is a source of income but it cannot be converted into capital, even temporarily.

The *badu* land in Pul Eliya in 1954 is shown on Map 5.1 by vertical shading. Although the regulations concerned date from 1935, they did not become fully effective until after the war. Thus, most of the *badu* holdings in Pul Eliya in 1954 had only been cultivated for a few years. It is obvious from the map that, because the Old Field and the freehold acre land between them had previously taken up all the best ground, all *badu* land is very disadvantageously sited. It seems likely that a good deal of it can never be properly irrigated, except at extravagantly uneconomic cost.

Irrigated Cultivation

The whole Pul Eliya cultivation area taken together, the Old Field plus the freehold acre land plus the *badu* land, amounted in 1954 to about 135 acres. This is approximately the same area as that covered by the main tank when full. This coincidence is not an accident. The Irrigation Department, which issues regulations on these matters, seems to have a rule of thumb that a village tank is capable of irrigating an area of land equivalent to itself.

The increase in total cultivation area since 1900 does not imply a corresponding increase in the productive capacity of the village, since there is insufficient water to supply the whole area during both cropping seasons. In recent years the Old Field has normally only been cultivated for the *Yala* harvest (September) and the acre land only for the *Mahā* harvest (March); in former times, when the total asweddumized area was less, much of the land was made to yield two crops a year.

The decision as to what fields are to be cultivated in any particular season is made at a public village meeting and formally declared to the V.C.O. The issue is a subtle problem of economic choice, since, if the water resources of the irrigation system are overextended, the outcome may be total crop failure. The village meeting makes its collective decision on the basis of the level of water in the tank and a gambling estimate of the prospect of rain in the weeks to come. In a normal year the *Mahā* rains will be much heavier than the *Yala* rains, and it is therefore logical that attempts to irrigate the outer fringes of the irrigation system should be confined to the *Mahā* season.

Tank-Bed Cultivation

To recapitulate, I have so far mentioned three main categories of ground. First, there is tank ground. When the tanks are full, this is covered with water, but at other times of the year it is soft pasture. In the dry weather the moist grassland in the rear of the tanks provides the main grazing grounds for the cattle and the buffaloes. At one time it was common practice to cultivate this ground for rice, but for many years now this has been prohibited. The reason for this is that the villagers often found it easier to cultivate in the empty tank bed than in the orthodox fields and ill-disposed persons were liable to break the bund of the tank on purpose. The prohibition on tank-bed cultivation applies only to government-owned tanks, and this form of cultivation is still practiced from time to time in the small *olagama*.

House Sites and Gardens

The second main category of ground is that on which the village buildings stand. Each dwelling house is in a small compound (*vatta,* plural *vatu)* in which there are a few coconut trees, plaintains, areca palms, and the like. In some cases there are several distinct dwelling houses within the one compound. There are also, in Pul Eliya at the present time, two garden compounds which contain no dwelling house at all.

The third main category of ground is the irrigated rice land.

"Highland" and Shifting Cultivation

Finally, those parts of the map which do not fall into any of these three main categories are, for the most part, rough scrub jungle. The ground surface is rocky and uneven and it is covered with a coarse type of forest, including much thorn and cactus. If we exclude the very rocky areas and certain temple ruins, nearly every part of this rough jungle has at one time or another been cleared for shifting cultivation. After more than a century of administrative indecision, offical policy on the use of such land remains very obscure.

Formal government opposition to shifting cultivation (*hēna,* anglicized as "chena") of all sorts goes back almost to the beginning of British administration. Colombo government officials have usually taken the view that forest trees are a valuable commercial asset which ought not to be destroyed. Forest lands should therefore belong to the Crown and be looked after by the Forest Department. Villagers should be confined to their irrigated farmlands. These general principles are sensible enough when applied to the ecological conditions which prevail in the wet southwest zone of Ceylon, but they ignore altogether the realities of the North Central Province.

In the first place, there is very little commercial value in the timber of the low-grade secondary forest which surrounds villages such as Pul Eliya. Secondly, it is a demonstrable fact that droughts and floods and other natural disasters are so frequent in the North Central Province that if the villagers did not from time to time resort to shifting cultivation, they would all have to starve.

There have been brief periods during the past century when shifting cultivation has been officially prohibited altogether. There have also been brief periods when it has been allowed without restriction. But for the most part the legal rules have been similar to those which apply today. Roughly, the present legal position is that if a village headman will certify that one of his villagers has a number of dependants and cannot be expected to support himself from the irrigated land which he works, then this man can apply for a license to clear a one-acre plot for shifting cultivation, for one year only, at a nominal rent. Such rules are unworkable and practice is very far removed from the legal theory. The principal actual effect of the present regulations is to ensure that all shifting cultivation is carried out in the most inefficient manner possible.

While the irrigated lands are devoted exclusively to rice, the shifting cultivation areas can be used for a variety of alternative crops. Traditionally the most important of these is a species of millet, known as *kurakkan*. In years when there has been a rice-crop failure, this may be a very important standby item of diet. But shifting cultivation can also be used for a number of important cash crops, such as gingelly and mustard. It is this which makes the problem so peculiarly difficult from the administration's point of view.

If all restrictions were withdrawn, many of the villagers would probably devote their entire energies to the cultivation of cash crops by shifting cultivation. They would then count on being able to purchase foodstuffs in the open market with only a proportion of the cash they obtained from their cash crop. At 1954 prices they would be better off than if they had stuck to their ordinary occupation of rice cultivation. Such behavior is sensible for the individual farmer but, if carried out on a large scale, might have quite disastrous consequences for the general economy of Ceylon.

So far as the individual villager is concerned, the position seems quite simple. He considers that, by ancient tradition, he has the right to clear land for shifting cultivation where and when he will. He looks upon this shifting cultivation as his main means of earning a cash income, in contrast to his ordinary activities as a rice farmer, which provide him with a subsistence living. For over a hundred years successive governments have expressed their disapproval of shifting cultivation; there is a constant threat that it may be prohibited altogether. The villager looks upon this attitude of government as merely vindictive. He sees it as a persecution designed to prevent honest men from earning a decent cash income.

Irrigation Ditches

The final map feature which I wish to consider is that of the irrigation ditches. As I have already stressed, the basic scarce commodity is water. Economic and political influence throughout the community is determined by control of water and it is this factor which has made the office Vel Vidāne (Irrigation Headman) so important. The tank water is conveyed to the house gardens and the irrigated fields by way of a sluice (*horovva)* and then by ditches (*vel*). Operation of the sluice is an exclusive prerogative of the Vel Vidāne personally. Every house-site garden in the village proper has access to a particular length of irrigation ditch, and the same basic principle governs the division of land in the Old Field. Under the traditional system, holders of land in the Old Field owned rights in a certain length of irrigation ditch rather than rights in a particular area of ground. A plot holder owned rights in water; he could cultivate as much or as little land as he chose.

Now so far as the Old Field is concerned, ditch maintenance is simple. The general obligation is that each plot holder must maintain that part of the main irrigation ditch which passes by or flows through his plot. Since nearly every villager has a plot in the Old Field, the obligation to maintain the original main irrigation ditch is, in practice, a widely distributed general duty.

But the development of "freehold acre lands" and *badu* areas has created a new situation. To feed these new lands several new main irrigation ditches have had to be constructed, but these ditches are private property. They belong to those particular individuals whose land the channel serves. These private rights in irrigation ditches are jealously guarded and are becoming of very special significance in the developing polity of the village. The position of the main irrigation ditches is shown on Map 5.1.

Individuals who work land served by the same irrigation channel have an inescapable obligation to cooperate. This fact is a most potent source of friendship alliances, but it is also a major source of hostility.

Calendar of Work

The pattern of labor organization varies greatly at different parts of the agricultural season according to the technical activity involved. The total sequence can best be considered in phases but first I should say something about the work calendar as a whole, which is shown in schematic form in Table 5.1.

Table 5.1.

Approximate calendar of activities, 1954–55

Lunar month	Date of full moon (new moon)	Season	Rites	Paddy-growing activity		Chena activity	Other activities
Bak	18 April	(Season between two years)	—	—		—	—
				Yala plowing	RAIN		
	(2 May)		New Year	—		—	—
Wesak	17 May	U	—	18 May: *Yala* sown		—	—
Poson	16 June	S	—	—	HOT	—	—
Asäla	15 July	N	—	—	DRY	—	—
	(31 July)	A	Mutti Mangalaya				
Nikini	14 Aug.	V		18 Aug.: *Yala* reaped		Chena clearing starts	Plowing by tractor
		A		Stack shifting			Tank low: cattle on *maha* stubble and at back of tank
		S		Stacks fenced			
Binara	13 Sept.					Chena burning	
		S				*Vi hena* sown	
		A	18 Sept.:	18 Sept.: Threshing			
		M	New Rice festival	Granaries filled			
Vap	11 Oct.	A	—				Tank repair work
		S					Tank fishing
		I		*Maha* plowing by buffalo			House thatching and building
		T		when rain starts			
		A		Temple field sown		*Kurakkan hena* sown	V.C.O.'s meeting 5 Nov.
Il	9 Nov.		—		HEAVY		
Unduvap	8 Dec.	—	—	—	RAIN	—	—
Durutu	6 Jan.		—	—		—	—
Navam	5 Feb.	S	—			*Kurakkan* harvest	—
		I		*Maha* harvest		*Vi hena* harvest	
Mädin	7 March	(Season T	—			—	—
Bak	6 April	between A	—	(*Mada* sown if *Maha* fails)	RAIN	—	—
		two years)					

Note on the Sinhalese months and seasons: The Sinhalese lunar year is usually held to begin with the month Bak (cf. Pieris, 1956, p. 92). Pul Eliya villagers seemed to regard Wesak as the first month. The three seasons are not precisely defined. Usna (hot) covers the period of the *Yala* crop growing and harvest; Samasita (cool) is the period of preparing the land and sowing the seed for both shifting cultivation (*hena*) and for the *Maha* crop; Sita (cold) extends over the main period of the *Maha* growing season up to the *hena* and *Maha* harvests. The interval between the end of the *Maha* harvest and the start of the *Yala* plowing is "the season between two years" and covers parts of the months Mädin and Bak.

Vas is a three-month period of Buddhist ritual significance. It ends at the full moon of the month Binara, which here synchronizes closely with the *Yala* harvest.

The Sinhalese, with their almost obsessional interest in horoscopes, have long been accustomed to reckon time by two alternative calendars—one solar and the other lunar—the "luck" of any particular day being dependent upon the conjunction of the two systems. Although the European "solar" calendar is not quite the same as the traditional zodiacal cycle, it is very similar, and the peculiar way in which the 1954 Pul Eliya villagers managed to combine lunar and calendar dates is very probably part of ancient tradition.

Whenever the Old Field is to be cultivated it is essential for the whole village to adhere closely to a predetermined program of work, for when the tank sluices are open the whole field can take water and when the sluices are shut the whole field must run dry. No plowing can be done on a dry field, but once the water has been let in to soften the earth, work must proceed everywhere simultaneously. Thereafter, to avoid loss by evaporation, the plowed field must be sown and the crops carried through to harvest with the least possible delay.

There must, therefore, be agreement about the dates on which the sluice will be open, the date at which sowing will be completed, the varieties of rice that will be sown, and the dates at which it is planned to have harvest ready and the field drained. Under rules in force in 1954 the Village Cultivation Officer held a village meeting at the beginning of each cultivation season and formally agreed on these various dates with the assembled villagers. Although this particular form of bureaucracy is a recent innovation, it has always been necessary for the villagers to agree among themselves upon such matters.

Since the prospect of rain during the *Yala* season is substantially worse than that for the *Mahā,* the usual practice is to grow low-yielding rapid-maturing varieties during the *Yala* season and rather slower-maturing varieties during the *Mahā*. The *Yala* varieties are known as "three-month rice" and the *Mahā* varieties as "four-and-a-half-month rice." In the *Mahā* season some land is cultivated from direct rainwater only, and in such cases the cultivator can use his own discretion. But a gambling element is always present. The higher-yielding varieties all take longer to mature and therefore run greater risk of being destroyed by drought.

The key dates to which the villagers agree with the approval of the V.C.O. are "lucky days" conventionally spaced. For example, in 1954 the first day of the month Wesak in the lunar calendar fell on 2 May. The Pul Eliya villagers considered this to be the New Year. The Wesak full moon fell on 17 May and they agreed with the V.C.O. that the *Yala* sowing should be completed by 18 May.

Since they were growing three-month rice, they planned to start reaping on 18 August, and in fact did so. They then held the New Rice festival on 18 September and threshing was allowed to start that evening.

A similar program required the *Mahā* season to be fitted in between the full moon of the month Vap (11 October) and the end of the lunar year (beginning of April), though as it turned out the winter rains that year were very erratic and the crop probably failed.

A similar mixed calendar applied to the chena cultivation. The clearing of chenas started around the full moon of the month Nikini (14 August) and proceeded spasmodically. The *vi hēna* (rice chenas) were sown around the middle of the month Vap (11 October) and the *kurakkan* chenas one month later. The rice in the *vi hēna* case was a variety called *dik vi,* a four-and-a-half-month paddy, which was expected to be harvested at the end of February, at the same time as the *kurakkan*.

Since chenas are not dependent upon a public water supply, the various cultivators have discretion about their precise timetabling. But there seemed to be a general tendency to pick sowing dates according to the moon while predicting harvesting dates according to the civil calendar. The reason for this is quite simple. In general, these villagers thought in terms of the months of the ordinary European calendar; but for any specific action in which luck might be concrned, or for religious festivals, they consulted an astrologer, who usually provided them with a date taken from the lunar calendar. Different astrologers seem to have different systems of calculation, and neighboring villages did not all adhere to exactly the same timetable.

The Theory of Traditional Tenure

The Classical System as Described by Ievers and Codrington

Ievers, in his *Manual of the North Central Province* (1899), describes an idealized type of land tenure which he represents as being the typical traditional system operating in all villages of the North Central Province. Codrington (1938) and Pieris (1956) have both accepted Ievers's account as an accurate historical reconstruction. All three authors seem to suppose that at one time this kind of tenure prevailed throughout the dry-zone Ceylon.

There is documentary evidence that tenures approximately to that described by Ievers existed over much of the northern part of the North Central Province in the latter part of the nineteenth century. This area has, even today, a consistent ecological pattern which is best summarized by the phrase "one tank—one village." But this ecological pattern is not general to the whole of the Ceylon dry zone, nor even to the whole of the two northern provinces, nor even to the whole of the Nuvarakalāviya District. It therefore seems to me unlikely that Ievers's tenure system ever prevailed over any very wide area. The system is certainly extremely interesting and it *may be* of considerable antiquity, but certainly, in part, it is a peculiarity of this particular region.

This ideal scheme specifies a very precise relationship among (*a*) the tank, (*b*) the field, and (*c*) the village. It further implies that the field shall be roughly rectangular, with its longest side lying at right angles to the bund of the tank. The field is in major sections, *bāga (bāgē)*. Maps 5.1 and 5.3 may be taken as illustrative of this account.

As Codrington recognized, this description fits very nicely with the ideal pattern previously described by Ievers, who wrote:

> Each tract or pota is divided into two or three portions called bāgē, viz. inhalabāgē, medabāgē and pahalabāgē. Each pota has two small strips, one at the tank end and the other at the opposite end which are called kurulupālu meaning "an allowance of extra land as compensation for damage by birds." Two larger strips at each end next the kurulupālu are called elapat. These are the property of the gamarāla, who is the hereditary chief cultivator of the village. The other portion in the middle of the pota is divided equally among the shareholders, including the gamarāla, and the divisions are called pangu. A panguwa is an original share but as the family increases it may become subdivided. . . . The pangu are divided across the field by ridges parallel to the bund of the tank and each contains one or two strips called issarawal. One issara is a range of beds between two of these ridges. . . . The gamarāla put up two watch huts one at each elapata and the other pangukārayō (shareholders) jointly build a watch hut for every three or four pangu and watch by turns. [Ievers, 1899, p. 172]

By the way of illustration Ievers published a diagram which is here reproduced (Diagram 5.1). The letterpress in the margin of the diagram has led to some confusion. Ievers states that his hypothetical field contains only ten *pangu,* but adds that "each *bāgē* contains an equal number of *pangu.*" Most readers would infer that his diagram shows five *bāga* in all, each with ten *pangu,* but with all five *bāga* controlled by one Gamarāla. I can only remark that such a field layout is in the highest degree improbable.

Since the Sinhalese term *gama* (pl.: *gam)* is commonly translated as "village," many writers have thought that a gamarala was simply a village headman; Ievers himself writes as if this were so. But, in this region, the normal holding of a gamarala is not a village but a *bāga*. The Sinhalese think of a village tank and its associated lands as constituting a single unified estate. The ultimate landlord of this estate is either the Crown or an absentee landlord or the priest of some local temple. Under the traditional system this ultimate landlord would always have sublet the estate to *one or more* primary grant holders—the Gamarāla. The position was accurately described by Codrington:

> The Sinhalese gama, plural gam, normally signifies a village but the word is applied to an estate or even to one field. . . . Of the gama, whether village or estate, the centre is the paddy land, of which the high land is considered to be the appurtenance. . . . The gama normally consists of paddy fields, gardens, and miscellaneous fruit trees and chena (hēna, pl. hēn) that is jungle land, cleared, burnt and cultivated periodically. . . . In the country where paddy cultivation is carried on by means of tanks, the houses are gathered together in the neighborhood of the tank bund. Such a village until recent times consisted of a number of families of the same caste and related to one another, presided over by one or two hereditary Gamaralas—whose holding was the gamvasama.
>
> The classes known in India as the village servants, the blacksmith, the washerman, the potter and others (tovilkārayō) live in separate villages of their own similarly organised. [Codrington, 1938, p. 1]

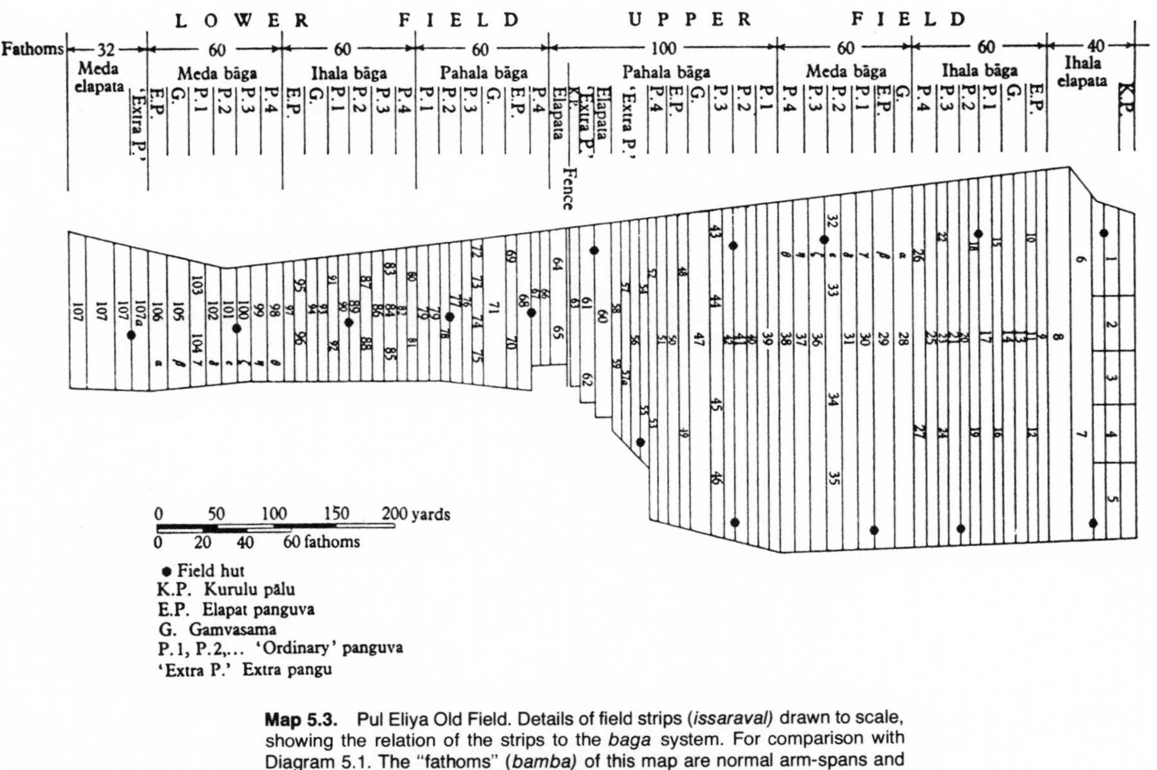

Map 5.3. Pul Eliya Old Field. Details of field strips (*issaraval*) drawn to scale, showing the relation of the strips to the *baga* system. For comparison with Diagram 5.1. The "fathoms" (*bamba*) of this map are normal arm-spans and are approximately 2 inches short of the English 6-foot measure.

The fields next the tank are called *Purampota*, or *Mulpota*, or *Upáyapota*.

The next range is called the *Hérenapota* or *Peralapota*.

The land opened in addition to the above two ranges is called Kaṭṭa Kaduwa or Alut Asweddumа.

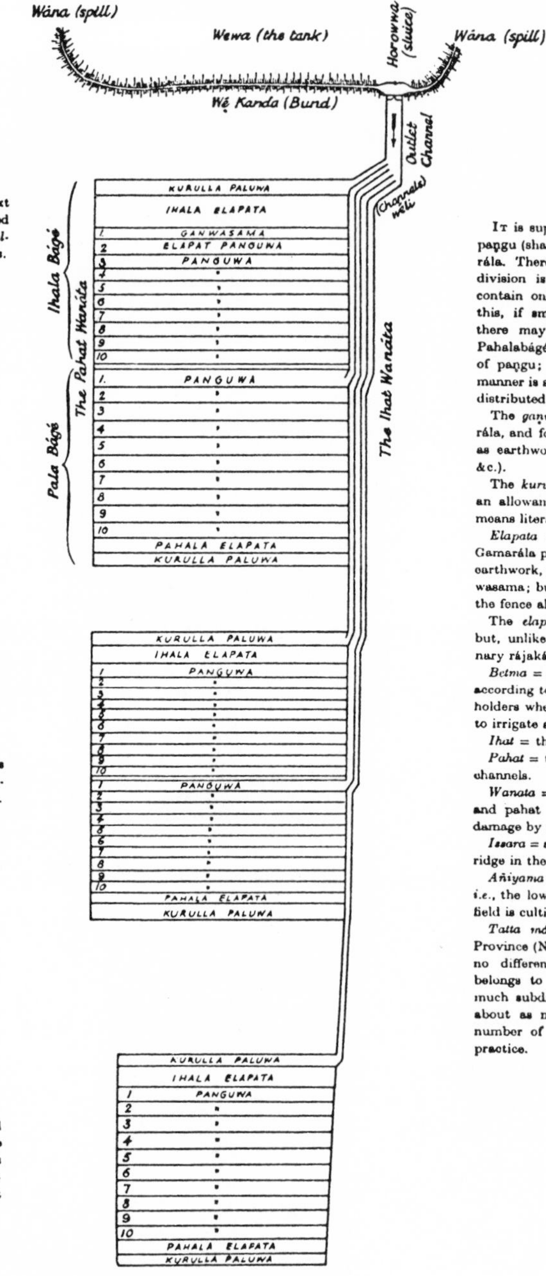

Lith Surveyor Generals Office Colombo N°350

Explanation of Terms

It is supposed that in this village there are ten paṇgu (shares) and a gaṇwasama held by a Gamarála. There may be any number of shares, but the division is in this manner shown. Some villages contain only one field called Purampota, &c., and this, if small, is not divided into bágé. If large there may be three bágé: Ihalabágé, Medabágé, Pahalabágé. Each bágé contains an equal number of paṇgu; and the reason of the division in this manner is so that good and bad land may be equally distributed.

The *gaṇwasama* paṇguwa belongs to the Gamarála, and for this he has no duty to perform (such as earthwork on bund, clearing channels, fencing, &c.).

The *kurulla páluwa* belongs to the Gamarála as an allowance for damage by birds, beasts, &c. It means literally "bird-damage."

Elapata means "a side." This belongs to the Gamarála paṇguwa, and is free of rájakáriya (either earthwork, pinpara, &c.), as in the case of the gaṇwasama; but the Gamarála must put up and keep the fence all round.

The *elapat paṇguwa* belongs to the GamaráIa, but, unlike the elapata and gaṇwasama, the ordinary rájakáriya must be done for it.

Betma = any portion of any field equally divided according to paṇgu, and cultivated by all the shareholders when the water in the tank is not sufficient to irrigate a whole field (pota).

Ihat = the channel side of a field.

Pahat = the side of the field opposite to the channels.

Wanata = the line of jungle yearly cleared on ihat and pahat sides for protection of the crop from damage by wild animals.

Issara = a paṇguwa, or, if it is divided in two by a ridge in the middle, half a paṇguwa.

Añiyama = a pahala elapata in a betma cultivation, *i.e.*, the lowest paṇguwa when only a portion of a field is cultivated. It is given to the Gamarála.

Tatta máru, Kara máru, Sóra máru. In this Province (North-Central Province) these words have no difference of meaning. When a paṇguwa belongs to several shareholders, or becomes too much subdivided, it is taken for cultivation turnabout as may be agreed upon according to the number of the shareholders. It is not a common practice.

Diagram 5.1. Ievers's schematic diagram of an ideal *bāga* system (for comparison with Map 5.3).

This description correctly outlines the general state of affairs prevailing even now in the vicinity of Pul Eliya. The unit estate associated with one particular tank may sometimes consist of one *gama* and sometimes of several. In the latter case each subunit is called a *bāga*. In any one village there are as many Gamarāla offices as there are *bāga,* though on occasion one individual may hold more than one such office at one time. The number of *bāga* in a village and hence the number of Gamarāla offices is permanent and unalterable.

The Formal Theory as Applied to Pul Eliya

In Pul Eliya and in all the neighboring villages the estates of the different Gamarāla are immediately adjacent to one another and are ordinarily fed from the same irrigation channel. The different Gamarāla must, therefore, cooperate.

The different Gamarāla who control a single field are not thought of as close kinsmen. They are heads of different *pavula* and essentially of exactly equal status. They may then be *massinā*—brothers-in-law—but cannot properly be brothers, since *ayiyā/malli* are always of unequal status. It is true that, on one occasion, I was given the hazy outline of a myth which started by saying that Pul Eliya was founded by three *sahōdara* ("brothers"), but when I pursued this and asked whether one of the brothers was senior to the others, my informant quickly changed the subject. The term *sahōdara* may, as a matter of fact, include persons standing in *massinā* relationship.

The Gamarāla was a grant holder, not an owner. In accepting a grant of land he also undertook a number of service responsibilities associated with the grant. The nature of the service depended on the grant holder's caste and various other factors. According to local tradition, the original service obligation undertaken by all three Pul Eliya Gamarāla was to provide annually a certain weight of water lily flowers for the use of the temples in Anurādhapura. This story is probably simply a rationalization from the village name, Pul Eliya, which means "flowery open space," but some service tenures of this kind are authentic.

The inhabitants of Pul Eliya are all members of the cultivator caste (Goyigama), which stands at the head of the caste hierarchy. The various grant holders in non-Goyigama villages took on much more menial service duties than that of picking flowers.

In addition to this traditional feudal service, the Gamarāla had an obligation to collect taxes, to entertain government officials, and to maintain the tank in good working order. Most of these latter obligations remained even after the coming of the British and the abolition of Kandyan feudalism. Some of these duties were carried out by the several Gamarāla jointly while others were worked on a roster basis. In the outcome there was no clearly defined village headman and it was this fact which led the exasperated British administrators to institute the dictatorial office of Vel Vidāne.

During the latter part of the British period the Vel Vidāne was the sole channel of communication between village and government while the rights and duties of the Gamarāla were of no practical account, but the independent authority of the Vel Vidāne was only gradually established. In Pul Eliya, down to 1926, the office of Vel Vidāne was always held by someone who was simultaneously entitled to call himself Gamarāla. The initial postulate that authority should be distributed among the different *pavula* of the village rather than concentrated in one place seems to have survived throughout.

But the traditional Gamarāla did not act on his own; he had feudal duties to perform and this implied that he must have tenants.

In every community each traditional *bāga* is subdivided into a certain specific number of traditionally established equal shares *(pangu)*. There might, for example, be two *bāga* each of ten shares, or three *bāga* each of twenty shares. In Pul Eliya there were originally three *bāga* each of six shares, making eighteen shares in all.

A share is primarily a share in tank water. The *gamgoda* land on which the house sites stand was originally laid out in parallel strips, *bāga* by *bāga,* in such a way that each *bāga* block had equal access to the irrigation channel running through it. Even today the house sites and their associated gardens conform approximately to this initial gridded plan (see Map 5.2). It is an arrangement of this kind that Codrington's informants were referring to when they told him that in a new village the house sites should be laid out "according to the *pangu*."

A similar arrangement prevailed in the irrigated field itself. The main irrigation channel was laid out in a single straight line, which was then divided up into major segments "according to the *bāga*." Then, within each *bāga* block, the subsections were divided up "according to each *panguva*." The details of this subdivision are complicated and will presently be described, but the net result is that each *panguva* has allocated to it an exactly equivalent portion of the main irrigation channel. Certain sections of the channel were not allocated to the *pangu* in this way, but were reserved as the perquisites of the various Gamarāla.

Once the sections of the main irrigation channel had been laid out and divided, each shareholder then proceeded to develop the land at right angles to the irrigation channel. How much land a particular man opened up in this way would depend partly on the lay of the ground, partly on the individual's energy, partly on the amount of water available. There were, however, no set limits. The shape of the Pul Eliya Old Field today simply represents the approximate shape it happened to have in the year 1900, when the government survey was made. Diagram 5.2 shows how the main irrigation channel runs in a straight line along the northern side of the field. The various strips are then extended at right angles from this channel toward the south (Map 5.3). But they vary considerably in length, and the irregularities were formerly even greater than they are now. In 1890 the strips embracing plots nos. 39–53

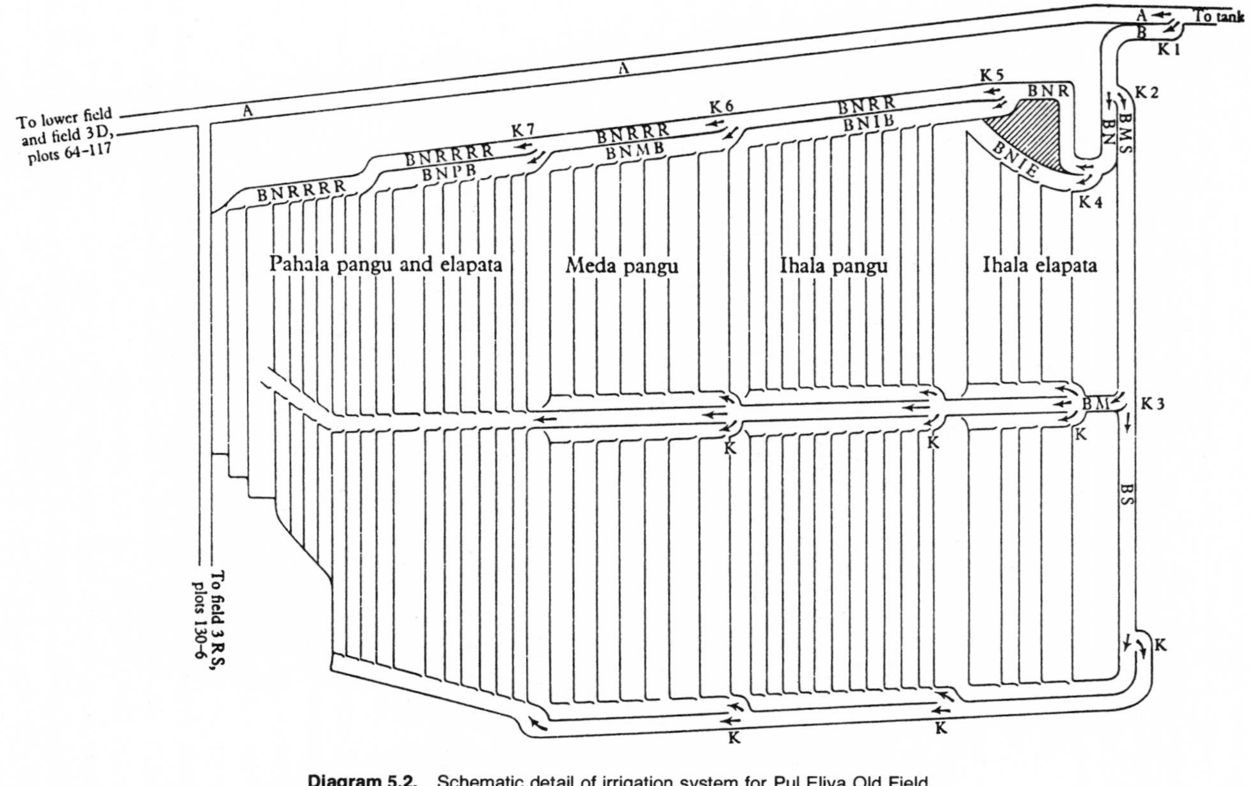

Diagram 5.2. Schematic detail of irrigation system for Pul Eliya Old Field.

extended considerably farther to the south, into a piece of ground that has now been abandoned.[2]

Since 1900 the shape of the field has been "frozen." Land tenure, as the British authorities understood it, necessarily consisted of rights to a particular piece of land rather than rights to a particular quantity of water. Consequently all *land* not actually under cultivation at the time of the survey was treated as Crown property. But it is the emphasis on rights in *water,* as opposed to rights in *land,* which explains the many peculiarities of the traditional system which are already apparent in Ievers's account.

It will be immediately obvious that in times of scarcity those parts of the irrigation channel which are close to the tank will receive a more regular supply of water than parts farther away. For the same reason, although the field is very nearly flat, and the field strips are supplied with water at both ends and in the middle,[3] the northern end of each strip, adjacent to the main channel, is slightly better placed than the southern end.

As an adjustment against these discrepancies we find that, within the field as a whole and also within each *bāga* as a section of the field, the land is subdivided into small allocations which are designed to minimize, as far as possible, the consequences of unavoidable inequalities in the distribution of water. In Pul Eliya this equalization principle has been carried to extremes.

From here on the reader should follow the description by reference to Map 5.3.

In the first place the field as a whole is divided into two sections, the Upper Field and the Lower Field. These are the units which Ievers describes as tracts *(pota).* The Lower Field has roughly half the area of the Upper Field. The three *bāga* are made up from sections in each field. Each field is divided into three main sections, but if we count the sections of the Upper Field as Ihala (A), Meda (B) and Pahala (C), then the corresponding sections in the Lower Field are arranged C, A, B. Each *bāga* also comprises an "end piece" *(elapata).* Thus the *bāga* are as follows:

A. Ihala bāga: upper end of Upper Field plus upper section of Upper Field plus middle section of Lower Field.

B. Meda bāga: lower end of Lower Field plus middle section of Upper Field plus lower section of Lower Field.

C. Pahala bāga: End piece made up of lower end of Upper Field plus upper end of Lower Field plus lower section of Upper Field plus upper section of Lower Field.

The general effect is that shareholders who own land in the lowest and least advantageous portion of the Upper Field also own land in the highest and most advantageous portion of the Lower Field.

2. See Map 5.1, empty area between plots 129 and 137. The ground was given up because of irrigation difficulties and poor soil.

3. This applies only to the strips in the upper field.

Pangu

Here it is necessary to digress and say something further about the exact nature of a *panguva* holding.

The translation of the word *panguva* as "share" is very appropriate, for it corresponds closely to the notion of share as it is used in the parlance of English company law. A *bāga* contains a fixed number of *pangu,* just as the ordinary capital of an English commercial company consists of a fixed number of "shares." It is possible for one individual to own several shares or alternatively for one share to be owned by several distinct individuals. Moreover, within one village community one particular individual may own parts of a number of different shares in different *bāga*. In theory, the number of shares is fixed from the start and should never change. But the number of individual shareholders is varying all the time. It follows that there is no precise correspondence between the number of strips (*issaraval)* in the field and the division into shares. On this point Ievers's original diagram and his written account are both misleading.

If we examine Map 5.3 in detail and consider only the central part of each field, we can see that each *bāga* consists of 60 fathoms in the Upper Field plus 60 fathoms in the Lower Field. Now there were originally 6 *pangu* in each *bāga*. Thus the original arrangement was that each *panguva* had 10 fathoms in the Upper Field and 10 fathoms in the Lower Field. Very probably, when the field was first laid out, the strips were actually laid out in this way, so that they were regularly at 10-fathom intervals all the way down the field. At the present day, however, the width of the strips varies considerably. There are strips of 4, 4½, 5, 7, 8 and 10 fathoms. This variation is shown to scale on Map 5.3. Even so, apart from minor discrepancies, the original general layout into 60-fathom major blocks has never been lost sight of. The *bāga* blocks as wholes have remained almost unchanged.

For every strip in the *pangu* portion of the Upper Field there is a corresponding strip of corresponding width in the Lower Field. This pairing is an essential feature of the system. It is not possible to own *pangu* land in the Upper Field without also owning a corresponding piece of land in the Lower Field. Land can be bought and sold or alienated in other ways, but if an owner wishes to dispose of one-third of his *pangu* land he cannot simply dispose of that worst portion of it, which is in the Lower Field. He will have to divide off one-third of his Upper Field strip and also one-third of his Lower Field strip. This fragmentation makes very good sense if it is remembered that what is being disposed of here is not really land at all, but rights to a proportion of the total water supply. The "fragmentation" that results is not an economic vice but a moral virtue!

The final detail which completes the logical perfection of the Pul Eliya system of equalized shares is that the order of the strips in the Upper and Lower Fields is reversed. What I mean by this can be seen from reference to

Map 5.3, where the details of Meda bāga are given for both fields. If the strips in the Meda bāga in the Upper Field are numbered α, β, γ, δ, etc., starting from the top of the *bāga,* then in the Lower Field they would have to be numbered in the reverse order starting from the bottom of the *bāga*. This feature, though not peculiar to Pul Eliya, is not found in all the villages round about, though in other respects the Old Field tenure in all communities is very similar. It seems evident that it is a feature which derives from an early phase in Pul Eliya history, for the present generation of villagers appear to be quite unaware that their strips are arranged in this fashion; all they know is that every Upper Field strip has its counterpart in the Lower Field.

Strips are sometines shared by more than one owner. For example, K. Dingiri Banda and A. V. Punchi Etani are the owners of plot nos. 15 and 16.[4] This holding amounts to half a *panguva* in the Ihala baga. The total holding consists of one strip in the Upper Field (plots 15 and 16) and another strip in the Lower Field (plots 91 and 92). Each strip is shared by the two owners. Division is effected by cutting each strip in half and reversing the holding in alternate years. Each owner works the northern half of each strip every other year.

All this sounds most improbable but, in fact, this part of the ideal scheme is adhered to closely. The generalization that every *pangu* plot in the Upper Field has a counterpart plot in the Lower Field owned by the same individual is very nearly true. Such small discrepancies as exist are due to the existence of "undivided holdings"—for example, plots 9–12 plus 95–97 make up a single holding shared by the heirs of a common estate; the same is true of plots 33–35 plus 103.

Down to 1954 it was still the general practice in Pul Eliya that no one could acquire a holding in the Upper Field without simultaneously acquiring a corresponding plot in the Lower Field.

The Allocation of Water

Finally we need to consider the arrangement by which proportionately equal amounts of water are fed into the field strips through the irrigation channels. There is only one main water channel leading from the *gamgoda* area to the Old Field (see Map 5.2). This channel is then divided into branches which feed into the Upper Field on the north and south sides and in the middle; the northern branch also extends along the northern edge of the Lower Field. These various ditches are not single channels, but multiple parallel channels which branch off and rejoin one another in a seemingly very complicated manner. However, once the principles involved are understood, these permutations and combinations turn out to be quite straightforward.

The distribution of the water is the responsibility of the Vel Vidāne. In theory he should make use of a traditionally established set of numerical ratios

4. Plot numbers refer to the numbers in the Vel Vidāne's paddy census return for 1954. The corresponding numbers appear on Maps 5.1 and 5.3.

which relate to the relative sizes of the different sections of the Old Field. The 1954 Vel Vidāne seemed scarcely to understand the rationale of the system, and much of my information on this point came from his predecessor.

The essence of the system of water allocation is that the flow of water in any particular channel can be subdivided into numerical proportions by making the water flow over a device called a *karahankota*. This is, in effect, a miniature weir consisting of a log of wood into which two or more flat-bottomed grooves of equal depth have been cut. The length of the grooves is proportional to the required ratio of division. For example, suppose the Vel Vidāne wishes to divide a moderate flow of water in proportions 2:1; he might make a *karahankota* with grooves of twelve-inch and six-inch breadth respectively. The flow of water over the twelve-inch "weir" is then assumed to be twice the flow of water over the six-inch "weir." The fact that the Vel Vidāne today makes his calculations in English inches does not necessarily mean that the system itself is recent. In pre-British times the Sinhalese possessed a carpenter's rule, *vadu riyana,* which was a cubit measure divided into twenty-four *angula* (finger joints). This could have been used in the same way, and even with the same numerical values, as the modern foot rule.

The traditional reckoning is that, if Upper and Lower Fields are both being cultivated, then each 60 fathoms of *pangu* land in the Lower Field is only entitled to one-third as much water as each 60 fathoms of *pangu* land in the Upper Field. The water rights of the three *elapata* seem to have been less precisely defined, though each was, in theory, 40 fathoms deep. The proportions of the different sections of the field are supposed to be as follows:

Field section	Fathoms	Numerical proportion of water allocation
Upper Field		
Ihala elapata	40	?3
Ihala bāga pangu	60	6
Meda bāga pangu	60	6
Pahala bāga pangu	60	6
Pahala elapata	40	?3 — 24
Lower Field		
Pahala bāga pangu	60	2
Ihala bāga pangu	60	2
Meda bāga pangu	60	2
Meda elapata	40	?2 — 8

When filling in "acreage" figures in the annual paddy census return for plots in the Lower Field the Vel Vidāne adjusts his figures so that the total acreage for the Lower Field (16 acres) is just one-third of that for the whole Upper Field (48 acres). Since the areas are quite fictitious, it seems obvious that this computation must have originated in the traditional rule about water allocation.

A complication results from the fact that the Upper Field is fed with water

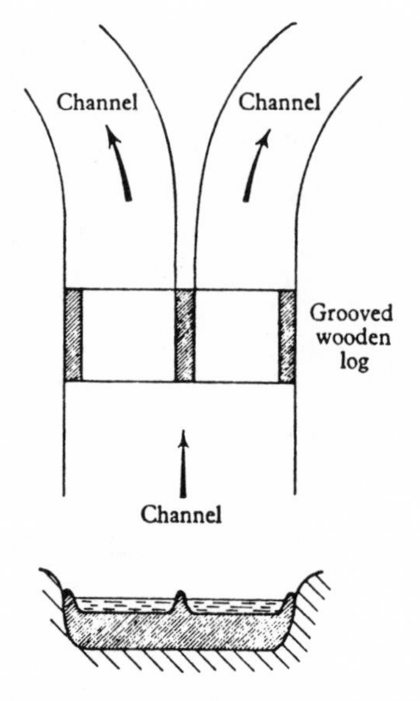

Diagram 5.3. *Karahankota.*

by three channels—one down each side and one down the middle—whereas the Lower Field is fed from one side only. If we distinguish the channels by letters as in Diagram 5.2, then the proportions of water flowing in each channel are supposed to be as follows:

Karahankota 1 (K1):
 Channel A direct to Lower Field, ¼ total supply.
 Channel B to Upper Field, ¾ total supply.

Channel B is then further divided as follows:

Karahankota 2 (K2):
 Channel BN to northern side of Upper Field, ¼ of channel B.
 Channel BMS to middle and southern side of Upper Field, ¾ of channel B.

Further along its course channel BMS is again divided:

Karahankota 3 (K3):
 Channel BM to middle of Upper Field, ½ of channel B (i.e., ⅔ of BMS).
 Channel BS to southern side of Upper Field, ¼ of channel B (i.e., ⅓ of BMS).

Channels BN and BS are then each treated in the same way, as follows: BN is first divided:

Karahankota 4 (K4):
 BNIE to Ihala elapata, ⅛ of channel BN.
 BNR to remainder, ⅞ of channel BN.

BNR is then ducted as far as the beginning of the Ihala bāga and divided:

Karahankota 5 (K5):
BNIB to Ihala bāga, ¼ of channel BN.
BNRR remainder, ⅝ of channel BN.

BNRR is then ducted to the beginning of the Meda bāga and divided:

Karahankota 6 (K6):
BNMB to Meda bāga, ¼ of channel BN.
BNRRR remainder, ⅜ of channel BN.

BNRRR is then ducted to the beginning of the Pahala bagā and divided:

Karahankota 7 (K7):
BNPB to Pahala bāga, ¼ of channel BN.
BNRRRR remainder, ⅛ of channel BN.

This channel BNRRRR then feeds the Pahala elapata. Within each *elapata* and each *bāga* block the water is ducted separately into each strip *(iscra)* but at this stage no special precautions are introduced to ensure equal distribution between each strip.

The subdivisions of channel BN on the north side of the field and of BS on the southern side of the field are virtually the same. Channel BM, which runs down the middle of the field, is somewhat different. It initially carries twice as much water as the other two, but feeds water left and right into the strips on either side.

The field as a whole slopes very slightly (*a*) from plot no. 1 downward toward plot 107, and (*b*) from north to south. The Upper Field as a whole, therefore, tends to drain off on the lower side of strips 55–63. This explains why channel BS is not carried on right to the lower end of the field. The division of water in the Lower Field follows the same general principles.

It must be understood that from the Vel Vidāne's point of view the ratios are established by tradition; they are not something which he has to work out for himself. Moreover, the ratios are not thought of as fractions but as *karahankota* of different sizes with the grooves measured in inches. The above seven *karahankota* should be:

	Lengths in inches	
Karahankota	Groove *a*	Groove *b*
1	6	18
2	6	18
3	12	6
4	3	21
5	6	15
6	4	12
7	12	6

In 1954 several of the *karahankota* were missing but those that existed were consistent with the possibility that some recent vel Vidāne had worked to the above numerical formula or at any rate had copied older *karahankota* which were constructed on this principle.

The complexity of the arrangement is itself relevant to my theme, since such a system is virtually unalterable. Although the present generation of Pul Eliya villagers are not at all clear about the inner logic of it all, they are keenly aware that the numerical formulae handed down from ancient times are very important. The general view seemed to be: We don't understand why things are arranged like this, but this is how they are, and we had better leave them alone.

The recent great extension of "acre land" cultivation—*sinakkara* and *badu*—has reduced the crucial importance of the Old Field holdings, and the owners are doubtless a good deal less fussy about the mathematical accuracy of their water rights than formerly. Nevertheless, it was quite clear that blocking or tapping a neighbor's water channel was still by far the most common cause of village quarrels and intercompound litigation. In the 1954 *Yala* season water was fairly plentiful. Had there been a critical scarcity, I might have observed the formal rules in much more rigid application.

Rājakāriya Duties

We see, then, that the use of both land and water in the Old Field is very elaborately enmeshed in a variety of institutional devices, all serving the same general purpose. The emphasis throughout this tenure system is on the principle that each *panguva* within each *bāga* shall have exactly equal rights to the total available water.

As against these equal rights, each *panguva* carries also exactly equal obligations with regard to the maintenance of the tank bund, maintenance of field fencing, etc. The precise obligations attaching to a *pangu* holding are today defined by government regulation; their substance has not changed since 1850. The duties are collectively known as king's work (*rājakāriya*). Briefly they are:

1. Each shareholder shall carry out a certain amount of repair work on the bund annually, proportional to the amount of his holding.

2. Each shareholder shall build and maintain that portion of the main field fence which is opposite the ends of any strip which he works.

3. Each shareholder shall maintain in good condition any irrigation ditches which go past or through the land which he works.

4. Each shareholder shall take his turn to sit up all night in one of the field huts to ward off wild animals which are liable to attack the field.

This last obligation, which is an extremely onerous duty, applies to the whole season during which the crops are in ear.

Perquisites of the Gamarāla

So far we have been discussing only the rights and obligations of shareholders in general, that is of all individuals who have rights to *pangu* within a particular *bāga*. We have not distinguished the Gamarāla's rights from those of his followers.

Today, indeed, there is no distinction. *All* shareholders, Gamarāla and tenant alike, are now presumed to hold their land freehold. But formerly the Gamarāla's legal status was quite distinct from that of the junior stockholders in his *bāga*. Ievers's original account here seems to be essentially accurate in all respects.

Of the total length of water channel allocated to a *bāga,* only part was originally divided up among the various *pangu*. The remainder was allocated as an end piece (*elapata*) and this was initially a perquisites of the Gamarāla. Moreover, out of the total *pangu* in any one *bāga,* two had a special status. One, the *elapat panguva,* was supposed always to be owned by the holder of the *elapata* and was, therefore initially a perquisite of the Gamarāla. The other, the *gamvasama,* must always be owned by the Gamarāla. This last is still the case. Always, within each *bāga,* there is one *panguva* which is described as the *gamvasama,* and the owner of this strip is automatically entitled to call himself Gamarāla, though today there is no advantage in doing so.

Formerly, as owner of the *gamvasama,* the Gamarāla had special obligations as well as special perquisites. Among other things, he had to contribute grain to the annual village feasts and he had various ritual duties at the associated religious ceremonies.[5] In Pul Eliya these obligations are still imposed upon the holder of any *gamvasama* plot, whether or not he chooses to lay claim to the title of Gamarāla.

As shown in Map 5.3, each of three *bāga* has one *elapata,* one *elapat panguva,* one *gamvasama* and four ordinary *pangu*.

According to Ievers the *elapata,* the *elapat panguva* and the *gamvasama* should *all* belong to the Gamarāla, but this represents only an ideal initial situation. When detailed Pul Eliya records began in 1886 the pattern had already diverged widely from this ideal. In that year, in each of the three *bāga,* the *gamvasama, elapat panguva* and *elapata* were in different hands.

Nevertheless, the theoretical association of the *elapata* with the *elapat panguva* provides yet another example of the principle of 'fair shares.'

Since the *elapata* constitutes the end of the field, it therefore carries with it the obligation to build and maintain the whole of the end fence. This is

5. For example, at the annual ritual known as *mutti manggalaya,* one Gamarāla tends the lamps in the *vihāra* and in the *kōvil*; one Gamarāla supplies the ritual pots (*mutti*); one Gamarāla pays for two shotgun cartridges, the firing of which forms part of the ceremony. The obligations rotate among the three Gamarāla in successive years.

about ten times more fencing than attaches to any ordinary *panguva* strip. Because of this extra fencing obligation, the owner of an *elapata* is excused from the duty of carrying out tank repair work. But the *elapat panguva* has no such privilege. Thus the idea behind the doctrine that the *elapata* and the *elapat panguva* should always be owned by the same individual is simply to ensure that no one wholly escapes from the unpleasant obligation of carrying out tank repair *rājakāriya* duty. This was felt to be particularly important since in the event of a breach in the bund all villagers must be equally responsible.

In a comparable way, while the owner of an *elapata* and the owner of a *gamvasama* must both pay for the building of watch huts, the latter, as Gamarāla, escapes the *rājakāriya* duty of night watchman. But, unlike the owner of the *elapata*, the *gamvasama* owner must do his share of bund-repair *rājakāriya* along with the other shareholders. In Pul Eliya this carefully differentiated system of rights and obligations has been rigorously maintained even though the status of the Gamarāla as a specialized class of individual is no longer formally recognized. The rights and duties attach to the land itself, not to the individuals who own it.

Contemporary Practice in Pul Eliya

So much then for the theory behind the tenure of land in Pul Eliya Old Field. Now let us consider the actual state of affairs as it existed in 1954.

According to present-day Pul Eliya tradition the Old Field originally contained eighteen *pangu*, six for each *bāga*, but at some unspecified date in the past two extra *pangu* were added to the Pahala bāga by reducing the amount of land allocated to the Pahala elapata.

The circumstances which brought this change about are not now remembered, so I was fortunate that among the few nineteenth-century documents relating to Pul Eliya which still survive there are two tax returns which appear to confirm the tradition.

The village Vel Vidāne still submits annually to the revenue administration a return purporting to show the exact amount of land cultivated throughout the village and the precise ownership of each plot. Today this return is compiled for the purpose of crop statistics, but its form is just the same as that of the paddy tax census of the 1870–90 period. It is, therefore, easy to correlate surviving tax census documents with the layout of the modern field.

Detailed analysis shows that the 1889 list is drawn up according to a scheme of eighteen *pangu;* the 1890 list on the other hand, fits the present-day arrangement of twenty *pangu*. The story of the "two extra *pangu*" must therefore be correct and the alteration must have occurred shortly before 1890.

Because of this satisfactory fit of documentary evidence with oral tradition, I feel confident that Map 5.3 (which fits the present-day arrangement of strips to an "original" system of eighteen *pangu* and three *elapata)* is justified and correct.

"Originally" the field consisted of three *bāga;* each *bāga* comprised a forty-fathom *elapata* and six *pangu;* each *panguva* comprised ten fathoms in the Upper Field and ten fathoms in the Lower Field. Such discrepancies as now exist result from the fact that shortly before 1890 two fathoms from the *panguva* 4 of the Pahala bāga together with twenty-two fathoms from the Pahala elapata were reclassified as forming "two extra *pangu.*" Since that date the Pahala bāga has been deemed to consist of eight *pangu* as opposed to the six *pangu* in each of the other two *bāga.* The principal effect of this reclassification has been to alter the type of *rājakāriya* obligation falling on owners of these plots of land.

Bethma

The arrangement of the irrigation channels together with the Vel Vidāne's assumptions concerning water allocation for the different parts of the field have the following implications:

1. The Upper Field consists of two equal parts—the north half of the field and the south half of the field.

2. The Lower Field is half the area of the Upper Field. Thus the field as a whole is divided into three supposedly equal areas, each of which contains the same number of strips of the same width, owned in the same way. One-third of every holding falls into each of the three main parts of the field. This symmetry has important consequences.

The North Central Province institution of *bethma* has received frequent comment. This is an arrangement whereby the shareholders in a field which is short of water may agree to cultivate only a proportion of that field and then share out the proceeds among themselves. The theoretical procedure, as described by Farmer, is as follows:

> The village has an admirable system, known as *bethma,* under which, if the whole extent of the paddy field cannot be cultivated for lack of water, as many of the tracts as can be irrigated are divided, regardless of their ownership, between the peasants in proportion to their several holdings, and thus cultivated as a compact block with minimum waste of water. [Farmer, 1957, p. 558]

The earliest reference to *bethma* in this form is in administration documents of the 1861–64 period.

I have studied these entries with care, but they are unfortunately ambiguous. It is evident that the government agent of that date imagined that the system was supposed to work in the way that Farmer has described, and he on several occasions records the fact that he had ordered reluctant villagers to carry out *bethma* division in this way. But it seems to me probable that this form of *bethma* was the unintended invention of the British government agent himself!

At the present time different villages seem to work *bethma* in different ways, and there is no means of ascertaining which, if any, of these methods is the ancient traditional system. But what is quite clear is that the Pul Eliya method is very much simpler than that described by Farmer. Furthermore it is *bethma* which provides the ultimate justification for fragmenting each individual holding in the complicated way I have described. For Pul Eliya the system is as follows.

If the villagers are to cultivate rice in the Old Field during the *Yala* (April/September) season they will decide from the start either to cultivate the whole of the field or two-thirds of the field (that is, the whole of the Upper Field only) or just one-third of the field (that is, the northern half of the Upper Field only). No pooling of proceeds or reallocation of holdings is necessary since the land is already divided up in such a way that each shareholder works the whole or two-thirds of one-third of his total holding as the case may be.

In my limited experience this is the most common form of *bethma* in all this area. The ideal scheme described by Ievers, in which the total field is divided into two or more tracts (*pota*), corresponds to the actual facts for all the villages in the Pul Eliya area. It is invariably the case that every strip or holding in the upper tract has a corresponding strip or holding in the lower tract, though the precise manner in which this is effected is not always the same. This fragmentation of individual holdings is always directly associated with the local practices regarding *bethma*. The relative size of the different tracts (*pota)* is such that when the water is scarce, cultivation of the upper tract only, or of half the upper tract divided longitudinally serves as a *bethma*.

Farmer's description, which is the orthodox one, implies that individual Sinhalese farmers get on so well together that they can readily agree to a reallocation of land in times of water scarcity. I can only say that this does not correspond to my experience!

References

Brohier, R. L. 1934–35. *Ancient Irrigation Works in Ceylon.* 3 vols. Colombo.

Codrington, H. W. 1938. *Ancient Land Tenure and Revenue in Ceylon.* Colombo.

Coomaraswamy, A. K. 1950. "Notes on some Paddy Cultivation Ceremonies in the Ratnapura District." *Journal of the Ceylon Branch of the Royal Asiatic Society* 18, pp. 413–28.

Farmer, B. H. 1957. *Pioneer Peasant Colonization in Ceylon.* London

Ievers, R. W. 1899. *Manual of the North-Central Province, Ceylon.* Colombo.

Leach, E. R. 1959. "Hydraulic Society in Ceylon." *Past and Present* (London), no. 15, pp. 2–25.

Pieris, Ralph. 1956. *Sinhalese Social Organisation: The Kandyan Period.* Colombo.

Wittfogel, K. A. 1957. *Oriental Despotism.* New Haven: Yale University Press.

6

Japanese Irrigation Cooperatives

Richard K. Beardsley
UNIVERSITY OF MICHIGAN

John W. Hall,
YALE UNIVERSITY

Robert E. Ward
STANFORD UNIVERSITY

Introduction

Niiike is a small rice-growing settlement in the western half of Honshu, Japan's Principal island. One hundred thirty people live in its twenty-four houses clustered at the base of a pine-covered hill. We are interested in these one hundred thirty persons not as so many individual Japanese but as members of a community. Niiike is a clearly defined, natural sociocultural unit, large enough to be the setting for most situations and events that are the basic stuff of a way of life, yet small enough to let us see this way of life at close quarters. As members of communities ourselves, we will recognize familiar conditions and problems among those that confront the people of Niiike. Certain other situations are peculiar to Japan or to this particular locality. But the solutions, even of those problems with which we are familiar, are often not what we would expect. Niiike's style of dealing with life's situations is what makes it distinctively Japanese rather than some other nationality, rural rather than urban, Niiike rather than some other settlement. Yet much of Niiike's way of life is also customary and normal in a great many other Japanese communities. Our study concentrates on the life of people in a rural community called a "buraku." In non-technical terms, the buraku is a small rural settlement, a social entity consisting of households grouped into a community. We retain untranslated throughout this study the Japanese word "buraku," because it has no appropriate counterpart in our experience and in order to avoid confusion with two other Japanese terms, "ōaza" and "mura," which refer to successively larger divisions of the rural countryside. From a purely geographical point of view these terms are generally so related that several buraku lie within each ōaza and several ōaza lie within each mura. But these terms connote more than a difference in geographic size. Outside the cities and towns, all of Japan is divided into mura: units of territory fitting border to border to furnish a blanket of government over the entire country, including hills and open fields as well as settlements. The mura has a full apparatus of government; it is the smallest legal administrative unit of Japan and may be considered the administrative village. The ōaza beneath it is also a territorial unit with a vestige of administration inherited from the period of Japanese history prior to 1872. Neither the mura nor the ōaza, however, is a community all of whose inhabitants have constant and face-to-face social interaction. The buraku is just such an entity; its government is largely unofficial, and the borders of its land are only hazily defined; but it is a clearly defined social community. Its inhabitants have a strong sense of group identity. For this reason, it is a useful and convenient unit of study.

To call this community by its Japanese title, Niiike buraku, unfortunately does not relieve us of all terminological difficulties. The same sort of community in some other parts of Japan is called an *aza,* while "ōaza" may be used to refer to only the principal community among a group of *aza,* all of them together composing a mura, or administrative village. Moreover, almost all over Japan, there is popular use of the term buraku in speaking of segregated settlements of the onetime classless people, the outcast Eta. Even in the vicinity of Niiike, this pejorative connotation may color the word in conversational use. But the usual meaning of "buraku" is the perfectly honorable one of "small rural settlement."

But we still must ask how appropriate Niiike is as an example of a buraku. It lies on the coastal plain of Okayama Prefecture in one of Japan's more prosperous farming areas. Among Japanese country folk, the farmers of this plain are regarded as generally up to date and forward looking. Niiike is run of the mill within the area. It is neither particularly old-fashioned nor progressive, rich nor poor, specialized nor diversified. It is a representative community of the Okayama coastal plain. But no one community can be fully representative of more than a limited area; its historical background, its present degree of prosperity, its size, its topographic and climatic setting, and its products, to name only a few features, differ slightly or grossly from those of other communities. Therefore, we have on occasion drawn comparisons, to show the variety of alternatives and necessities from which the people of Niiike and the Okayama coastal plain have shaped their living. These comparisons also make us aware of the diversity of rural Japan and of the importance of a region's history to the character of its communities.

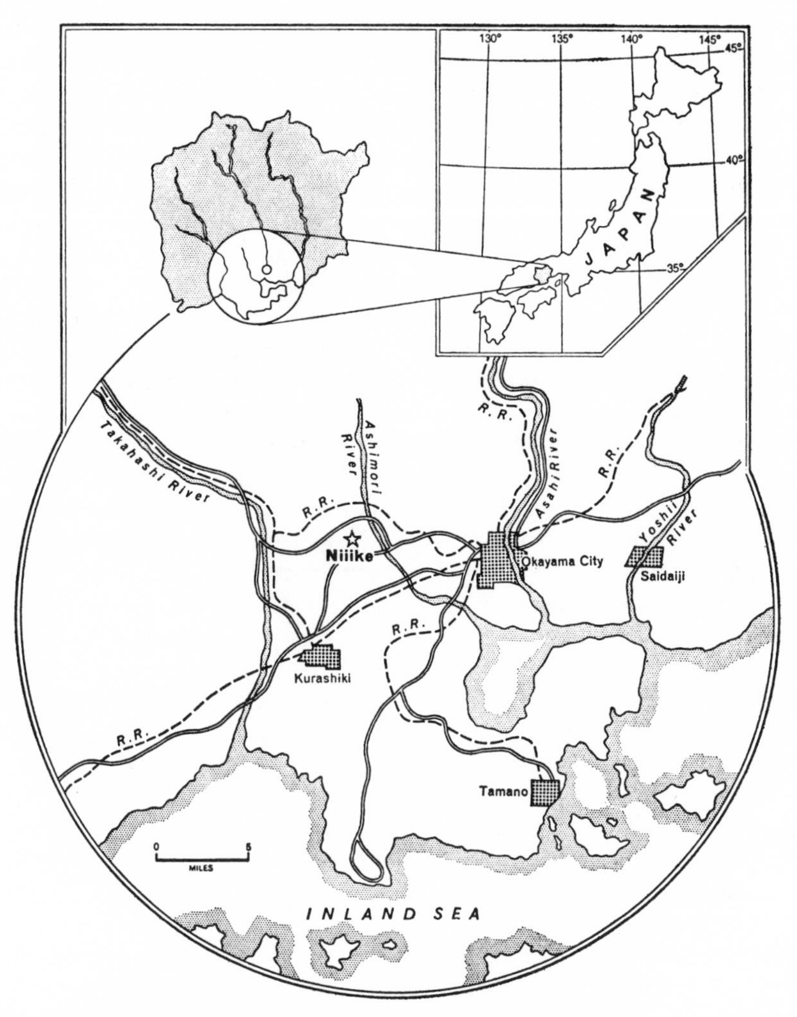

Figure 6.1. The location of Niiike in Okayama Prefecture, Japan.

Land and Water

The serried paddy fields lying beneath the pine-garbed hills of the Okayama Plain give an impression of changeless antiquity. Even the unhurried, deliberate movements of the farmers as they work among their crops or in their drying yards contribute to the aura of timelessness which lies over the landscape. There are evidences of the past all about. Yet the scene itself has been constantly changing. This landscape contains few of the natural features as they were a thousand years ago. They have been modified over the centuries and are still being reshaped by man. Over the centuries men have effaced the sluggish, estuarine streams that once drained these valleys. They have diked the Ashimori River to curtail its floods and have dug ditches to drain the swamps. Men have made ponds to hold irrigation water and later have erased them to set fields in their place. Men have cut down the ancient forests, terraced the hillsides for crops, and then once more covered the hills with trees. At a pace measured in generations, from the time of first settlement through the medieval projects of drainage, clearing, and canal-building to the modern period when revision of the ditch system transformed and rationalized the entire pattern of rice fields, the farmers of this valley have transformed the landscape. These many generations of planning and reconstruction were unified by the consistent effort to feed an increasing population. Flood control, new varieties of crops, alteration of soil conditions, and the adoption of new agricultural techniques all contributed to this effort. The wasteful prehistoric chieftains, it is true, placed their great tombs in the middle of the fields, but every other move affecting the appearance of the valley floor and hills, or even the distribution of houses, can be interpreted in terms of more efficient planting and management of fields.

Rice cultivation has been the central theme in the development of Okayama agriculture. From the dawn of agriculture in Japan, irrigated rice has been the overwhelming important crop. Dry fields have been used largely for supplemental purposes; even the dwindling forests and meadows have been used to support rice growing, providing green cuttings to fertilize the paddy fields. Today 56 percent of all Japan's cultivated land is in paddy. In Niiike this figure is 85 percent. Wet rice cultivation has been a potent force shaping communities such as Niiike. Its implications must be understood before we can adequately comprehend the problems of land and water faced by the people.

Even at a simple level of technology, rice grown with ample water provides a maximum yield per unit of cultivation. Water acts as a retardant to weeds and certain insects and in other ways aids the cultivator. In particular, nitrogenous bacteria that flourish in the still water covering the fields constantly replenish the soil and make possible its repeated use. Thus, apart from the high yield, wet growing conditions minimize the need for crop rotation. These natural qualities provide a secure and constant base for an agricultural population, contrasting in important respects with the technique of dry farming familiar in Europe. Europe's traditional agricultural technology centers around grains that grow well under dry soil conditions: wheat, barley, rye, and oats. It puts a premium on diversity of land use and rotation of crops. It has fostered close integration between animal husbandry and agriculture, since the constant need for fertilizer has made the Western farmer depend on manure from his animals.

Thus, starting with cereal crops that grow well under different conditions, two distinct agricultural technologies have grown up on opposite sides of the Eurasian continent. Farmers of the Orient increasingly have emphasized intensive cultivation of land, perfecting irrigation systems and organizing their communities to manage them. They have become able to grow two or more crops annually in the same field. As units of cultivation have grown small, maximum production per unit has been achieved by lavish expenditure of human energy. In contrast, most occidental farming has more land available, so its techniques are prodigal of land, e.g., fallowing and the grazing of large animals. As acreage expands, these techniques permit less manpower per acre and lower unit costs; so the system encourages extending the acreage of each farm. Substitution of machines for man and animal power fits the logic of this *extensive* farming, which is probably most highly developed today in the American Middle West. Instead of more crops per man-hour, however, *intensive* farming develops the logic of more crops per acre; it reaches a peak in Japan's small, double-cropped farms. In their version of the equation, MANPOWER UNITS + LAND UNITS = CROP UNITS. Land units are necessarily tiny, so they boost manpower in order to produce peak crops.

Intensive cultivation of rice tends to impose certain conditions on the society it supports. Its high yield stimulates dense population; its heavy requirements of manpower in turn depend on dense population. Paddy technology is complex and demanding. Preparation of paddy fields and the planting and transplanting of rice are laborious operations little amenable to types of mechanization devised to date. Rural Japan thus has long used its heavy concentration of population as a work force. Problems arise from the fact that this population is usefully employed only through a limited season; but the workers live the year around on the farm, so they often face long periods of underemployment or, as it is called in Japan today, "concealed unemployment." This lack of year-round work, however, does not contradict the reciprocal relation between rice and population; as population rises to take care of cultivation, the methods of cultivation must be intensified to produce food for the increased number of months.

Paddy culture is self-perpetuating in yet another way. The amount of labor and capital put into the creation of paddy land provides a continuing incentive to maintain the system. The construction of paddy fields is laborious and costly, hard for those familiar only with "natural" agriculture to comprehend. The fields must be constructed to hold water an exact inch or two deep over the entire surface. In preparing them the farmer must frequently start several feet below the surface and build up with carefully graded soils. Moreover, each paddy plot must be linked to an irrigation system, which in itself is often a major undertaking. The same irrigation system must also provide proper drainage of rain and ground water during the seasons when the fields are dried. This complex system of terraced fields and interlacing irrigation and drainage ditches is not easily assembled, nor is it easily modified. It becomes the cherished heritage of an entire community generation after generation.

In the final analysis this type of technology leaves a strong imprint on the community which engages in it. Each household, as a unit of production as well as ownership, needs internal cohesion and cooperation to keep and make the most of its landholdings. The household, in turn, must look to the local community, both as a source of cooperative labor at crucial times and as the management unit for irrigation. Close association among households exist in most areas of Japan today, even though the forms of association vary sharply according to local conditions.

Beyond the level of the family the social implications of paddy culture merge with many other molding forces, yet the effects of this technology are clearly visible. Intensive cultivation under these conditions tends to limit returns to a subsistence level. The great supply of labor tends to limit the mechanization of agriculture. In so far as the workers must be fed from their limited land, food crops cannot readily be supplanted by ones that have high market value or shifted to meet variable market demands. Farmers with low income on scarce and high-priced land tend to lose their independence and, in any case, lack income for improvement. Agricultural improvement frequently depends upon government or the wealthy and powerful. Land reclamation and water management tend to lie beyond control by the individual farmer and rest with the community, the landlord, or the feudal lord. In Japan, as in other Asian countries with similar economic features, political power has been closely related to the control of land and water.

Important though land is to contemporary Niiike, it is no longer the sole reliance for every household or for the community as a whole. Recent measures have pushed back the threat of landlordism, and the circle that leads from rising population to more intensive cultivation and back to newly increasing population has been broken by outside economic factors as well as ones internal to the community.

Land Types and Their Use

Every square rod of land in the Okayama Plain is precious. The average Niiike household owns somewhat less than three acres of productive land, which must provide not only food for the household but the produce sold to the city as well. Because the premium is high on land that can be made to produce heavily and constantly and because rice outproduces every other crop, irrigable rice fields are most prized, particularly if they will produce a second crop. In ordinary conversation people take irrigated land as the measure of a family's holdings. Yet other types of land are listed on the tax records. In the productive category, two other kinds are important: dry field and wood lot. Uncultivated land, of some importance elsewhere for pasture and other purposes, is inconsequential in the vicinity of Niiike. Other private holdings include lots for farmhouses and outbuildings and privately owned ponds or canals.

Communally owned land, although minimal, also exists in Niiike. It comprises the cemetery, the lots upon which the village assembly hall and fire-equipment shed are built, and one patch of wood lot. The last is owned in common by all buraku of upper Shinjō ōaza, which use it only as a source of wood for funeral cremations. Roads, paths, canals, and streams are public property and not taxed, although responsibility for upkeep belongs to the buraku or larger entities, depending upon their location and importance. The space given to these is restricted. Cultivation is the primary goal.

The importance of paddy land can be illustrated in two ways. In 1956 in the Okayama Plain, paddy land was selling at prices ranging between $1,100 and $2,300 per acre. (The figures here approximate the real price rather than the nominal price fixed by government regulation.) At this price, most Niiike farmers have holdings worth $4,000 or more, equal in value to ten or twenty acres of farmland in the American Middle West. This comparison is inadequate, not only because the prices of other things such as housing differ, but also because there are aspects of landholding that defy calculation in monetary terms. Cultivated land is so hard to get that those who possess it often are unwilling to sell at any price, to the vexation of public officials who must then confiscate land required for public or industrial developments. The fact that cultivation can compete in cash profits with other land uses helps to explain why plots of rice may be seen growing well inside the fringes of a city. In non-economic terms, the value of paddy land can also be illustrated by a distinctive system of place names. Peaks and promontories, streams and valleys very often have no standard names. A river or stream may change its name every few hundred yards along its course, and the same mountaintop will be known by different names to the persons who live north and to the persons who live south of it. By contrast, almost every little parcel of paddy land has its distinctive name, which it has carried through generations, sometimes from as far back as the Taika Reform of A.D. 645. Dry fields and wood lots rarely have acquired names in this fashion.

Yet every household, for healthy economic balance, needs to own wood lot and dry field as well as paddy land. The wood lot, at the minimum, supplies kindling for winter warmth. Dry fields provide much of the household's food. Once minimal needs have been met, few households strive to increase their dry-field holdings. These remain nearly constant while either paddy land or forest holdings are built up. New or impoverished households, who must worry first about subsistence rather than cash production, are apt to have a high proportion of dry fields. Some conception of the balance of landholdings in Niiike is given in Table 6.1.

Table 6.1.

	Average Holding		Largest Holding		Smallest Holding	
Type of land	Percentage	Acres	Percentage	Acres	Percentage	Acres
Paddy field	58.0	1.57	33.1	2.88	28.5	0.18
Dry field	10.4	.28	3.9	.34	30.8	0.20
Wood lot	31.6	.86	63.0	5.49	40.7	0.26
Total	100.0	2.71	100.0	8.71	100.0	0.64

Dry-field acreage varies much less from the average than does the area of paddy land or of wood lot, as the acreage figures show. The smallest owner, Iwasa Tamaichi, one of the evacuees who took up life in Niiike on whatever land he could get during the war, is short in all three categories. He has to make up his deficiency even in dry-field produce by money earned from his side trade as a barber. Note, however, that even the largest landowner has not a great deal more dry field, for such land has little value either for current income or as investment. Dry fields are gauged to meet subsistence needs within the household but are not increased beyond this point if paddy land or wood lot can be purchased. Land reform did not touch dry-field holdings, so it has not altered this situation.

The distribution of land types shown in Table 6.1, while valid for the average household today, differs markedly from the balance prevailing in Niiike until a generation ago. The farmers of Niiike have shifted in recent years to an increasing emphasis upon paddy farming for a number of reasons, among them the increased importance of cash-crop farming, the collapse of the market for locally grown cotton, and the improvement of hitherto marshy paddy land. To make clear the reasons for this shift, we must briefly review the recent history of land development in the vicinity of Niiike. Although this is the history of a local change, it typifies changes which have brought about an increase of paddy land throughout Japan.

Niiike is a young settlement compared to its neighboring buraku, most of whose beginnings go much further back than the several centuries Niiike can claim. Niiike's first settlers struggled with drainage problems, for much of their land was backmarsh. To judge from the shapes of fields and their names, drainage problems first were solved nearest the river, probably because river silts deposited there raised the land level. Historical records show that cultivation near the river antedates the Taika Reform which reworked some of these fields into the regular *jōri* pattern. During the following centuries, fields closer to the head of the subsidiary valleys were brought under cultivation, and the land at the head of the side valleys was reclaimed last. Niiike, pocketed at the very head of its little valley, was the last area to be drained and cultivated. Piecemeal attacks on the ever present drainage problem up to 1925 had resulted in meandering ditches that crosscut the fields into tiny, irregular shapes. A large gash was cut across the valley floor by a diked ditch that led run-off water from the hills on the north across the valley to the prehistoric moat at the base of Tsukuriyama and thence drained into the Ashimori River. Although this ditch was built to protect the acres lying between it and the Ashimori River, floods recurred, and one tract—the Shinjō Depression—although cultivable, remained so swampy that wry jokes were made about outsiders who, blundering through its footpaths, fell in and were never seen again. The dike of the large ditch, moreover, prevented the river water from reaching the land in front of Niiike in dry periods. Irrigation water had to be supplied by a good-sized pond immediately west of the present village. The presence of this pond, according to one local conjecture, accounts for the name Nii-ike, or "new pond." Thus the Niiike settlement grew to much its present size by combining intensive use of the dry fields on the hill behind the settlement with the greatest possible use of the unsatisfactory paddy fields in front.

The worst features of this situation were corrected through a large-scale program of land improvement begun in 1925. At that time the central government, seeking a monumental enterprise to commemorate the reign of the Taishō emperor, made special funds available for irrigation and land improvement loans. All over the nation a wave of projects ensued which resulted in "Taishō ponds," "Taishō canals," and "Taishō dams." The project centering on this valley affected not only Niiike and other buraku of Kamo village but the buraku of neighboring villages as well. It never was fully completed because its cost outran the Y240,000 fund raised to finance the project. Instead of seventy-two additional acres of fields, only twenty-four acres were created, and some land still lies too low to produce two crops each year.

Niiike, however, profited immensely from the improvements. Leveling of the entire valley floor between 1925 and 1930 made possible parallel ditches for drainage and irrigation between large, rectangular fields over most of the area; the earlier, irregular pattern, preserved where ancient tombs were not completely leveled, makes a striking contrast (see Fig. 6.3). A uniform grade of feeder-drainage ditches was established to enlarge and straighten the fields; a better drainage ditch carried run-off water from the hills on the north to the Ashimori without impeding irrigation from the river; in place of the large local pond, which shrank to a fraction of its former size, water became available from Taishō Pond in the hills of Yamate village. Also, a new road built from the highway across the valley past the settlement was wide enough to admit motor vehicle, thus vastly improving the village's communications.

Before this improvement of paddy land, most households in Niiike acquired their cash income from cotton grown on the hillside fields behind the buraku. Cotton was then a major product of the Okayama Plain, where some of Japan's largest spinning mills had been built after the Meiji Restoration. The mills later turned to the cheaper and more plentiful cotton imported from

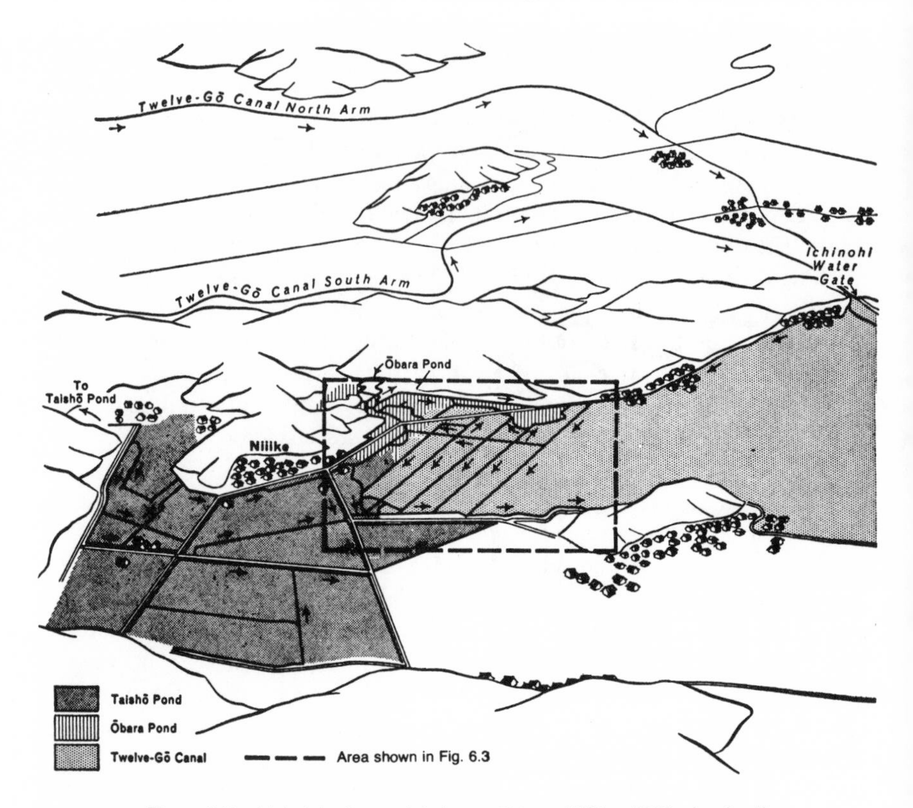

Figure 6.2. Main irrigation and drainage ditches of Niiike, 1950, showing the flow of water and the fields served by each water source.

the United States and southern Asia. The market for domestic cotton thus declined sharply after 1910. The Niiike farmer turned to raising mat rush, the basic material for the covering of floor mats. A swamp plant, this rush could be grown as a second crop even in hard-to-drain paddy fields. Intensive use of chemical fertilizer began about the same time. Household labor was needed now in the paddy fields or could be profitably employed in weaving the mat rush into *tatmi* covers. Gradually cultivation of the hillside dry fields was abandoned, and pine trees were planted. Today the major part of Niiike's cash income is derived from the cultivation of rice and mat rush. Niiike became a paddy farming village.

All parts of the valley floor low enough to lead water to are utilized for paddy. As has been said, such land is not merely cleared, it is constructed, beginning with the subsoil. Where decomposed granite from the hillsides forms the subsoil near Niiike, drainage is adequate; but where the base is muck derived from the original stream or swamp, proper drainage can be insured only by adding loads of gravel to lighten the soil. Over this base soil is

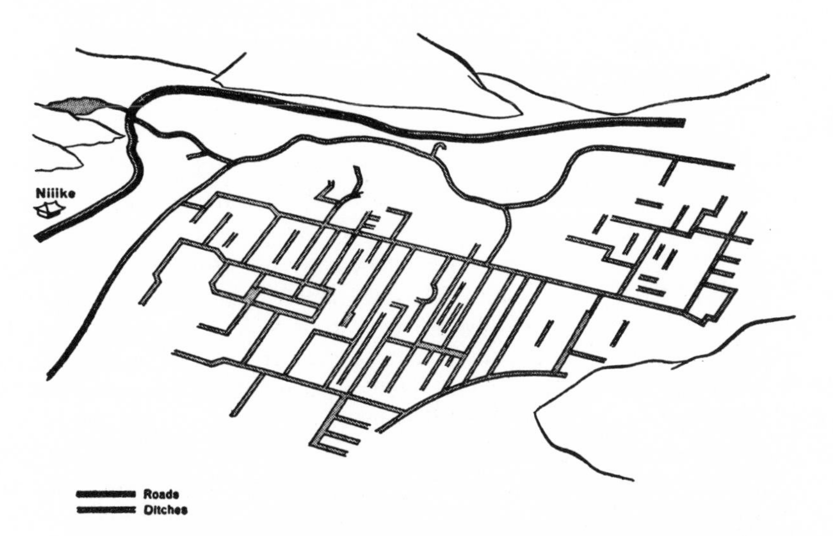

Figure 6.3. A portion of Niiike's fields in 1892, before the revision of the ditches. Compare the crop area with the fields in the area demarcated by broken lines in Figure 6.2.

the surface mud, made black by tons of organic material added through the centuries. This is carefully leveled and surrounded by a low dike. Even after such careful construction, fields are not identical in their characteristics. Some are overwatered when rains are heavy. Others go dry during a drought. The wetter fields can grow only one crop a year or three crops in two years; most desirable are the fields which bear two crops each year. The most swampy fields of all, in the Nishinuma sector, are rarely planted to anything other than vetch in winter, for they can never be drained, as is required one year in every three for growing mat rush as a winter crop. But even here the unending pressure for more crop land has led to such experiments as planting wheat in wide-spaced furrows, each furrow piled extra-high to raise the roots above the standing water in the troughs.

Dry fields lie above the practical limit of irrigation. In much of the Okayama Plain, particularly the eastern half, dry fields are valuable for their orchards. The relatively dry summer climate of Okayama is beneficial for fruit. But in Niiike only two households in recent times have grown fruit for outside, city markets. Niiike's hillside fields are more important for their crops of dry grains, white and sweet potatoes, beans, and root and leaf vegetables that serve primarily for family consumption. Most of the dry fields are terraced or sloped on the hillside behind the housing area and along a strip that extends around the upper slopes of Ōbara Pond. Chemical fertilizer and compost are used chiefly in the paddy fields, but night soil is used in the dry fields. This permits their intensive use, both through cultivation of successive crops throughout the year and through intercropping or "multi-story" cropping, the system under which one sees in a single field rows of vegetables planted between rows of grain, the whole overtopped by a grape arbor or fruit trees.

One major contrast in the economic function of dry and wet fields appears in the division of household labor. Adult men, women, and young people past school age all work in the fields. The men are invariably found in the paddy fields, not merely because of plowing and other heavy work, but because these fields produce the primary cash crops. Women and young people also work in the paddy fields, particularly at periods of peak labor, but the cultivation of dry fields on the hillside is almost exclusively their concern. The household head wields a hoe or other small hand tool in a dry field only if he raises fruit, melons, or some vegetables for sale.

There is wood lot on the hillside behind the settlement which was planted when cotton-growing was discontinued forty years ago. This wood lot meets the regular needs of each household. But most families also keep some plots of trees on the mountain slope across the valley for lumber and for winter firewood. Fast-growing species, particularly red pine, predominate; these are the trees that are massed in dark green profusion on top of nearly every hill rising out of Okayama Plain. The original, native vegetation of hardwoods such as oak and maple was cut centuries ago in most localities, to provide fuel and building lumber for the dense population of the coastal plain. Even the fast-growing species must be husbanded carefully to prevent denudation and consequent erosion, a danger of which the farmers in this area of torrential rains are keenly aware. The potential commercial value of trees has led a few households to acquire additional wood lot as an investment. During the war, higher prices and government exhortation led some households to sell patches of pine which were cut, sent away, and used in an ill-starred attempt to produce aviation fuel from resin. Such patches of land have been replanted to peach trees and edible bamboo.

Most of the smoke haze that hovers over the valley from the evening bath and cooking fires comes from rice straw, the primary cooking fuel. Charcoal, produced in the mountains and bought through stores, is used in small handfuls to provide most of the winter heat. Unlike many areas of Japan, compost made from pine needles, grass, and brush swept from beneath the trees is only of minor importance in Niiike. Nonetheless, the wild grasses which luxuriate through the summer are cut in the fall and the ground beneath the trees raked clean to make compost piles.

As in villages throughout Japan and the Orient, the land owned and worked by any one household of Niiike buraku is fragmented into many small, scattered parcels. This fragmentation and dispersal of landholdings has been a matter of concern to agricultural economists on account of the limitations such a system imposes on farm management. The factors leading to fragmentation and scattering are real and enduring.

Roughly speaking, the buraku and its lands form a unit centered upon the housing area (see Fig. 6.4). Most of the fields of Niiike lie together, distinct from fields belonging to people of other buraku. Until the land reform of 1946–47, anyone was able to buy and cultivate land wherever he could obtain it. Various land reform regulations now make it difficult for a person to obtain land unless he is an inhabitant of a nearby community. However, it is the convenience of having fields near one's house, rather than any legal necessity, that tends to concentrate a buraku's landholding. The unity, however, is far from being perfect. Various households in Niiike own plots of one or another type of land relatively distant from their own community. Conversely, the farmers of nearby buraku own and work a certain number of fields close to Niiike. The fringes of each buraku's holdings are made decidedly ragged by the way they are interlocked with fields of other buraku. Habits of cooperation with people of one's own buraku make it, on the whole, somewhat more convenient not to have an outsider working a neighboring field, but there is no strong feeling against outside ownership. Thus, it is only personal convenience and a general sense of the fitness of things that brings most land of the buraku together as a unit.

The pattern of distribution of land owned by any single household, however, is quite different. The acreage held by each house is made up of patches

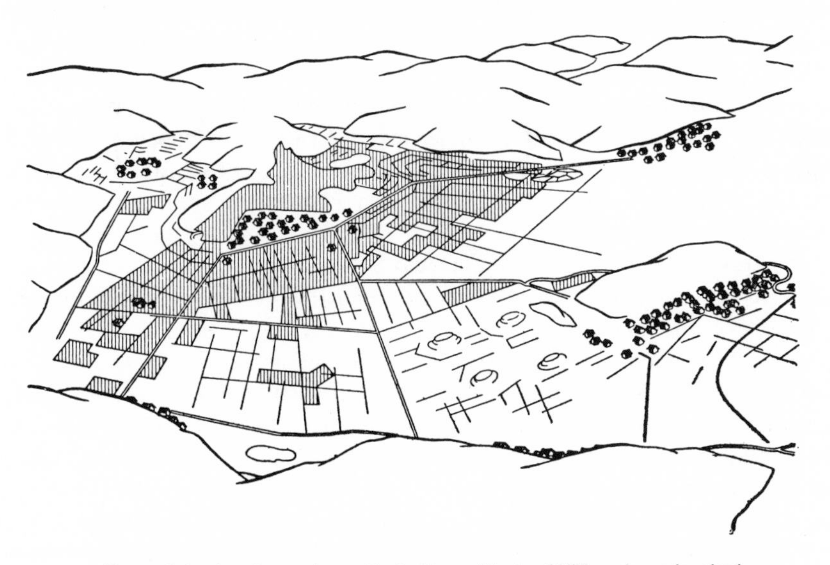

Figure 6.4. Land owned or cultivated by residents of Niiike, shown by shading, is concentrated near their houses. Wood lots, not set off here, cover the hill behind the houses and part of the hill on which we stand.

and plots scattered here and there, the number of such plots ranging from a minimum of eight to a maximum of thirty in the case of Niiike. Except for wood lots across the valley, Niiike is fortunate in having most of its fields close to the settlement. Yet for any given household, two fields seldom lie together or even on the same path. This maximizes the time spent carrying tools, seed, and fertilizer to fields and in bringing the harvest back. The same time and effort are required whether the field be a quarter-acre in size or only a few dozen square yards. Well-established households with more fields have greater carrying distances than newer dependent households with fewer plots. But the "portal to portal" distances range from 145 to 945 feet. Actually house-to-field distance in Niiike is trifling compared with that of farmers in the foothills and mountains, who measure their house-to-field distances in terms of hundreds of yards rather than feet. But in either case a great deal of effort is consumed nonproductively as a consequence of the scattering of fields. Because of the miniature scale of transport vehicles required to negotiate the narrow paths between paddy fields, several trips to a single field are often necessary for completing one operation, such as the scattering of fertilizer. This wastage of manpower is only one consideration that makes field scattering disadvantageous for farm management. Fragmentation poses great difficulties for mechanization or a possible trend toward market production of one main crop. Each field must be cultivated as a separate and distinct operation, raising tractor fuel consumption per cultivated acre to an almost prohibitive figure and canceling most of the advantages gained from working a single crop. It requires little imagination to see how fragmentation developed. Family holdings are seldom split among heirs but are passed on as a unit to a single heir. However, small plots are acquired, exchanged, and sold piecemeal as family fortunes fluctuate, new fields are developed, or the balance of cultivation is shifted. Older households, especially head households, tend to show more clustering of fields than younger or branch households, especially in areas of double-crop paddy. This is evidence of their recognition of the advantages of having fields close together. Nonetheless, the continuance of fragmentation testifies to the centrifugal tendencies built into the traditional land tenure system.

Yet, it should be noted that there are certain advantages in the scattering of fields. To the extent that each household is self-sufficient but makes its living from a very small total acreage, this acreage must be scattered in order to incorporate the different sorts of land required for different crops. Scattered holdings also provide some local security against natural disasters. If the crops are ruined in certain fields by flood, drought, or blight, others may escape. Finally, the appraising eye of the tax collector, who is traditionally regarded in the same category with natural disasters, may miss one or more fields when he visits the village to compare actual holdings with the tax records. Consequently, scattered holdings do provide security, both in terms of satisfying

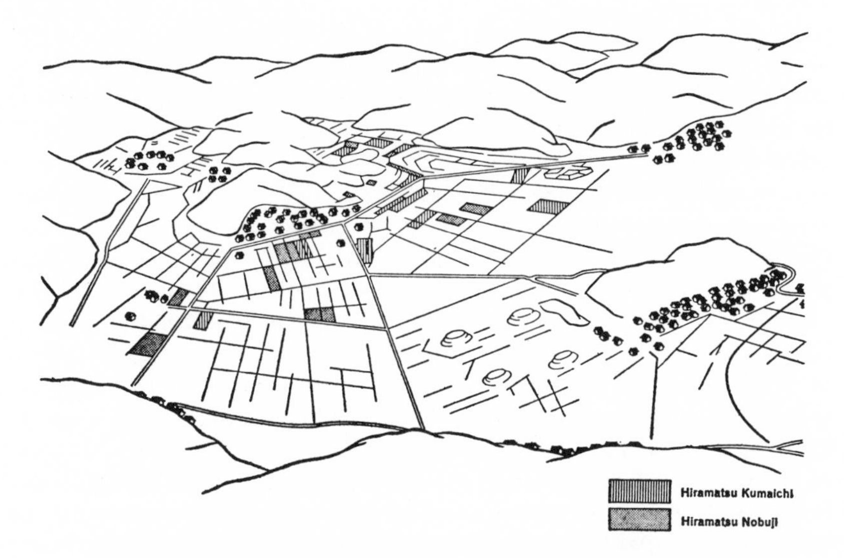

Figure 6.5. Dispersal of landholdings; the fields belonging to Hirmatsu Kumaichi and those belonging to Hiramatsu Nobuji (separately shaded) include patches in every sector of Niiike.

the diverse needs of the household and in terms of offsetting natural calamities. The farmers themselves, more concerned with the lessons of past experience than with promises of future development, have not until recently shown any serious discontent with the situation as it is.

Water Control: Irrigation and Drainage

Let the farmer labor ever so hard and manage ever so craftily to acquire land for his family's security, this security is valueless unless he has access to water. It is meaningless to own rice land without water. In recognition of this vital connection, legal water rights automatically go with the ownership of irrigation land. Despite such interdependence, the use of land and the use of water differ in one vital respect. Land is operated by each household as a unit. However, single households do not control irrigation water, except for very small ponds. Water is managed by communities. The households of Niiike participate in water-use communities of varying sizes, all of which are larger than the buraku itself. Community management of this vital resource has extremely serious implications for social and economic organization. Land management unites the household; water management unites the community.

Water-control problems are not limited to irrigation. Drainage and protection against flood are also of constant concern. Every measure taken to prevent drought must work also against inundation. These contradictory needs, together with the conflicting demands of the various members of the irrigation community, make delicate and complicated the arrangement of water courses, water gates, ditches, dikes, and ponds. One cannot but marvel at the resourcefulness which, over the centuries, has provided for these conflicting needs. By the same token, one comes to understand that the inadequacies that remain cannot be settled by easy means.

The water needs of crops are so well controlled through irrigation that the day-to-day fluctuations of direct precipitation concern the farmer only when they pose a drainage problem. Ultimately, rainfall over the river drainage basins in the hills is more important than rain falling directly on the fields. Local rainfall is only a small part of the larger problem of controlling run-off to prevent or reduce floods, and this in turn is but one aspect of the problem of water control. Water control means holding back the water in times of overabundance, saving it for use when it is critically needed, and distributing it equitably to all who have a call on it.

The days when water problems were met by the buraku or its component households are lost in antiquity. Today these problems transcend buraku limits. They are handled at the lowest level by the various ōaza and, beyond that, by organizations of larger scope. Irrigation, in fact, involves the prefectural and the central government; national measures for river control, dam-building, and land reclamation have direct impact upon the buraku. But, in contrast to many phases of life which the central government absorbs completely once it touches them, irrigation problems are too complex and variant to permit higher government to take over and mold the situation at will. Water control is vital to the national economy—the rice harvest as a whole may fluctuate 20 percent or more between wet and dry years. Higher levels of government are therefore deeply concerned with irrigation and drainage. But they must rest content with sharing this responsibility with local units of government.

Among Okayama coastal plain communities Niiike has a somewhat below average proportion of irrigated land. Moreover, because of poor drainage, even this land is below average in its ability to support two crops a year. Whereas 85 percent of Niiike's cultivated land is in paddy, Kamo village as a whole has 95 percent in paddy. For all of Okayama Prefecture, which embraces mountain land up to the Chūgoku Divide as well as coastal plain, irrigated land amounts to 57 percent of all cultivated and pasture land. A large sector of rugged mountains bordering Hiroshima on the west, where all villages have less than 50 percent and sometimes as little as 20 percent paddy land, is responsible for the low prefectural average. Everywhere else, the proportion of paddy land is above 50 percent and rises almost to 100 percent in some villages of the coastal plain.

The figures for irrigation facilities suggest the complexity of irrigation development in the prefecture as a whole: there are 10,000 irrigation ponds and somewhat more than 5,700 inlets from streams. Despite this, drought and flood are perpetual threats. The people of Okayama remember the severe droughts of 1925, 1945, and 1946, when pond-watered land lost up to 60 percent of its crop. They also recall 1953, when typhoons brought floods from one end of Japan to the other. Nonetheless, Okayama Prefecture is relatively well favored. Floods are seldom as severe as elsewhere, and the folk tradition of *amagoi,* a communal gathering around a great fire built on the hillside to call for rain in time of severe drought, is less frequently encountered than in other parts of Japan. Okayama's advantage over other prefectures is partly due to its natural setting, its screen of mountains south and north that protect it from typhoons and winter storms but provide a good rain shed. Beyond this, Okayama Prefecture escapes severe damage because its water-control facilities are far advanced.

Niiike has access to almost every important type of irrigation and drainage resource. The resources for irrigation include small ponds owned and managed by individuals, ponds managed by communities, canals owned and managed by irrigation cooperatives, and rivers in the public domain. Individually owned and managed ponds are of minimal importance in Niiike, although in areas where other water facilities are limited they may be instruments of economic coercion for the benefit of their owners. Counting even the tiny ponds used primarily for raising fish or planting lotus for its edible roots, there are sixteen. These are scattered about Niiike's fields, mostly around the uphill fringes of the main pond, Ōbara. Most are the property of individual Niiike households, but two are owned jointly by a man from Niiike and a man from the buraku of Horen. One pond, remnant of the original main water resource of the buraku Einoike, serves for little save a water reserve for fire-fighting. Title to the two larger ponds providing irrigation water for Niiike is held by the central government, so that no one pays taxes on them. This arrangement at the same time precludes their being manipulated for power purposes by an individual owner. The smaller one, Ōbara, is managed by farmers of Niiike buraku alone, on a volunteer basis, a system which exemplifies the ancient tradition of water management. The second and larger pond, Taishō, supplies five buraku of Kamo and various buraku of two other villages, Misu and Yamate.

Ōbara Pond, over two acres in area, lies in an elevated hollow around a shoulder of the hill from the housing area of Niiike. Water from this pond reaches fields totaling somewhat over 10 acres (4.11 *chō),* almost all of which are owned by Niiike residents. It is operated jointly by the sixteen households whose fields receive its water. In groups of four, the heads of participating households take turns as water guards, rotating variously over a cycle of sixteen years. Those who benefit least, through having the smallest acreage involved, work the least. But the duty is not onerous. It comes principally during the one hundred days of rice growing in the summer when one to two inches of water are constantly needed on the fields. The period of transplanting mat rush during January and February again brings a brief tour of duty. This system of management exemplifies the older tradition: service is rendered by the persons who benefit, on a pattern of cooperative, rotating responsibility carried on without written rules and without a formal director. The farmers on water-guard duty receive pay for their services but at a token rate. The money comes from an assessment per acreage and is paid to the guards in two ways: a direct payment from the landowners concerned, and a payment from water taxes transmitted through the ōaza.

Formally, the ōaza is the unit in charge of Ōbara Pond and other ponds within its domain. Although it may leave the operation of the pond to the discretion of the buraku that benefits chiefly or entirely, the ōaza in effect is responsible for the system of water guards and is in charge of maintenance and repair. Each pond may be thought of as having three parts: the pond basin and dam, the lead-off ditches to the fields, and the feeder ditches from the hills. Maintenance of small lead-off ditches is invariably the responsibility of the buraku, whatever the source of water. But the ōaza, using a portion of all irrigation assessments paid it, employs ''engineers'' (who often are part-time farmers) to inspect and do minor maintenance of the ōaza irrigation system and to submit to the village office estimates of costs for major repairs.

Maintenance and repairs take place for the most part quite inconspicuously. However, each three or four years every pond must be drained in order to clean out weeds and water grasses and to inspect the ladder-like rows of outlet plugs. This becomes a festive occasion for the entire neighborhood, since the draining ordinarily takes place in the slack season while the rice fields are dry for ripening. The water guards set a convenient date and spread the word around the neighborhood. Old and young gather from all directions. Men and boys make great sport of wading through the shallows of the larger ponds to net fish and stab the slippery eels, and the spectators delightedly watch them getting splashed and muddy. The fish harvest is a respectable one, and it would be unneighborly of Niiike to take it all for itself. Anyone is welcome to join the festivities and share the catch, whether he uses the water of Ōbara Pond or not.

Taishō Pond lies to the west of Niiike in Yamate village. Pocketed in a group of higher hills, it is larger and deeper than Ōbara Pond and has twelve levels of drains as against three for Ōbara. Its water irrgates about 120 acres, divided among adjoining parts of three villages: Kamo, Misu, and Yamate. Built thirty years ago during the ditch-revision program of the 1920's, its management follows a new tradition of professionalism. Money collected on an acreage basis from all farmers who receive irrigation water goes from the respective buraku to their various ōaza and from there via the village offices to the office of Yamate. Yamate uses some of this money for maintenance and with the remainder hires a guard at a token annual salary of Y987, less than $2.75. The present system of a single guard is a change introduced in 1947.

Prior to that there had been five. One may well wonder why five guards should have been needed to do the work that obviously makes only a part-time job for one man. The original five guards were chosen from the three villages in approximate ratio to the land irrigated in each village. We may suppose the reduction was made after a generation of use had given some assurance that no disputes were likely to arise over the distribution of water and that there was no need for each village to protect its rights by having official representation.

As compared with canals, ponds occupy relatively large areas of otherwise productive farmland. But pond water has the advantage of being easily conducted directly into gravity-flow ditches whereas canal water must always be pumped to the ditches or fields. As many as thirty outlets may pierce the pond embankment at regular vertical intervals between high and low water levels, each outlet having its own sluice to lead the water to the regular ditch system below. Yet a field directly below a pond is not necessarily free of pumping costs. In Niiike, for example, the allocation of water from Ōbara Pond changes as the pond level drops. It is reserved during low-water periods for higher fields, not for lower ones, on the ground that to force the owners of high fields to pump water all the way up from canal level while the owners of lower fields enjoy the pond water without cost would be an unfair expense to the former. Under this low-water allotment, everyone pumps; water is lifted to the high fields from the pond, to the low fields from the canal.

Water-lifting devices, accordingly, are an essential part of everyone's farm equipment. Round about Niiike, some of this work of raising water is done by various picturesque well sweeps and water wheels. These devices now do very little of the great amount of day-to-day water-lifting, yet they are worth noting as examples of methods of older times. The sweeps, which in certain places stand tilted against the sky in clusters like a flock of alarmed wading birds, function to raise and lower buckets in wells. Except where the only water to be had comes from such wells, more efficient methods are current today. Self-acting water wheels are sometimes used at the edge of a stream or canal where a steady current flows. They cannot raise water over a high dike, but they are an ancient device, and their construction is ingenious. Around the outer rim of a large double frame of wood or bamboo, sections of bamboo that form natural tubes are lashed at just the proper angle to catch up water from the stream on the start of the upswing and empty it, on the downswing, into a trough leading to a field. The fields so served are usually ones developed temporarily (and illegally) on the government owned bottom lands inside the dikes of a stream. Water wheel, crops, and all risk being swept away when a hard rain fills the bottom with rushing floodwater.

Throughout the irrigating season every household puts to frequent use a small, foot-powered water wheel. It can easily be dismantled into several sections, each one light enough to be carried from one field to another. When in use, the wheel, composed of paddles radiating out from the hub, is seated

Figure 6.6. Water-lifting devices of Niiike. A foot-propelled, portable water wheel is shown assembled and in use; part of a motor-driven cylindrical pump is cut away to show the Archimedes'-screw moving part in the interior. The large, automatic water wheel used on the river and main ditches is not shown here.

into a wood frame sheathing one quadrant. The frame is lodged solidly against the ditch edge, its short trough projecting over the border of the field. Two poles rise from sockets in the frame sides to support an armrest on which the operator steadies himself as he steps from blade to blade, rotating the wheel so that each blade in turn catches water and lifts it up within the sheathing to the trough. When the field is adequately watered, he dismantles his contrivance and carries the wheel, sheathing frame, armrest, and supporting poles to the next small field.

Motorized irrigation is preferred for larger fields, provided that a path is nearby. A metal cylinder, which is open at the bottom and incloses an Archimedes'-screw core, is set into the water at the edge of a ditch so that a trough from the top of the cylinder projects over the low dike into the field. From a wheel at the top, a belt is stretched to the flywheel of the household's 4.5-horsepower kerosene motor, set nearby on its carrying sledge, and the motor spins the screw to send water gushing into the field. Powering these irrigation devices is one of the principal uses for the all-purpose motor. Weighing well over one hundred pounds at the lightest, it cannot readily be used in fields that have no nearby path or firm ground. Hence the wooden foot-turned wheel continues to be used for small or distant fields even though the motor-driven screw pump has become standard equipment on the Niiike farm.

Among water-lifting devices, there should finally be noted several large, stationary electric pumps, owned by the ōaza rather than by individual households, which help in drainage rather than in irrigation. Those that concern Niiike are installed at the dike of the Ashimori River, where they work constantly throughout the irrigating season, discharging over the dike all the water that has been pumped or led onto the fields in the valley from its ponds or canals.

Canal water supplies Niiike with about half its irrigation water. Niiike's access to this water is through its membership in a large irrigation canal organization known as the Twelve-Gō Irrigation Association (Jūnikagō Yōsui Kumiai). Its canal runs eastward from the Takahashi River. Niiike and other burkau have no independent status in this organization but participate only through their respective ōaza and villages. The organization, despite its names, includes thirteen villages and towns and contains various suborganizations at different levels. More exactly, perhaps, we should describe the whole system as being made up of two coordinate organizations, the Twelve-Gō (Jūnikagō) and the Six-Gō (Rokkagō) irrigation associations, with subordinate units in each. Niiike is a very small part of this organization indeed, but to understand the significance of the organization to the life of people in Niiike we must step back for a large-scale view of the whole.

Eight of the thirteen villages and towns of the Twelve-Gō Canal System, including Kamo, lie in the drainage basin of the Ashimori River four to five miles east of the Takahashi River. Their access to the Takahashi River water relieves them from dependence on the capricious Ashimori, which has a dangerously torrential flow after a heavy rain but carries little water during most of the year. The Takahashi River has provided summer water to these distant villages for more centuries than to villages close by; not until recent times was a second canal system, the Eight-Gō (Hakkagō), created for the benefit of villages and towns close to the Takahashi River. The Twelve-Gō Canal, older and larger than the Eight-Gō Canal, branches from the river at Tatai, where the river first breaks out of the hills, and moves eastward across the plain, following what may once have been the bed of the Takahashi River. It soon splits into two arms. The northerly arm follows the hills on the upper edge of the plain until it spills into the bed of the Ashimori River; take-off ditches serve nearby fields en route. The southerly arm bends down to the edge of the hills along the south and, serving as a transit canal, retains all its water until just before it reaches the Ashimori River. There, at the Iwasaki water gate, it splits into three arms. One drains into the bed of the Ashimori River, mingles with the excess water of the northern arm, and flows south to be used by villages farther downstream. Of the remaining two arms, one bends around the hills on the west side of the Ashimori and eventually reaches Niiike after having been split once more into an Upper and a Lower Shinjō Canal. The remaining arm dives under the bed of the Ashimori River to re-emerge via a siphon on the east side, where it waters the lands of eastern Kamo and its neighbors. As Figure 6.7 shows, Niiike gets its share of Takahashi River water only after it has been diverted by two water gates (the Tatai water gate at the Takahashi River and the Iwasaki water gate near the Ashimori River) and channeled through several branch and subbranch ditches. Niiike's lands are only a fraction of the 13,600 acres watered by the canal.

The Twelve-Gō Canal, an exceptionally important factor in the agricultural development of coastal land, is reputed to have been constructed along

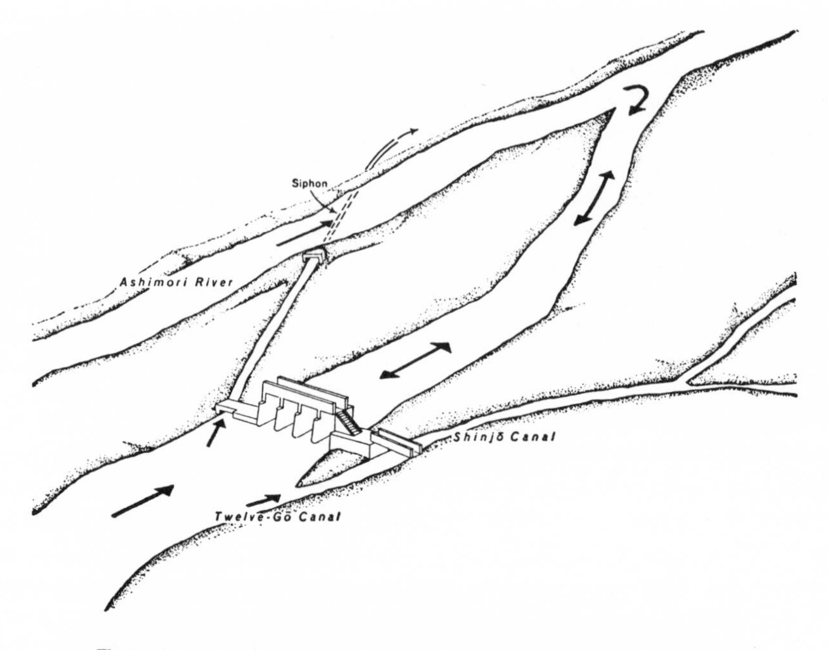

Figure 6.7. The Twelve-Go Canal at the Iwasaki water gate. Canal water flowing from the left divides here, part crossing under the river to the far side, part going through the First Gate of Shinkjo (Ichinohi) (lower right) and part falling into the Ashimori River bed for use downstream. In flood, the river water backs into the canal and may inundate fields between canal and river or beside the canal upstream.

its present general course about A.D. 1220, serving twelve of the village-like units of the time known as "*gō*." Local tradition has it that the course was laid out by Kageyasu Senotarō, who set out from the Takahashi River on horseback and had the canal dug along the twisting route he followed across the plain. Centuries afterward, as new land was created from the tide flats of Kojima Bay, the villages established there hooked onto the canal and became incorporated as downstream members of the Twelve-Gō system, which today is among the five largest irrigation systems of Okayama Prefecture.

As one follows the route of the canal and its branches today, one comes across a few control and diversion structures of recent date and modern aspect standing side by side with very humble, primitive arrangements. Prefectural loans have helped to build concrete water gates at the Tatai entrance and, since 1950, at the Iwasaki juncture with the Ashimori River. En route between the two gates, sections of concrete or rock retaining walls alternate with heavy piles or stakes interwoven with bamboo strips to support the banks. Some diversion ditches that water the fields in the vicinity are no more than holes in the bank. At other places there are crude diversion devices made of piles of submerged rocks and pine trunks laid across the canal, of calculated thickness and depth beneath the surface to control the amount of water diverted. These simple arrangements are readily displaced by accident or high water, and their reinstallation becomes a matter of sometimes acrimonious arbitration; it is difficult to replace them with more precisely measured modern devices, not only because of the expense involved, but because of jealous fears that one side or another will come out on the short end.

Tight conservatism preserves many archaisms in the administrative organization of the canal system. It derives in large part from the tensions and jealousies over the husbandry of this important resource. All the participating villages in the several canal organizations linked under the Twelve-Gō system work on a common problem: the provision and distribution of adequate water in ways which will not jeopardize flood control. But there is no single ideal solution, since some fields are disadvantaged by conditions that are ideal for others. Each unit of a canal organization is in a sense a separate sovereignty treating with others for a satisfactory working compromise. This uneasy confederation is more apt to follow historical precedent than laboratory theories of hydraulic engineering. Although cooperation is mandatory, it is not achieved without a certain amount of hard feeling or even open conflict. Serious tensions prevail at every level of the rather complex organization.

In order to use canal water from the Twelve-Gō Canal, Niiike is a member of three organizations of successively larger size, each involving it with portions of the canal farther upstream. It belongs to the smallest through its ōaza, Upper Shinjō, which joins with Lower Shinjō to clear and repair the feeder ditch as far as the water gate at Iwasaki. The two ōaza form an organization, the Shinjō Ichinohi (Shinjō "First Gate"). Similar minimum-level organizations composed of one to three ōaza exist all along the system. The Ichinohi ditch, it will be remembered, is only one of three ditches that branch at the Iwasaki water gate. Hence the Iwasaki gate affects not only the Shinjō Ichinohi fields but also those of the ōaza of Kamo east of the river and of the six remaining villages and towns downstream. All these, accordingly, form a cooperative, in which the several ōaza of Kamo, including the two that use Shinjō Ichinohi ditch, speak through Kamo village as their administrative unit. This is the Six-Gō organization, responsible for the Iwasaki water gate and downstream ditch maintenance. Finally, the villages of the Six-Gō cooperative, because of their interest in the head water gate at the Takahashi River and in the transit ditch from there to Iwasaki, are confederated with upstream villages as members of the Twelve-Gō organization.

Formal charters cover the structure and operation of the Twelve-Gō and the Six-Gō organizations. Several of the lower echelon organizations also have charters, but most do not. All but a few of these date from 1889, the year when present village organization was formalized, but some revisions took place about 1925, when grand plans for ditch revision focused attention on irrigation matters. Chartered associations do not differ basically from the others, for the charters did not establish them or, in essence, change them except to recognize modern village units in place of ancient ones. Each charter merely codifies tradition that has long been accepted. For example, each ōaza, as a basic participating unit of the Twelve-Gō Irrigation Association, is charged a share of the costs and allotted a percentage of water fixed to the fourth point (Upper Shinjō gets 1.6071 percent). The ratios are fixed on the basis of acreage served at some time in the distant past and have not taken into account later changes in the extent of fields. Even though charters, in a sense, merely crystallize tradition, they are nevertheless convenient and even necessary for the larger groups. Because costs vary from year to year, their apportionment in an annual budget is much facilitated by the formal organization of the irrigation group under a written charter.

The mayor of Kamo is fairly prominent in the councils of the twenty-four-man administrative board of the Twelve-Gō organization, because, by tradition, he is chief administrator of the Iwasaki water gate, located near the Kamo village office. The fact that Kamo is farthest upstream and the oldest participant in the Six-Gō system also gives him seniority.

Irrigation taxes rise as the years go by because cash expenses have gradually increased. Villagers themselves once took care of most maintenance and repair, contributing their free labor; unpaid labor is still contributed but extends only to maintenance and repair of the smallest ditches within each buraku. Cash is now needed to hire guards for each water gate, to hire workers for repair of the main canal, to pay for major improvements, and for incidental organizational expenses. These costs are met by an earmarked irrigation tax in each village budget.

Water Control: Tensions and Disputes

The theoretical rights to a certain percentage of water are clearly established by charter or by tradition for each participating unit. The problem is to implement these rights in practice. The condition of the canal itself makes this difficult. At certain points, facilities are precisely and firmly fixed; for example, the dimensions of the outlet from the main canal to the Shinjō Ichinolhi branch, 1.1 feet high by 3 feet broad, were fixed by arbitration and made permanent by being cast in concrete. As the people say, every man knows these figures as well as he knows his own house number. But no one has measured the precise volume and rate of flow of water either in the canal or in the river from which it comes, nor does anyone know what happens to the water as it passes over and around crumbling rock dams and rotting diversion logs before reaching the Ichinohi outlet.

Equally important as a source of friction is the tradition of unequal water rights. In time of drought the villages on the reclaimed land in Kojima Bay get shortchanged, according to their view. Their claim to this water goes back only a century, at most, and they are at the tail end of the canal. Hence, the older villages upstream claim prior right to the water. In any case they have first access to it, and it is their own people who decide, as water guards, how much to let through. Frictions thus are inevitable. At a still higher level, the Eight-Gō organization has gone to court against the Twelve-Gō organization in a dispute over division of the Takahashi River water. The dispute concerned the question of how high the Twelve-Gō organization ought to be allowed to build its diversionary dam. Because there is no codified law concerning water rights, the case stagnated until both parties, appalled at the mounting expense of lawyers and litigation, compromised out of court. Such friction occurs at every organizational level down to the ōaza and, in some cases, among individual farmers within the ōaza. The details of the following dispute show how difficult it is to reconcile conflicting interests.

Rice fields near the Six-Gō water gate at Iwasaki but upstream from the gate are not entitled to water from the transit canal bordering them but instead draw their water from the canal arm running along the edge of the hills to the north. The tangle of river and canal channels near the Iwasaki gate is arranged to prevent flooding at that point but, at the same time, to water fields on the downstream side of the gate. Ordinarily, the trickle of water in the bed of the Ashimori seems much too small to justify the high dikes that border the river; the river bed serves merely as a convenient channel to carry canal water. After a hard rain, however, the river rises twelve or fourteen feet. The canal level fluctuates only slightly, owing to the fixed size of the opening from the Takahashi River. Thus, the flooded Ashimori may back up along the canal and, since no high dikes border the canal, may flood the low fields there unless the water gates are closed in time. Although the canal may serve as a safety valve for the river in flood and prevent its bursting its dikes elsewhere, farmers owning fields at this point suffer if the water gates are opened. Shortly after a heavy rain, we saw the gates standing open, the rice in low fields nearby completely under water, and a crowd of angry farmers clustered on and around the water gate. The mayor of Kamo, summoned posthaste, splashed up through the mud to learn that unknown persons, presumably those with fields at a higher level nearby, had used a bootleg crank to raise the gates and water their fields. The cultivators of the low fields, upon seeing the water rise over the top of their rice, had stormed up to the water gate and were now demanding that the gate be closed and they be compensated for damage. The mayor's decision, finally, to close the gates did not relieve him of having to hear a half-hour tirade launched against the suspected culprits, the system, and his own management, delivered with violent gestures by the disputants on both sides.

Although there is no visible end to disputes of this sort, progress can be made toward solving one of the basic irritants, periodic shortages of water. Small ponds and ditches already exploit localized ground water so thoroughly that additional water needs cannot be met except by major projects in the upstream drainage basins of large rivers. This requires action quite beyond the domain of single villages or canal cooperatives. Prefectural and national dam-building is required. Such action has been started in the postwar period. The prefecture has pushed plans for construction of major, multipurpose dams on two important rivers, the Takahashi and the Asahi. The irrigation community to which Niiike belongs is to benefit directly from the smallest of these, the Osakabe Dam. This, besides supplying 3,000 kilowatts of electricity and impounding water to diminish the flood threat along the Takahashi River will provide irrigation water during the low-water season. A still larger dam downstream, not yet in the blueprint stage, will provide even more water than the Osakabe Dam and will further alleviate the perennial water shortage of villages on the downstream end of the Twelve-Gō Canal. The central government, which advances the entire cost to begin with, is eventually repaid 40 percent of this amount, 15 percent by the prefecture and 25 percent by the participating irrigation cooperatives. This postwar dam-building, involving both local communities and the highest levels of government, is something new in water control in the Okayama Plain. It is occurring in the normal course of events rather than as a specific outcome of war or occupation. As such, it seems a token of a future in which communities such as Niiike will be linked ever more intimately into an economic and financial framework of prefectural and national scale.

7

Irrigation Societies in the Northern Philippines

Henry T. Lewis
UNIVERSITY OF ALBERTA

Ilocano irrigation societies were reported as early as 1914 by E. B. Christie, who provided the first substantial ethnographic reports on the Ilocos area. The *zangjera*,[1] or cooperative irrigation society, is a special development of Ilocanos in Ilocos Norte. Keesing referred to Ilocano irrigation societies and, on the basis of early Spanish reports, dated their origin at about 1630 (1962:145; 305–307). Although they may date even earlier, perhaps predating Spanish contact,[2] they are quite old and their age suggests an early population density of some significance. What relationships and significance the irrigation societies may have had for population density and why they occurred in

1. The name apparently derives from the Spanish word *zanja*, an irrigation ditch or conduit. At first glance the word *zangjera* (sometimes spelled *sanghera* or *sanhera*) might suggest that it was derived from *sangre*, or blood, but I was informed, on good authority, that this is not so. At the time, the suggestion of a link to the word *sangre* seemed pregnant with meaning for the kinship system!

2. Keesing's evidence, from the Blair and Robertson collection (1903–09: vol. 7, p. 174; vol. 12, p. 210) is based upon references by Spanish priests who were developing irrigation systems at "mission-created settlements." There are no comments, unfortunately, as to whether or not such irrigation projects existed elsewhere but there is no reason why the Spanish should have limited such success to the northern Ilocos area. The use of a Spanish-derived word, *zangjera*, is not necessarily solid evidence for origins since there is an Ilocano equivalent, *pasayak*. Also, none of the technical or operational terms associated with zangjeras are of Spanish derivation; all are Ilocano words—*puttot*, a dam for stopping water; *padul*, a diversion dam across a large stream or river; *kali*, a main canal; *aripit*, a small ditch; *sayugan*, a flume; *bingai*, a share or membership; *gunglo*, working sections of land; *kamarin*, a community meeting place; etc. We are not, however, considering etymology here.

Ilocos Norte and not elsewhere are questions which cannot be answered here. Demography, climate, and topography are undoubtedly important factors and to these one might add the Ilocano personality type ("hardworking, thrifty, industrious"), except that there are no irrigation societies in Ilocos Sur or La Union.

The manifest function of irrigation societies is simply to procure a stable, reliable supply of water, which can increase crop production in some cases by more than half. Given this goal, the zangjeras employ a wide variety of organizational means and methods. It should be stressed, however, that these variations do not reflect different "types" of irrigation societies; they simply reflect various solutions to different technological problems. For instance, some zangjeras are restricted to membership from within a single barrio. Such "restrictions" are incidental, however, and zangjera membership is invariably independent of barrio membership. The point is that zangjeras are not designed to correspond to the members or geographical boundaries of the barrio; they are designed only for obtaining water for particular field areas.

Most zangjeras have members from two, three, and more barrios with some of the largest societies having members from ten or twelve. In some zangjeras the members are all landowners; in some, landowners and tenants; and, in several, all members are tenants. In a few, formal ownership of the land is vested in the zangjera itself with members owning only use rights to the land. Some societies are dominated by one or more family groups while others have no suggestion of extended family control. In one instance, in a barrio near the poblacion of Piddig, an irrigation society is an independent group which sells or leases water for a percentage of the crop. The members of this "professional" irrigation group farm no land (at least as members) themselves. In the final analysis, membership is "decided" by the hydraulic engineering employed by Ilocanos to get water, and a wide variety of social ties and relationships becomes involved. As Leach has noted for the village of Pul Eliya in Ceylon, the "inflexibility of topography" and the "crude nursery facts . . . that water evaporates and runs down hill" are inescapable conditions for the social organization of zangjeras (1961:9).

Within the municipality of Bacarra there are twenty-six zangjeras ranging in size from less than six hectares to more than one hundred hectares. In the whole of Ilocos Norte there are reported to be 185 societies in all.[3] Among the individual zangjeras there are various "levels" of intersocietal cooperation. The complex of dams, canals, reservoirs, and drain-off systems has resulted in the need for a wide number of verbal and written agreements between

3. Some of the more general figures and information of zangjeras outside the Buyon area were graciously provided me by the Bureau of Public Works, Office of the District Engineer in Laoag. Francisco T. Tamayo, the senior civil engineer, was especially helpful. However, neither he nor other members of the District Engineer's Office are in any way responsible for the interpretation given the data here.

zangjeras in any given area. Different zangjeras may share the use of a main canal or even a single diversion dam. In other instances, where there are several *padul,* or diversion dams, located along a desirable section of a river, a number of zangjeras will have agreed to cooperate on the repair of dams damaged or destroyed during the monsoonal flooding. Different societies may have interconnecting canals by which water from one system can be diverted to another system which has become temporarily inoperative. Drain-off water from one system may be used to supplement the basic water supply of a nearby or adjacent society. Usually one irrigation society will be involved with several other societies in various sets of mutually cooperative relationships. As a consequence, the interdependence of various zangjeras tends to moderate any conflicts which might arise among them.

In the Bacarra area, forty zangjeras (including several from the adjacent municipalities of Vintar, Laoag, and Pasuquian) form the *Federation of Communal Irrigation Systems for Bacarra,* an organization which acts to settle disputes between its members and acts on behalf of the membership on matters of political importance, especially those relating to government irrigation developments. The federation operates in situations, internal and external, which threaten one, more than one, or all members, but it exerts no direct control over the "internal affairs" of any single zangjera. Individual societies may join or withdraw as they so choose. Leadership in the federation is elective, and the president of the group is an important and influential man in Bacarra politics. He is a member of two zangjeras and president of still another.

The area of land occupied by all zangjeras in Ilocos Norte is estimated to be in excess of 17,000 hectares. Christie's (1914:99) report from over fifty years ago gives 15,000 hectares, which, as far as figures go, is not inconsistent with current estimates. Unfortunately, however, both figures are probably considerably excessive. Estimates in many instances are confused or exaggerated by the fact that most irrigation societies sell water to nonmembers whose lands border the systems, and such lands are often included in government estimates as being within the zangjera. In several instances the area of nonmember lands actually exceeds those formally within the system. Individual zangjeras also tend to exaggerate the size and importance of their organization, especially when government aid may be involved. In any event, the development of cooperative irrigation in Ilocos Norte is impressive, especially by comparison with other parts of the Philippines.

The above description of zangjeras hardly seems to apply to Isabela, where there are so few of these organizations. Although immigrants in central Isabela were quick to obtain the water rights to their lands, few groups followed this up by developing cooperative systems. Most of the water used to irrigate the rice-growing fields of Isabela comes from a single source: the 30,000-hectare government-constructed Magat River Irrigation Project, located southwest of the town of San Mateo. Beyond the limits of the Magat system there are a few privately developed irrigation systems, usually the work of one of the large haciendas, or, very occasionally, an Ilocano zangjera. The actual number of cooperative systems outside the Magat River project is difficult to ascertain, as records on these are much less complete. Some are but paper organizations, a few have ceased to function, others were absorbed by the Magat River project, and a few were appropriated by individuals. In all, there are probably no more than six effective irrigation societies (four were actually examined) in the whole of Isabela. They exist for the same reason that their ''parent'' organizations in Ilocos Norte exist: to provide irrigation water. They also exist because the migrants from Ilocos Norte had a history and tradition of irrigation cooperatives and possessed the knowledge and the incentive to construct them—something which the indigenous Ibanags and other immigrants lowland groups did not. The few zangjeras in Cagayan Valley were all constructed by Ilocano immigrants from Ilocos Norte.

Yet, the same ecological conditions which encourage the operation of zangjeras in Ilocos Norte were absent (and, for the most part, still are absent) in Isabela. Zangjeras in Isabela ''suffer'' as a consequence of the relative prosperity of the people and agriculture in Cagayan Valley. The few zangjeras there function quite differently than do those in Ilocos Norte because there is less need for irrigation and because the social factors relating to irrigation societies are so different. This is especially evident with regard to the derivative functions of zangjeras in the two provinces.

Irrigation Cooperatives in and near Buyon

The Laoag District Engineer's Office lists two zangjeras as being located in Buyon: the Zangjera de Camungao and Zangjera San Antonio. Most members of the Zangjera de Camungao live in Buyon, and a large number of those in the Zangjera San Antonio also live there. The land for both cooperatives is located generally within the formal boundaries of Buyon. Eight other zangjeras also have members living within Buyon, the most distant of these zangjeras being about three kilometers away. The bamboo and rock diversion dams for both Camungao and San Antonio are located on the Bacarra-Vintar River just south and west of the town of Vintar (see map 7.1). The main canal for Camungao is approximately three kilometers long; the main canal for San Antonio is about one kilometer. About one kilometer northwest of the town of Vintar, and for another one-half kilometer downriver, there are four dams representing five irrigation cooperatives: the first and farthest upstream belongs to the Zangjera Narpayat; farther downstream is the dam belonging to San Antonio; the third dam is used by Camungao and the Zangjera Dibua; the fourth and farthest downstream belongs to the Zangjera Curarig. For specific

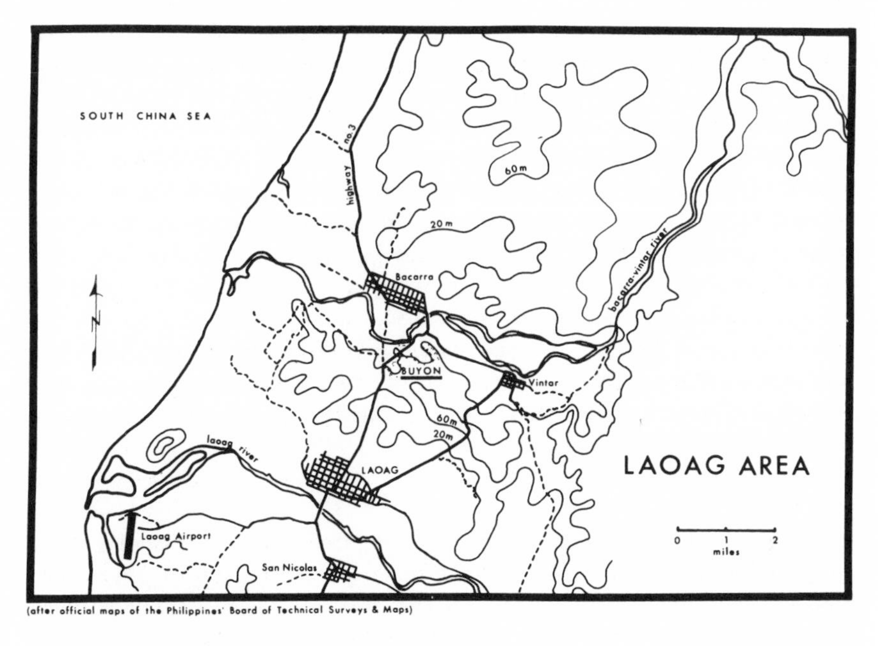

Map 7.1.

problems associated with the dams, these five zangjeras will act as a single, essentially cooperative, group. For instance, when any single dam of the four is damaged or when excessive silting has occurred at the entrance to a main canal, all five zangjeras will act together as a cooperative body to undertake repairs.

In addition to the arrangement among these five zangjeras, there are several sets of relationships within this group of five, plus special arrangements by one or more with still other zangjeras affected by their operations. For instance, the main canals from all four dams pass through the lands irrigated by Narpayat, San Antonio, and another zangjera, the Zangjera Bangbangkag. Camungao and Dibua share a single main canal which continues past Camungao's lands and through the lands of Zangjera Curarig. The main canal of Curarig passes through Camungao, and some of the overflow from Camungao is emptied into the Curarig system. Where the main canals of the upstream systems pass over (via stone or brick aqueducts) the canals of the more downstream systems, arrangements have been made for diverting an emergency supply of water from the higher into the lower system. In addition to the primary source of irrigation water supplied by the diversion dams, there are secondary water sources—creeks, springs, secondary diversion dams, water drained off from other systems—which are utilized by one or more of the cooperatives. All of these arrangements involve special agreements between two, and often more, of the zangjeras in any given area, and such agreements may involve either oral or written contracts.[4] The diversion, passage, and drainage of water involves a complexity of arrangements much greater than the simple bamboo dams would seem to suggest. At the time of the field research, the Zangjera San Antonio was faced with problems of major proportions. A flood in late 1960 altered a main channel of the river and destroyed almost all of the member lands, and many of the nonmember lands, irrigated by the system. Some of the land which was not completely removed or buried under rock and gravel was being reconstructed. An attempt was being made to bring new members into the zangjera and some suggestions were even made that San Antonio should join with Camungao. Despite the seriousness of the situation, which involved the loss (since their lands were washed away) of working members, the Zangjera San Antonio remained a small but viable organization.

Camungao lands have suffered almost as badly as a consequence of a more prolonged process. The main channel of the Bacarra-Vintar River abruptly alters its east–west course below the main highway bridge at Bacarra; at that point it turns directly south to hit and turn again against the main lands of the Zangjera de Camungao (see map 7.2). The riverbank has become undermined at this point with chunks of land regularly dropping into the river.[5] Members of the zangjera claim that over three-fourths of the original lands have been washed away. Out of perhaps an original thirty hectares, the member lands of

4. A written contractual arrangement exists between the Zangjera Camungao and the Zangjera Dibua with regard to sharing the diversion dams and main canal. There are, however, no written agreements between the Zangjera Camungao and the other zangjeras regarding the various other intersocietal arrangements mentioned. Whether such "contracts" are written or simply verbal should not be taken as an indication of the relative importance of one agreement over another. All such contracts have relative and immediate importance with regard to the particular needs and problems solved in any given agreement, and the existence of a written contract can be related to the specific circumstances and the particular individuals concerned at a given time and place.

The written agreement between Camungao and Dibua was made in 1937 and is written in Spanish. I was somewhat bothered by the fact that Spanish was used because, as far as I could ascertain, none of the members of either zangjera spoke any Spanish. The officers of Zangjera Camungao were good enough to loan me the contract which was duly translated and I subsequently provided the members with the English translation. Two of the officers expressed their thanks for copies which they could read and both agreed that it said, in fact, what everyone thought it said. My question as to why it was written in Spanish instead of Ilocano, or even English, was answered by the older members. They informed me that it was drawn up for them by a local lawyer who felt that Spanish was the appropriate "legal language." The written document was, in effect, a symbolic, essentially decorative, gesture involved in a *real* social contract.

5. This was instrumental in bringing about the sharing of the same canal and dam by Camungao and Dibua. The main canal of Dibua was located along the northern border of Camungao but was lost to erosion by the river. In 1937, the two zangjeras agreed to share both dam and main canal plus all construction and maintenance work. Because Dibua is considerably larger, Camungao was "compensated" for having to widen the canal by acquiring three times the labor force it could normally muster for itself.

Camungao now number no more than six hectares. More land, thirteen hectares, is irrigated for nonmembers in Buyon than for members.

The organization of San Antonio is structured along more general lines: members own their own lands or are tenants for landowners. Each member contributes work on the repair of the dam and main canals. Individual members have to contribute small amounts of money to buy bamboo and other materials used in dam and canal construction. Nonmembers pay 10 percent of their production as the fee for using water, and, if they so desire, individuals may join as full participating members by contributing one-third the total amount of land to be irrigated (or an agreed-upon equivalent) to the zangjera. Given the importance of land in Ilocos Norte, it is obvious that new members are not readily or regularly forthcoming, the 10 percent fee being the more usual agreement.

The Zangjera de Camungao has a unique system of organization which discourages potential members from joining. Instead of a group of individuals, each with his own or his landlord's land, the Camungao cooperative itself owns the land and controls the water rights. Individual members own only the use rights to that land, rights which are invariably inherited, and, when

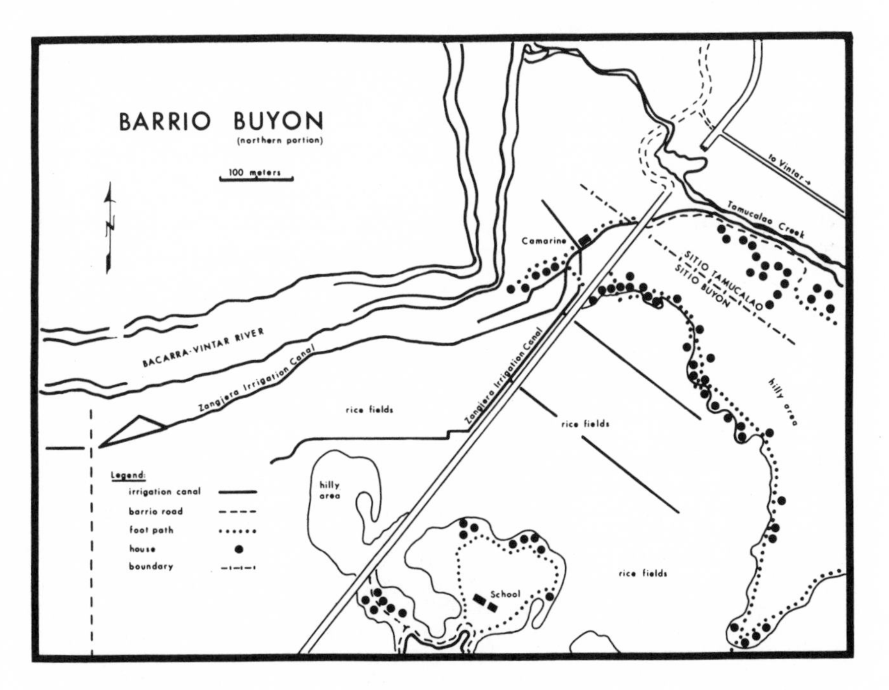

Map 7.2.

available, the shares of which can be purchased for 250 pesos, or about one-fourth the local land value. The shares of land are about one *uyon* in size, or, translated into land measure, about one-fifteenth to one-tenth of a hectare (.16 to .25 of an acre). In some instances there are half, and even quarter, shares, these being the result of adjustments made over past years because of the continuous erosion of the land. In addition to individual member shares, there is a communal section of land representing two *uyons* (an area equal to two shares), which is set aside for the use of the zangjera as a whole. This section is cooperatively worked and provides the means for obtaining the necessary supplies used in the maintenance of the irrigation system. An additional half-share is also set aside for the head man, called the *pangulo* or *maestro,* as payment for the extra work he must perform. No one in Buyon knows the original basis for such an organization. The original "constitution" no longer exists but the written agreement, mentioned above, between Camungao and Dibua, dated 1937, states that the Zangjera de Camungao was founded in 1793. The name derives from a local family, and a few people with the same name still belong to the organization. Whatever the historical origins for this kind of organization (and an extended kin group ownership seems the most reasonable guess), it is now virtually impossible to attract new members since the individual must give up formal ownership to his land. This, together with the continuous wearing away of lands by the river, means that the Zangjera de Camungao is faced with the possibly of losing all of its member lands.

The amount of work which an individual must do to maintain and repair a system in any given year is considerable. Damage to the bamboo and rock diversion dams can occur a number of times each year, especially during the monsoon and typhoon seasons.[6] Heavy silting occurs in the main canals, requiring considerable yearly maintenance throughout the system. As a consequence, each member must contribute from forty to sixty full working days a year. Fines are levied for work absences, and, if an absence continues, the

6. A comment is perhaps necessary about the dams. One of the first statements made by Christie (1914:99) concerns the "crudely constructed dams" which are "either completely destroyed each year or require considerable repairing." These rock and bamboo dams are not, however, as impractical as Christie suggests. Bamboo is relatively cheap, and rock and cooperative labor involve no formal expense; in addition, the dams are comparatively easy to repair. Without government aid and assistance, a concrete dam would be prohibitively expensive to construct, and, except on the smaller streams, impossible to maintain. Such a dam was constructed for the Zangjera Curarig at the lower end of Tamucalao Creek (see map 7.2) and, during the flooding of 1960, the dam was broken. After the flood Curarig rebuilt their bamboo diversion dam on the river. The traditional bamboo and rock dams, being so easy to repair and replace, are more "reliable." The danger from flooding comes at the middle of the rice-growing season when there can be no delays such as waiting for the dry season or waiting for government assistance. A concrete dam is practical only in a society which can afford the time, money, and delay involved in construction, maintenance, and repair. The people in Ilocos Norte can afford none of these.

loss of land may result. In Zangjera San Antonio, where members own their own land, repeated failure to attend work sessions can result "only" in the loss of water, but considering the importance of water and the thin margin of subsistence, the individual simply cannot afford to neglect his obligations to the zangjera.

In addition to the subsistence and technological demands of cooperation, there are certain social rewards, valuable in an environment where there are few such benefits. Camungao, like most zangjeras (though not San Antonio), has a meeting place, or *kamarin,* where, once each year following the harvest, a feast is held. Usually a religious personage is invited—it may be a Catholic priest, an official of the Aglipayan Church, or a Protestant missionary. Considerable amounts of food and drink are provided. Various kinds of meat (beef, pork, goat, sometimes dog) are eaten with rice, and large quantities of *basi,* or sugarcane wine, are consumed. Sometimes food offerings are made to the local spirits, though care is usually taken not to offend a priest or minister. Local and provincial politicians are often invited to these events. In the absence of barrio fiestas, these yearly feasts constitute the only community-wide celebrations in the area—the "community" being the zangjera, however, not the barrio. Yet, even work projects are occasions for some degree of sociability; food, such as that provided at the annual feast (items not normally eaten at everyday meals), is provided the working group, and here, too, the favorite drink is *basi*. Money derived from the fines assessed against absentee members is usually insufficient to cover either the annual feast or the workday expenses since normally very few members are ever absent in a given year. During 1963, a total of fifty pesos (the usual fine being two and one-half pesos for each day's work missed) was collected with no fines outstanding. Funds for covering social expenses largely come from the income derived from selling water to outsiders.

Irrigation Cooperatives in and near Mambabanga

There are two zangjeras in the general area of Mambabanga: the Society Mambabanga and the Union Bacarrena.[7] The fact that people from Mambabanga belong to two zangjeras is certainly atypical for Isabela—especially since there are probably no more than six operative systems in the whole province. The zangjeras have not made Mambabanga more cohesive, however. The Society Mambabanga may be used as an example of what has

7. Only the older men and more recent immigrants to Isabela refer to irrigation cooperatives by the term zangjera; invariably the Ilocano term *pasayak* is used. I never noted the use of Spanish words, except for *sombra,* or sluice, used with reference to a *pasayak*. The honorific term *don* was used in some instances in Ilocos Norte as a term of address to the zangjera officers, but it was never used in Isabela. Ilocano terms were always used.

happened to at least some of the early ideas and plans for developing irrigation which were based on the traditions brought by Ilocano immigrants into Cagayan Valley.

Water rights to the original land grant were acquired by the founders at the time the barrio was settled in 1918. However, no action was taken to develop an irrigation system until more than twenty years after this and only then because of the efforts of a single individual, whom we have called Mr. Cruz. During that twenty-year period Mr. Cruz had become the single largest landowner in Mambabanga, acquiring just over half of the original two-hundred-hectare land-grant purchase. From the very beginning, his holding of over forty-three hectares was the largest in the barrio; in the process of acquiring land from other less fortunate or less successful individuals he created considerable enmity and jealousy. At the same time, whether it was his original design (as some people contend) or not, his lands were strategically located further upstream on Macanao Creek, and consequently, nearest the dam. According to a number of persons in the barrio, he attempted to use this geographic position for his own personal gain. Even more animosity developed against him when, through a legal maneuver, he obtained the formal title to all water rights. A number of families in Mambabanga claimed that he used these rights and his location upstream to create difficult circumstances for the families more removed from the dam by withholding amounts of water at crucial times. By creating a severe hardship, he attempted to buy up the lands of the more hard-pressed, small farmers. If this was his intention he was unsuccessful, for no new lands were added to his total holdings after the construction of the irrigation system. Along with the development of his irrigation system, Mr. Cruz brought new tenants in to work his land; all of them settled in the western section of the barrio, or what is now Harana (see map 7.3). Most of Mr. Cruz's tenants were Ilocanos newly arrived from Pangasinan and considerably poorer than the people of Mambabanga, who considered them rude and aggressive—as the Ibanags had considered the people of Mambabanga a quarter of a century earlier. Partly because of the increasing animosity toward him and partly because of his age, Mr. Cruz sold the land and moved to a town located a considerable distance away.[8]

The sale of land took place in the mid-1950s after Harana had become a

8. Shortly after my arrival in Mambabanga a legal case which developed from the sale of this land was completed in Luna. Mr. Cruz was taken to court by his brother, a Hawaiiano and a resident of the United States, who contended that he had financed Mr. Cruz for the initial purchases and the subsequent additions to that land. Mr. Cruz claimed that this was not true and, since no formal agreement existed between the brothers, the claimant from the United States lost his case. Although quick to state that this was very un-Filipino-like behavior, the people in Mambabanga were not unduly surprised, for the incident corroborated what they already "knew" about the character of Mr. Cruz. My own opinion is that such behavior between close kinsmen, in this case siblings, is not as uncommon as anthropological literature would suggest.

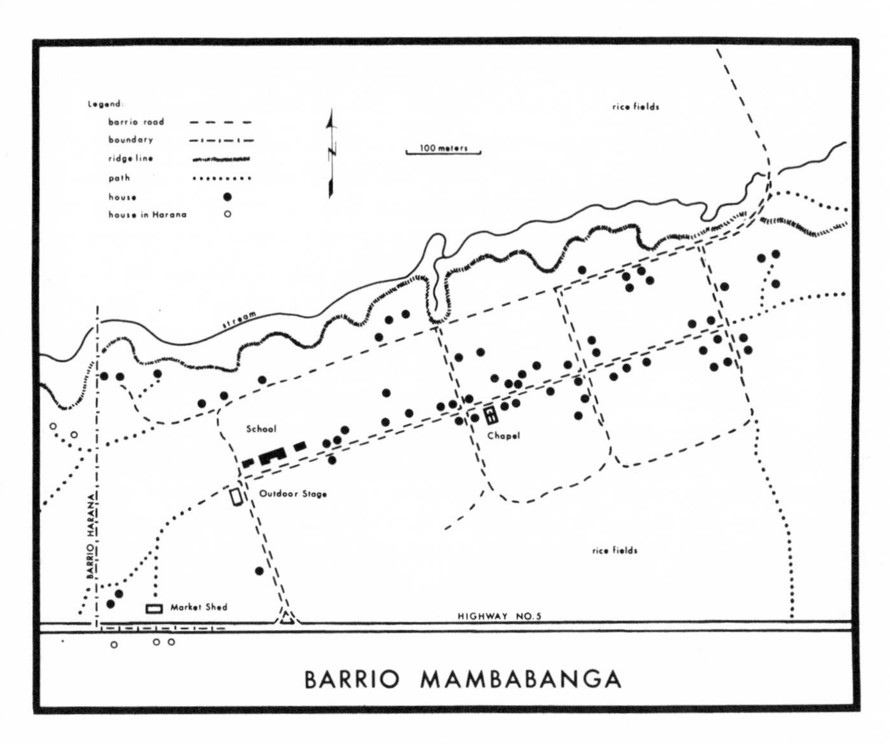

Map 7.3.

separate barrio. The new landowner was a Chinese mestizo who owned a large tract of land in Pangasinan, and many of the more recent tenants were attracted from his holdings there. This individual went out of his way to settle some of the problems created by the former landowner, and water is now distributed equitably to all members of the society. All people within the system, the tenants in Harana as well as a number of families in Mambabanga, are required to contribute work on the dam and on the main canals. The officers of the organization are almost all tenants of the new landowner, the reason being given that their residences are much closer to the dam and they are thus in a position to regularly check and control the flow of water. An annual feast is held each year following the harvest, and the large landholder makes the major contribution of food and drink on these occasions. Dominated as it is by a single, important individual, the Society Mambabanga is much less a communal effort than the other irrigation cooperative, the Union Bacarrena. The members in Mambabanga are happy with the system as it operates despite the fact that it was used quite differently by the former landholder, Mr. Cruz.

The Union Bacarrena covers about half again as much area as the Society Mambabanga, but not all of its over 300 hectares is regularly irrigated. Named after the Bacarra area, from which many of the original members had come, the Union Bacarrena was originally a group of immigrants who, like the settlers of Mambabanga, had applied for and purchased a land grant from the Friar Lands Estates. All of the Bacarrena land lies north of Macanao Creek (see map 7.4). These settlers first established three different barrios: Puroc, where today almost a third (twenty-seven) of the members reside; the nearby barrio of Luna (now the poblacion of Luna), which has nineteen members; and the barrio of Concepcion, currently with eleven members. Through intermarriage and inheritance, memberships have spread to other barrios as well. The actual barrio sites having members in the Union Bacarrena are located on less fertile or higher spots of ground within or bordering the system of irrigated fields; Mambabanga, however, is located well outside the system. There are eleven members in Mambabanga. Several families moved there from Concepcion following a flood in 1937, and it is these people and their offspring who are the members from Mambabanga.

In terms of formal organization, the Union Bacarrena is like the Zangjera San Antonio in Buyon: members own their own land; the society controls the water rights; individuals must contribute work on communal projects; water is

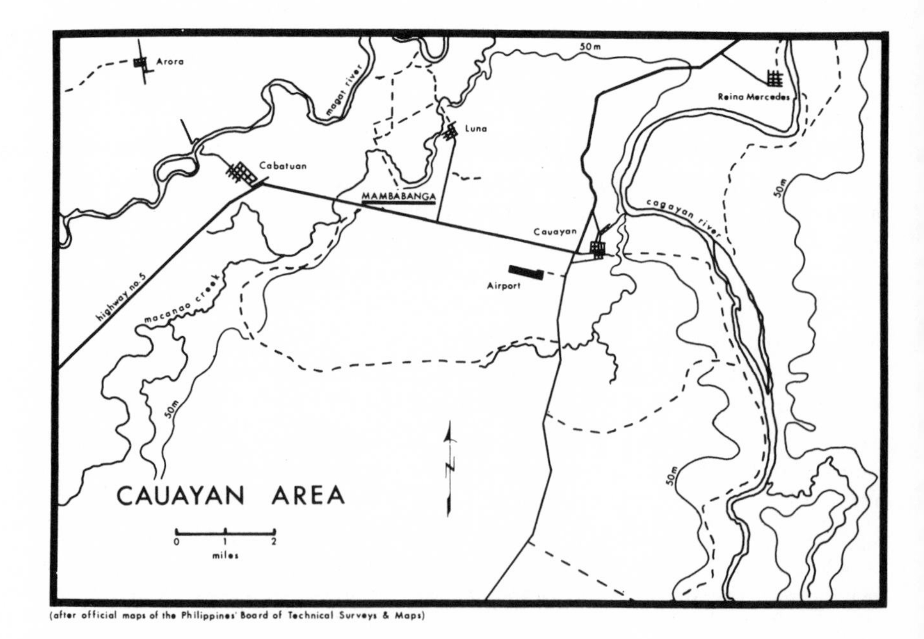

Map 7.4.

sold to nonmembers; a feast is held each year. Both the Society Mambabanga and the Union Bacarrena take their water from Macanao Creek, where Bacarrena has a concrete and log dam which backs the stream up to form a reservoir. Although this dam and the one of Society Mambabanga are subject to some damage by flooding, the threat of high water is never as serious as it is on one of the major rivers. Partly because of this, only one-third to one-half the number of workdays are required by Union Bacarrena members as are required of the zangjera members in Ilocos Norte, where, as was mentioned, water is diverted from the Bacarra-Vintar River.

Even though members of the Union Bacarrena are less often called upon to work, they are also less willing to work. Some members regularly fail to appear for labor details, and fines assessed against them are often impossible to collect.[9] Many complain that the system itself is inadequate and that it only provides sufficient water for those fields nearest the water sources. Several individual members farthest removed from the dam, those most commonly affected by water shortages, withdrew entirely from the society. New members were solicited from among the outsiders buying water from the Bacarrena, but most prefer to pay the required 10 percent of their crop rather than contribute work on the system. New memberships were opened simply on the basis of participation, but no new members were forthcoming. The national government has twice provided financial assistance and without this the Union Bacarrena would probably have had serious, if not insurmountable, maintenance problems. The stated hope is that the system will be eventually incorporated into the planned extension of the Magat River Irrigation Project, which ends only a few kilometers away.

It is not greater motivation which makes zangjeras in Ilocos Norte work more ''efficiently'' than those in Isabela. Zangjeras (like varieties of rice, the cohesiveness of extended families, or the amount of money saved in banks) differ according to the circumstances. Although the physical environment has

9. During my field research period, the officers of the Bacarrena met to take formal, legal action against recalcitrant members. The amounts due in outstanding fines were totaled up and a list was made of all persons in arrears. This caused some embarrassment since 90 percent of the members, including most officers, had outstanding debts! The list was shortened to include the ''really'' outstanding debtors, those owing P1,000 or about $250, and even this amounted to over 30 percent of the members. A local attorney in Luna who acts as their legal representative informed the officers that their ''constitution'' had no formal, legal status so there was no legal way to enforce the fines. The attorney was well aware of the social importance of a zangjera, and informed them that ''to be a cooperative you must cooperate!'' The concerned members then asked the municipal mayor to speak to the delinquent members. This action was considered by the older members to be especially bad policy. It would have been understandable in dealing with another organization, they said, but it was inappropriate for outsiders to settle an internal problem. The mayor called a special meeting in Luna, where he admonished the delinquents to live up to their responsibilities; characteristically, however, most of the very delinquent members were absent and most of the fines remained unpaid.

a direct effect on the operation of the zangjeras in two areas, there are important social factors as well. One of the most important of these is the relative significance, the variable latent function, which the zangjera has in the municipal and provincial political scene.

Irrigation Systems and Political Systems

Like many other barrios involved in the pioneering development of Isabela, Mambabanga was formed out of common purpose and maintained out of shared needs and fears. A socioeconomic basis for barrio communalism still persists in Mambabanga, through a complex of overlapping and interrelated ego-centered work and landowning groups. Thus, both historical continuity and social interdependency are involved in what constitutes the community of Mambabanga. And the annual barrio fiestas reflect the importance of the barrio as a socioeconomic unit in most of Isabela; the virtual absence of fiestas reflects the poverty of the barrio as a socioeconomic unit in Ilocos Norte.

In Buyon the historical conditions of barrio communalism are long forgotten. Because shared ammuyo work groups are virtually nonexistent and landownership is invariably a real or potential source of conflict, existing social ties are abbreviated and relatively ineffectual. In Ilocos Norte, only the zangjera forms a meaningful socioeconomic community of any size. Partly because the irrigation societies are effective social units and partly because of the nature of political activity in Ilocos Norte, zangjeras have assumed an importance and function considerably beyond the manifest purpose of providing irrigation water. A brief reexamination of the class and landowning situation is necessary to illustrate these relationships.

A traditional occupation for the Philippine elite has been to manage farm lands, sometimes directly, sometimes as absentee landlords. For what are essentially geographic and demographic reasons, the Ilocano elite (like the Ilocano non-elite) has been much less endowed with land than the upper class of other cultural groups in the Philippines. Because of this the Ilocano upper class has more often turned to politics. Beyond the relatively narrow limits of shared interests and individual loyalties, the most important component in maintaining alliance systems is wealth. This becomes especially important in the complex of alliance systems making up the larger alliance system and the following of a political person. Although wealth is so often the limiting factor in Ilocos Norte politics, it is to politics that so many of the elite want or need to turn. Intensification of political activity has developed in recent years with a growing number of non-elite persons turning to politics, individuals just as politically ambitious and just as wealthy—or poor—as the elite. The economic rewards of political life in the Philippines are always an inducement to seek elective office; and, corresponding to the relative impoverishment of all classes in the Ilocos area, political office has become that much more attractive and competitive.

Politically active persons have to involve themselves personally and directly with an ever growing number of associates who, in turn, have limited influence and political ambitions of their own. Thus, the promise of support in this election may be partly based upon supporting others in future elections. The consequent lack of continuity in office adds still more to political instability. Though often owning little land, the landowning elite attempt to make the widest use of their resources in property to influence a large number of people. Consequently, the landowner breaks his two- or three-hectare holdings into fifty to seventy-five tenth-of-an-acre plots to acquire as many tenants and, hopefully, as much support as possible. Other "political landowners" do likewise, so that a single tenant may be subject to the demands of several landowners, not one of whom has an exclusive "right" to his support. In the face of increasing competition and decreasing reserves of wealth and influence, the politically ambitious individual builds what he can on political promises, commitments, and social obligations—a precarious and often hazardous alliance system. Pressures and the potential for conflict continue to mount, competition increases, personnel shift between alliances, promises are made and broken, animosity grows, and the final complication of political life in the Ilocos emerges: political warfare and feuding. In fact, political killings and terrorism have themselves become an important means to the realization of political ends—and not simply when all else has failed!

Within the Ilocos area itself there are significant differences in the intensity and instability of political life. This is most pronounced when comparing Ilocos Norte to the southern Ilocos area, particularly Ilocos Sur. During the election year of 1963 (a non-presidential off-year election) more than 100 "political" killings were reported for Ilocos Sur and twenty-five for Ilocos Norte.[10] The reason often given in Ilocos Norte for the difference is that Ilocos Sur has a particularly large number of *Bagos,* or "new Ilocanos," the new or recently Christianized Tinguians from adjacent Abra Province. The assimilation of Abra people may indeed play some part in this inasmuch as some of the most politically explosive areas border Abra, but just how or in what way is impossible to say here. The significant difference is in the presence of irrigation cooperatives in Ilocos Norte and the particular agricultural economy which exists in Ilocos Sur.

To illustrate, tobacco has been grown as a commercial second crop on the

10. It is difficult to distinguish between "political" and "nonpolitical" killings. Authorities in Ilocos Norte and Ilocos Sur are understandably reluctant to add more publicity to that which they already receive. Consequently exact figures are difficult, if not impossible, to obtain. Though the newspapers perhaps tend to exaggerate the number of shootings, it is also true that not all shootings are reported to the newspapers.

Ilocos coast since Spanish times and, in fact, in the late eighteenth century a revolt was narrowly averted in the Laoag area over the imposition of a tobacco monopoly by the Spanish colonial government. The varieties established early in that period, now called ''native tobacco,'' are used for cigar production. A major change in tobacco production occurred shortly after World War II with the introduction of Virginia leaf tobacco, which is better suited for cigarette production. In 1952 the national government established import duties and a local price support program to aid the growing industry and to protect the growers against foreign competition. Partly to keep the Chinese business interests from gaining control of the market and partly to gain political patronage, the program—grading, purchasing, payments, jobs, etc.—was administered by local political officeholders. This resulted in an intensification in the political alliance systems. First, it provided a means by which those in power could not only increase their rewards to the political faithful and attract new support, but could punish the opposition in terms of low gradings, delayed payments, etc. Second, it has made the attainment of political office increasingly more popular and correspondingly more competitive. Finally, it has made the individual farmer more independent and more difficult to reach by the traditional social and economic means. A study of the effects of Virginia tobacco in Ilocos Sur and La Union noted the following:

> The economic prosperity of the farmfolk has somewhat lessened the tenant's subservience to his landlord in the matter of choosing national as well as local officials.
>
> Political leanings of the Barrio folk are more influenced by material aid. . . . [Garcia, 1962:10]

Except in the southernmost section, relatively little Virginia leaf tobacco is grown in Ilocos Norte. This is apparent by the absence of tobacco-drying sheds north of Batac. The most important second crop in central and northern Ilocos Norte is either garlic or onions or, less commonly, the traditional ''native tobacco.'' The sale and marketing of these are more subject to changes in the marketplace and not directly to changes in the local political system. The highly arid and extended dry season of Ilocos Norte, which makes the growing of Virginia leaf unfeasible, probably influenced the early development of communal irrigation societies. Also, because of the longer growing period required for bayag varieties of rice, artificially irrigated lands are much less commonly planted to tobacco than are nonirrigated or naturally irrigated lands.

The irrigation societies constitute the second condition distinguishing the relative political stability of Ilocos Norte. The decline in affluence and influence of the upper class, the increasing isolation of the individual, the various interrelationships of these factors and the resulting weakness of the political alliance systems have already been outlined. Only the communal irrigation systems constitute relatively large and, at the same time, stable social groupings with shared community interests. These groupings generally must protect their water rights and obtain materials (especially cement and other building materials) to improve upon and maintain the irrigation works. The zangjeras, in the absence or incapacity of other social forms, particularly the barrios, are an important political resource. The backing of several irrgation systems can widen and extend the scope and effectiveness of a politician's alliance system. Because of the presence of zangjeras, and because of the corresponding absence of the highly political Virginia tobacco "industry," the political situation in Ilocos Norte is somewhat less intense than is the case further south. Competition for the zangjera vote is intense, and political violence is by no means abrogated; but the influence of the irrigation societies in Ilocos Norte provides at least some block support together with a higher degree of voter predictability. The extremely individualized, intense house-to-house search for votes which is found in the southern Ilocos area is tempered and less frenetic. The difference is only relative, however, not absolute.[11]

Politics and the Barrio

Political feuding does occur in Isabela, but it is by no means as pronounced or intense as it is in Ilocos Norte. It is normally the landed, elite class in Isabela, as elsewhere, which occupies itself with politics. The builder of a personal political alliance system there is wealthier, and has a relatively large and stable social group, the barrio, to which he can appeal. Though tenants in Isabela are relatively independent and can usually find other lands to work, nonetheless, landlords do exert considerable influence over the voting of their tenants. It is, after all, the landlords who ultimately control part or all of the tenants' means of livelihood. Yet, as mentioned earlier, this superordinate-subordinate relationship is not without its rewards for both parties, and the tenant does not feel particularly coerced or forced into voting against his will. In fact, except where the charisma or ethnic position of a candidate intervenes, the barrio voter will make his choice according to the immediate and practical considerations relating to his voting—his landlord's wishes, his ties with barrio mates, the promises made by a candidate, an obligation to a friend, relative, or compadre, the need to sell a vote, and so on. Because barrios are effective social units, they can often be influenced to vote for one party or alliance system. The older barrios, such as Mambabanga, vote along

11. In November 1963, just before an election, an ambush occurred outside the poblacion of Bacarra which resulted in the death of four persons, one of whom was the Liberal Party candidate for mayor of that town. The attack was explained as being the result of a long-standing blood feud between the candidate for mayor and the leader of the ambush group. The complex of relationships within each group (the attackers and those attacked), between the two groups, and in the web of sociopolitical systems and subsystems in Bacarra was fantastically involved. It stands as a classic—and very tragic—example of the highly involved and unstable nature of alliance networks in the Ilocos area.

traditional lines which have been maintained and nurtured by both voter and politician. Although particular barrios are often described as "Liberal" or "Nationalist," what is meant, in fact, is that these barrios have traditionally supported particular alliance systems in Luna which are associated with one of the national parties.

Mambabanga differs from some barrios in that it is split into Liberal and Nationalist factions; yet the split is along traditional, essentially predictable social lines. The original settlers of Mambabanga have traditionally given support to certain alliance systems which are now identified as Liberal. This is also true of Harana, in large part a result of Harana's having been a sitio of Mambabanga, so that all of Harana and most of Mambabanga identify with Liberal Party candidates. Of the sixty-one houses in the barrio, thirty-six identify with Liberal Party candidates; nineteen claim to have traditionally supported the Nationalists; and two claim to be "nonaligned." These latter two families are very recent arrivals, one having ties established in the town of Cabatuan and the other not having yet established itself. The nineteen supporters of Nationalist candidates are all members of the families which moved to Mambabanga from Concepcion in 1937. The thirty-six families which support Liberal Party candidates are the descendants of the original settlers in Mambabanga plus several "converts" from the later arrivals. Most of these converts from the Nationalist group are second-generation families or families from Concepcion who intermarried with the original Mambabanga families. Although most of the late arrivals from Concepcion still own land and have important economic and social ties in the old area, they have become increasingly involved and interdependent with the original Mambabanga families. Mambabanga now holds a rather special position in the municipal political scene for it can deliver a pivotal block of votes in an election. The barrio people, as a consequence, have profited from both sides.

Irrigation societies in Isabela, on the other hand, are almost nonfunctional with respect to political life. While politicians seldom miss the opportunity to speak at a gathering of either the Society Mambabanga or the Union Bacarrena, it is to the barrios that they direct most of their political efforts. Besides the fact that barrios are socially and economically important, the irrigation societies in Isabela are simply too few and too poorly organized to constitute an important political focus. In Isabela, because of the interdependency of various socioeconomic ties, the barrios are significant politically. In Ilocos Norte, because of real social and economic poverty, the zangjeras partly fill a political vacuum from which they are able to profit. Thus, communal, cooperative efforts to obtain water by the same cultural group have been applied in two environmental settings with strikingly different results. All this points up the fact that the behavior of a given group does not result from social arrangements or cultural tradition or environment alone. The behavior of Ilocanos in both Buyon and Mambabanga is a consequence of the shared social and cultural traditions as those traditions relate to and are interrelated with their respective natural and social environments.

References

BLAIR, E. H., and ROBERTSON, J. A., eds. 1903–09. *The Philippine Islands, 1493–1803*. Cleveland: A. H. Clark.

CHRISTIE, EMERSON B. 1914. "Notes on Irrigation and Cooperative Irrigation Societies in Ilocos Norte." *The Philippine Journal of Science* 9:99–113.

GARCIA, NATIVIDAD V. 1962. "A Study of the Socio-economic Adjustments of Two Ilocano Villages to Virginia Tobacco Production." Abstract Series no. 15. Quezon City: University of the Philippines Community Development Research Council.

KEESING, FELIX M. 1962. *The Ethnohistory of Northern Luzon*. Stanford: Stanford University Press.

LEACH, E. R. 1961. *Pul Eliya: A Village in Ceylon*. Cambridge: Cambridge University Press [Chapter 5 in this volume.]

8

Mountain Irrigators in the Philippines

Albert S. Bacdayan
UNIVERSITY OF KENTUCKY

This paper deals with the irrigation system and social organization of the Tanowong people in the western portion of the Mountain Province, in the Northern Luzon highlands of the Philippines. An analysis is made of the organizational impact of the expansion of the preexisting traditional Tanowong irrigation works, and of the problems that ensued in intervillage conflict and enforced cooperation, as well as the Tanowong's emerging relations with the national government as they sought the sanction of the state as the ultimate guarantee of their desperately needed water rights. I propose that this irrigation expansion served to reinforce traditional Tanowong social organization and to integrate their contemporary relationships with other villages at the same time that it enhanced the articulation of the relatively isolated Tanowong community with the outside world, in particular with the structure of the Philippine national state.

The Tanowong Villages: The Social Organization of the Traditional Irrigation System

The Tanowong people occupy four villages located at varying altitudes from the bottom to the summit of a mountainside in the northern portion of the municipality of Sagada. Tanowong, the mother village, is closest to the river

This paper was originally presented at the Symposium on Irrigation and Communal Organization during the annual meeting of the American Anthropological Association in New Orleans, November 1973, and was entitled "Securing Water for Drying Rice Terraces: Irrigation, Community Organization, and Expanding Social Relationships in a Western Bontoc Group, Philippines." It is reprinted from *Ethnology* 13 (1974):247–60. Footnotes have been renumbered.

bed at the bottom of the mountain, Kadatayan stands midway to the top and is the second oldest, while Nadatngan and Madongo, on the plateau at the summit of the mountain, are the youngest villages, having been founded by migrants from the two older villages in 1915 and 1939 respectively. The total population of the four villages is estimated to be roughly 1,000 people. Although living in discrete villages, the people see themselves as one sociopolitical group separate and distinct from their neighbors. Their religious ceremonies are closely coordinated, and before the implementation of the Philippine Barrio Charter in the area, they had only one set of political officials. In spite of the recent split of the group into two different barrio governments, there is still much coordination in the social, religious, and political life of the entire group.

The Tanowong people are Bontoc culturally,[1] although most of their extravillage interactions are with the Lepanto groups, particularly the Sagada group to the southwest of them. Like the other Igorot of the area, Bontoc and Lepanto alike, the Tanowong cultivate terraced rice fields. In this endeavor they expend most of their man-hours and their religious efforts, although more than 50 percent of their year-round diet consists of sweet potatoes rather than rice. As is typical of the Bontoc and Northern Lepanto groups, the Tanowong are organized into different *dap-ay* groups. *Dap-ay* is the Tanowong designation for the men's house, which is better known in the anthropological literature by the term *ato*.[2] A *dap-ay* group consists of the families belonging to a particular *dap-ay* which in Tanowong would number up to thirty. The number of *day-ay* in a village vary according to size and age of the village. There are four in Tanowong, two in Kadatayan, and one each in Madongo and Nadatngan.

The *day-ay* are the religious, social, and political centers of village life,[3] where major decisions are made and through which the villages are mobilized

---

Grateful acknowledgment is made to the University of Pittsburgh for an Andrew Mellon Postdoctoral Fellowship in Anthropology, during which this paper was written. Being a native of the Pedlisan and Tanowong area, I am familiar with the Tanowong irrigation expansion from its inception. I took a very active part in the concluding negotiations between Tanowong and Agawa during my field work in the area in 1971–72.

1. The diagnostic Bontoc culture traits that distinguish the Tanowong people from their Lepanto neighbors are use of the basket hats called *kinaw-it* or *ballaka,* the singing of the song called *alassan,* and the dancing of the war dance called *tallib,* which is also called *talliberg* or *ballangbang*.

2. See Jenks (1905) and Keesing (1949). Although in former times *dap-ay* membership was a function of locality or contiguity so that *dap-ay* group corresponded with neighborhoods within a village, this is no longer strictly the case, although it still tends to be that way.

3. It is also an educational center for the males. Boys sleep in the *day-ay* from about the ages of six or seven until they are married. The men go there in the evenings, early mornings, and during the rest days to socialize with one another and with the young men and boys. At such times the young are trained to obey through the performance of chores assigned them by the men, like stoking the fire, scratching the soles of their elder's feet, or massaging their legs and backs. Also, the boys are instructed in village law and morality.

and grouped for communal action. While there is explicit competition among *dap-ay,* they always coordinate their efforts for the welfare of the community as a whole. The competition between the *dap-ay* is always within the framework of a "loyal opposition," with the welfare of the village remaining uppermost in everyone's thinking. Community problems are discussed in individual *dap-ay* gatherings as well as in village-wide meetings held in one of them. There is much consultation between the different *dap-ay* prior to any decision being made for the entire community at one of the meetings. Decisions binding upon the whole community are implemented through the *dap-ay* system. While each *dap-ay* theoretically has an informal council of old men who make the decisions, in actual fact, especially at present, every mature man participates in the deliberations of the council.

The *dap-ay* usually take turns assuming responsibility for the standard calendrical activities of the village. For example, a different *dap-ay* is responsible each year for performing after the planting season the sacrifice to the village sacred tree for the sake of all the village fields. Another *dap-ay* is usually assigned the performance of the main socioreligious ceremony called *begnas* for the village as a whole. While the other *dap-ay* of the village also perform ceremonies at these times, only the appointed one is considered to be the performer of the ceremony, with the others playing an active supporting role.

The *dap-ay* serve as focal points of mobilization and accountability where decisions require implementation, such as the collection of materials or money, and the procurement of labor for community trail and irrigation repairs. The *dap-ay* are important in the discussion and dissemination of information about decisions reached in a major meeting in other *dap-ay* or in meetings outside the village, such as in the municipal government centers which village officials attend. Often each *dap-ay* is assigned responsibility for specific segments of regular projects like trail and irrigation repair. In such a case, year in and year out, the members of each *dap-ay* know exactly where to go at the appointed day or days of work. Temporary assignments are similarly made. With this division of labor and responsibility, it is easy for the Tanowong to evaluate *dap-ay* performance and to spot those which do not fulfill their communal duties.

These convenient groupings within the community provide a healthy medium for competition in the performance of civic duties, since they often try to outdo one another in cooperativeness, promptness, and thoroughness in their assigned work. The *dap-ay* that lags behind is taunted by the others. In order to avoid this, each one tries to keep abreast of the rest. This spirit of competition directly benefits community concerns, such as the maintenance and repair of the irrigation works.

While the *dap-ay* is the unit of mobilization and accountability in community-wide endeavors, the nuclear family within the *dap-ay* is the unit of assessment for whatever is required by a community project. A *dap-ay* divides the assessment of goods, money, or labor among its families and enforces their delivery.

The rice terraces of the Tanowong people are mostly located in the immediate environs within roughly three kilometers north and south of Tanowong and Kadatayan, the two oldest villages in the group.[4] There are a few springs in the immediate area, but they are so small as to be insignificant in terms of the volume of water needed for the fields. The required water must be obtained elsewhere. For a long time it was supplied by two streams northwest of the fields in the upper reaches of the plateau on the summit of the mountain on whose eastern side of the fields are located. Because of a history of an egalitarian exploitative pattern of their territory, and an equally long history of village endogamy which is now breaking down, all the rice fields in this area are traditionally owned by Tanowong people. More important, all families in Tanowong own at least one rice field in the area. Given the tremendous effort needed to construct and maintain the irrigation ditches and the dams at the water source which divert the water, and given the fact that everybody in Tanowong is involved because everyone owns terraces, irrigation among the Tanowong is a serious communal affair.

The water from the two streams is conveyed to the fields by a ditch approximately four kilometers long and on the average about two and one-half feet wide and three feet deep. This ditch is repaired annually in late December after the planting season by all the villagers, mobilized through the wards in the manner discussed above, after the date for the work is decided on by the men at a meeting, either in one of the ward houses or in one of their innumerable house-to-house group drinking sessions.

The repair crew is composed of men and women of all ages and sometimes includes young boys and girls, depending upon the available labor in each family. While there is no discrimination regarding sex and age of the participants in this undertaking, if the job requires hard labor, as when typhoons have been especially destructive and ditches and stone retaining walls must be rebuilt, then men if possible should represent their families. Maintenance of the ditches consists of cutting the growth on the banks of irrigation ditches, removing earth that has accumulated and any other obstruction to the water flow, plugging leaks, repairing retaining walls that have given way, sometimes making and installing wooden flumes and bridges across the ditches where such are used as paths, and repairing the dams diverting the water.

Apart from major reconstruction, these jobs usually take one day, at the end of which the water flows freely toward the fields. The working group,

4. A considerable number of rice terraces are found outside of this area away from the villages farther upland, but the terraces involved in this study are the most productive, most valuable, and most important.

particularly the men involved, then split into smaller groups and go to the different Tanowong villages to collect fines from those who failed to participate. The fines take several forms: money (one peso), rice (five bundles), or a good drink and meal for the group.

Once the water is flowing, there are eight to twelve water distributors who take over the task of systematically distributing the water as fairly as they can to the different fields. This practice of having water distributors seems unique to the area. It is convenient and efficient in that it saves the individual owner from having constantly to check his fields and sometimes to sleep by them to see that they receive water. Perhaps more significantly it prevents conflict among the different owners, since they cannot accuse one another of stealing or of taking more than their share of water. In neighboring areas where there are no water distributors, suspicion and 24-hour vigilance is standard practice, punctuated with not infrequent altercations and even violence at the height of the dry season and the peak of irrigation need.

The water distributors are informally selected by the people on the basis of dependability, diligence, and fairness. Men who are thought to be good as water distributors are urged to take the office, or else are mentioned to the leading men of the village as able candidates. Sometimes men themselves apply for the job by indicating to the people their desire; if they are thought fit for the job, eventually they are told to give it a try. No woman so far has been made a water distributor. Distributors are paid at the end of the harvest in kind at 5 percent of the harvest: five bundles of rice for every 100 bundles harvested.

The routine of the water distributors is a demanding one. They rise very early and by 5:30 A.M. are on the job. Some follow the ditch every morning and late afternoon to check that there are no breaks, no obstructions, and no leaks. The rest spread themselves among the terraces, working the water through narrow ditches between the terraces and through each terrace outlet (a cut in the mud dike) to those fields that had no water the day before, or to those that are drying up. By 9:30–10:00 A.M. the first part of the job is finished and the distributors often gather on a knoll at the top of the mountain overlooking the terraces, where they exchange information and from where they can observe which part of the area will need water the most that evening. They then may go home to have breakfast and do chores. At about 5:00 P.M. they begin again and work until dark. They stand on call all day, however, in case of emergencies.

Because adequate water is an absolute necessity for the rice to mature, any source of water is important, particular during the driest months of February, March, and April, when the rice is growing rapidly. Between late February and early March sacrifices are made to the small springs by those whose fields are nearest them. This sacrifice consists of killing a chicken near the spring, cooking it there with salted pork, and having a meal to which the spirit of water is invited and prayed to in order to increase the flow. The performance of these sacrifices by the few whose fields adjoin a spring is viewed by them as a sacrifice for the sake of all the people in the community. Should this obligation be neglected, it is viewed by the people as selfishness, requiring some form of censure, usually taking the form of a "talking to," if not a fine, by the old men.

After the rains really set in about early May, the irrigation system is no longer needed and the ditches therefore are left untended. From this time until the next season, irrigation is individualistic and not communal. Those who need water, perhaps for the repair or expansion of their fields,[5] must go alone to clean the ditches for the water to flow from the streams without obstruction.

Although in the past irrigation was not an overwhelming concern of Tanowong village organization, it has been one of the most important concerns in the recent history of their villages. The fact that it was not considered critical was due to the adequacy of the water supply. This complacent situation changed radically in 1954, when the growing insufficiency of the water supply led the Tanowong people urgently to seek out hitherto untapped and remote water sources to expand their irrigation, a venture to which we now turn.

The Bwasao Stream Diversion Project

Over the years the inadequacy of the original irrigation sources became more and more of a problem for three reasons: (1) the expansion and increase in the number of terraces on the original site, (2) the construction of new terraces along and below the irrigation ditch which necessarily diverted water permanently, and (3) the denuding through careless cutting and frequent fires of the pine forest of the mountains in the environs of the streams which served as the source of irrigation water. Gradually, therefore, more and more of the original rice terraces, particularly those located in the lowest tiers, were not adequately watered and thus became increasingly unproductive, leading to their conversion to the growing of sweet potatoes (*camote*). At first the water distributors were blamed for the shortage. They were thought to be lax in checking leaks from the ditches, and unfair in their distribution of the available water. The water distributors became even more strict with the water, looking with extreme disfavor at those people close to the source who had vegetable gardens and who had heretofore diverted water at least once every two days to water their plants. These moves were all to no avail, since there simply was not enough water.[6]

5. Water in such cases is generally used for the process called *gobogob,* to erode soil from one place to another. In this way an area can be leveled by the removal of earth through the use of water power.

6. Although it can be argued that the water supply per se did not decrease and that the cause of the shortage was the increase in the area to be watered, the older people also claim that the amount of water at the source had become much less than in the past.

On realizing the actual water shortage, the Tanowong regarded it as a serious blow to village life, if not a calamity. It was not that they thought it would lead to starvation, because as already noted, sweet potatoes constitute a major portion of the diet and do in fact grow well in the dried-up terraces. Strictly speaking, since rice does not grow very well in this mountain area and its cultivation is so time-consuming, the *camote* or sweet potato is a more economical crop. Rice, however, is considered to be the best food there is. It follows that rice terraces are one of the most highly valued kinds of property for inheritance.[7] Given this fact, plus the fact that the religious ceremonies, which are very significant kinship group and communal rites of intensification, are keyed to the rice cultivation cycle, the ownership of a rice terrace is like a badge of citizenship and of continuity, rootage, or identity in the group. This cultural significance of the rice terrace is the background for the intense concern shown over the dwindling water supply. In view of the seriousness of the problem, the Tanowong people struck on the bold idea of tapping Bwasao, a fairly sizable stream deep in the forest in the mountains north of them. As Bwasao lies within their hunting, mushrooming, and bamboo-gathering range, and directly on their path to the southern Ikneg territory where they go for trade and employment in building terraces, every male of Tanowong was familiar with it. They knew that water there is steady, that there is no great variation in its flow between the rainy and dry seasons, and that it therefore is a dependable year-round source. But they must have been intimidated by the distance of Bwasao to their fields. It is at least a three-hour fast walk from Tanowong to Bwasao, using the shortest route. Since the proposed ditch would have to follow the contours of the mountains, it would have to be much longer than the path, perhaps up to 25 kilometers long. But everybody agreed that the project was worth the try, for should it succeed, preservation of their precious rice terraces would be ensured.

For some time the Tanowong discussed the technical difficulties and other problems presented by the project. Finally, during the slack season (February, March, and April)[8] of 1954, the people, mobilized, and organized through the *dap-ay* system, began working at the water source. Their strategy was first to build the dam at a point which was high enough to allow the water to flow

7. One of the most important activities during a Tanowong wedding celebration (also true among the neighboring Bontoc and Lepanto villages) is the announcing of the couple's inheritance by their parents. This takes the form of a "sing" on the night of the celebration by both kin groups involved in the marriage, with each side announcing the inheritance in the form of a song and challenging the other to match what it has given. Rice terraces are the main subject of these songs.

8. During these months the most serious agricultural activities are the *kames,* removing weeds between the rice plants and evening the plants by replanting with shoots from established plants those spots where the original seedlings died, and *agabat,* weeding the banks of the terraces. Since these are ordinarily women's tasks, the men are free and usually go to the neighboring Ikneg (southern Tinguian) area of Maeng, or else to the mines in the province of Benguet to the south to seek employment.

with adequate grade, and then to build a ditch leading from the dam to Tanowong. This remarkable engineering feat, which would appear overwhelming for a strictly gravitational flow through the mountains with no sophisticated surveying equipment, was made possible only through the people's detailed familiarity with the territory from years of hunting and foraging.

Since the Tanowong appreciated that the project would be a long one, they assessed each family five days of labor, with the understanding that after those days were expended, another cycle would begin, and so on until the irrigation system was completed. Because of the importance of the fields to everyone, there was strong commitment to the project and enthusiasm on the part of all.

The Tanowong people were fortunate that at about this time a man from the municipality of Sagada won a seat in the now disbanded Philippine Congress. The Mayor of Sagada at that time was from Tanowong and through his efforts the Tanowong people had supported this winning Congressional candidate. After seeing the serious self-help effort of the Tanowong people, the newly elected Congressman included the Tanowong irrigation expansion project among those to be supported out of his "pork barrel" funds. It then happened that after the donated labor of the people began to lag, compensation became available from the government for the additional labor needed to complete the irrigation works by the following year, 1955. After these government funds were expended, subsequent grants were obtained from the government for further work on the irrigation system for widening and straightening its course.[9]

Maintenance of the Bwasao irrigation system is one of the most important annual activities of the people of Tanowong today. As in the original irrigation system described earlier, the Bwasao ditch is repaired every year in January after the original ditch has been repaired. Because the Bwasao ditch is nearly 25 kilometers long, different *dap-ay* are assigned specific segments of the ditch as their charge. Unusually heavy damage to a segment, however, such as a massive landslide, calls for labor from all the people. Groups from the different *dap-ay* work their way to the forest area early on January mornings. As soon as possible, water is fed to the ditch where part of the maintenance task is tamping (*detdet*) the moistened soil at the bottom and sides to prevent loss through leaks or seepage. As soon as the water is safely flowing through the ditch, the water distributors take over as in the preexisting irrigation system. In all aspects of the maintenance of the newly

9. It has become a matter of competition for the Tanowong village officials to obtain money from the government or other agencies and to spend it in a most significant way on the irrigation system. The latest development along this line was the procurement of a food assistance grant in September 1971 from the Food for Peace Program through the Church World Service Agency in the Philippines in the form of about 600 sacks of bulgar, flour, and rolled oats to feed workers and their families while they worked on improvements in the irrigation system.

expanded irrigation works, the traditional social organization and patterns of maintenance established for the pre-existing irrigated terrace system prove adequate.

Intervillage Conflict over Irrigation Water

Control of water has been a sore point between a number of villages in this part of the Mountain Province. No sooner had the Tanowong begun work on their expanded irrigation works than another village, Agawa of Besao muncipality, made claim to the water source. The upper reaches of Agawa's eastern boundary is by-passed by the Bwasao ditch and Agawa is somewhat more closely situated to the source. But the major basis of Agawa's claim was that Bwasao lies within her territory and that at some time earlier in her history, she had already begun work to divert the stream toward Agawa. The former is a false claim, for in fact Bwasao is an open hunting territory for both Tanowong and Agawa, confirmed by long usage. The latter portion of the claim was based on the fact that early in the 1930s a gold mine, which soon failed, was started in the area; part of the mine's plan was to divert the Bwasao stream to obtain water for a proposed mill and compound three kilometers from the source. The laborers employed to divert the stream were from Agawa. Thus, while Agawa could show they had started a ditch more than twenty years before, it was not part of the Agawa plan to divert the river for themselves. The Tanowong were quick to point this out. After much negotiation between the leaders of both villages, the Agawa people reluctantly agreed not to interfere further with the Tanowong project. Both groups rationalized that such an agreement was in the best interests of all concerned since there was a growing number of intermarriages between Agawa and Tanowong, and fields watered by the irrigation works were as a result owned by people from both groups.

The Agawa people asked, however, that any creeks flowing toward Agawa territory across the path of the irrigation ditch should not be diverted or interfered with. Although this was an inconvenience to the Tanowong, as it meant the construction of two water aqueducts across streams, they conceded. An agreement on paper was drawn up to this effect. In addition, the agreement provided a fine for its violation, the amount of the fine to be determined at the time of the violation.[10] Also, the Agawa people later claimed that the agreement included an understanding not put down on the paper that if at any time in the future Agawa should need water, the Tanowong were to share some with them.

This last claim was a source of contention between the two groups

10. In March 1972 the Agawa people accused the Tanowong people of a violation of this agreement. Although the accusation was not upheld, it provided an occasion for discussing the importance of the agreement, thereby refreshing the memories of both sides.

between 1971 and 1972. In 1970 the Agawa people were given a large contract to install the waterworks for the Besao municipal seat, on condition that they themselves locate or provide a water source. Their solution was to tap one of the springs above the Tanowong dam at Bwasao. This was rumored in Tanowong and was confirmed when Tanowong hunters came upon Agawa men and a government engineer in the environs of the dam. Shortly thereafter, the Agawa leaders and some of the municipal officials of Besao were seen in the provincial capital of Bontoc at the Engineer's Office, purportedly there to draw up the plans for the waterworks.

The Tanowong people were upset by this turn of events; and after waiting in vain for some time for the Agawa people to inform them of their intentions, Tanowong sent some of their officials to Agawa to find out what was going on, fearing that their irrigation works might be in serious jeopardy. The Tanowong officials on their trip to Agawa ascertained what the Agawa people proposed to do. This led to a series of internal meetings in Tanowong to determine how to counteract the Agawa threat, and to a series of meetings with the Agawa leaders as to how to solve the problem between the two groups. The Tanowong people let their determination be known to fight any incursion of their water resources. The Agawa asserted their traditional claim to the water and their recollection that there was an understanding that they could share the water with Tanowong at some future time, and now was the time. Threats of force were made by each side. Cooler heads prevailed through effective appeals in the negotiations to the unity of Tanowong and Agawa as a result of the increasing number of intermarriages and the resulting joint ownership of rice fields.

The Tanowong people proposed that the Agawa tap another stream in the vicinity but one not flowing to their dam. The Agawa countered that the stream was not large enough and that it dried up during the dry season. The government engineer was taken to that proposed source by representatives of both groups and while he agreed that it was small, he pointed out that with sufficient storage tanks it had good possibilities. After much bargaining, the Agawa countered that they would accept the Tanowong proposal if the Tanowong people would carry the pipes from the Besao municipal center to Bwasao (a distance of some fifteen kilometers) and if they would install the pipes from the proposed source for a distance of about two kilometers toward Besao. These conditions were stoutly resisted by Tanowong, but they finally agreed to them in order to eliminate the threat to their water source. A contract was drawn up to the effect that the Tanowong would help Agawa carry and install the pipes, but in return Agawa was never again to interfere with the Tanowong irrigation system in any way. Tanowong immediately mobilized to carry out the agreement and after one month they fulfilled their commitment in the contract. In doing so, they now hope that there will be no further irritants arising from the Bwasao irrigation system to disrupt the relationship between Tanowong and Agawa.

In the course of dealing with Agawa's challenge, the Tanowong people became convinced of the necessity of obtaining government sanction for their irrigation system as a means of ensuring permanent control of the water source. An application to the government for a water right, accompanied by a petition from the community, was filed early in 1956 to initiate the proceedings. Officials of the Bureau of Forestry were then sent several times to inspect the area. A long period ensued during which other requirements were fulfilled. Finally, in April 1972 Tanowong was notified that they had been granted a temporary water right certificate by the secretary of the Department of Agriculture and Natural Resources.

During the filing and follow-up period the officials of the Tanowong community made periodic trips to the provincial capital and to Manila, first to file the application formally and then to follow it up. These trips necessitated contributions by the people to pay for the transportation and food of the village officials sent. The people also contributed to the required filing fees. In addition, gifts in kind in the form of coffee beans and even a live deer were given to government officials as this was thought to facilitate matters. Every occasion on which the people had to contribute necessarily focused attention on the irrigation system. The need for contributions was discussed publicly in the *dap-ay*, and finally the officials went from house to house to collect the required amount.

Before receipt of the notification of their temporary water right further anxiety among the Tanowong people was caused by a rumor that the Agawa people were in fact also applying for the water rights to Bwasao. This rumor spurred the Tanowong to send another delegation of officials to Manila, which necessitated yet another collection from the people. Two such trips were made during the three months preceding their notification in April 1972. The notification quieted their anxiety.

Unlike the conflict it caused between Agawa and Tanowong, the Bwasao irrigation project had a unifying and integrating effect with respect to Tanowong's relations with her northern neighbor, Pedlisan. Pedlisan is also a Bontoc village and the two are therefore very similar culturally. Because they are closely situated about five kilometers from each other and because they are traditionally friendly, there has been a growing number of intermarriages between the two villages. They also share much territory, the Bwasao area being one of them. As a result of intermarriage, the bilateral pattern of inheritance, and neolocal residence, the ownership of the terraces in both Tanowong and Pedlisan is becoming increasingly shared between the two communities.

Tanowong's problem with its water supply therefore affected many families in Pedlisan. Pedlisan was also experiencing water shortage and irrigation problems and this likewise affected Tanowong families with fields in the vicinity of Pedlisan. It was natural therefore for the two communities to realize the necessity for close cooperation. Whereas they had historically cooperated in matters such as bringing home the body of a member of either community who had died abroad or searching for lost animals, the present critical need to support each other with regard to water is an especially articulated and verbalized reason for more general cooperation between Tanowong and Pedlisan.

Many meetings between the two communities have taken place, in which the theme has been cooperation over the water problem. The meetings are usually held in a hamlet midway between the two, where the Episcopal church[11] has a mission establishment in the form of a church, parsonage, and schoolhouse. Cooperation is expressed in varying forms of support for each other in their relations with an outside party: in applying for water rights from the government, in opposing concessionaires who want to exploit the forest resources around both communities, or in contributing signatures to petitions or funds for travel expenses of village officials. In the spring of 1972 the people of both villages united strongly in opposition to the owners of a paint manufacturing firm in Manila who wanted to gain concessions to tap for oleoresin in the pine forests surrounding the villages.

This opposition to the exploitation of the pine forest resources is due to the realization, perhaps belated, of the correlation between the density of trees and the amount of available water. As a result there is a strong movement to have the surrounding forest areas declared a watershed or communal forests by the government, which would make them immune to outside exploitation for lumber or oleoresin. But to gain such protection, the villagers must deal with the national government structure, often a remote, complex, and frustrating entity for the uninitiated to approach.

Discussion: Organizational Impact of the Expansion of the Irrigation System

It is noteworthy that the traditional Tanowong social organization of irrigation focusing on the *dap-ay* proved adequate for the construction and continuing maintenance and distribution of water from the Bwasao water works, a tribute to the vitality and capability of this traditional framework of communal action. However, more significant aspects of the irrigation expansion project lie in: (1) the manner in which it provided and still provides the entire political community with a central issue with which all members identify very strongly at a time when other serious socially disruptive forces are at work; (2) a new consciousness and favorable attitude toward the normally remote national

11. The church is increasingly referred to by the people as the "common *dap-ay*" to which the people of Tanowong and Pedlisan all belong. It therefore has the potential of serving as an intervillage uniting agency, so far as the Pedlisan and Tanowong people are concerned.

government; and (3) a new dimension of experience and education in leadership, particularly in the skills of negotiation with other villages and in dealing with the modern government bureaucracy.

To serve as a stimulus for cooperation and solidarity in contemporary Tanowong community life is no mean contribution of the Bwasao irrigation project. For in these days even the relatively isolated Tanowong villages are racked with all sorts of forces of change: religious change (pagan animism versus Christianity); education and contact with the outside world; old values versus those of the young; and election politics of the barrio, municipality, province, and national government. Any of these singly and especially in combination seriously strain the village structure. Both the old and the newly expanded irrigation systems have been a boon for community integration and for the continuation of traditional communal organization.

The capacity of the Bwasao irrigation works to galvanize community action rests on the cultural significance of the rice terraces to the Tanowong people that was noted earlier in this paper. For this reason, any threat to the terraces, such as the dwindling water supply and the Agawa attempt to prevent Tanowong from constructing the Bwasao project and later to divert one of the springs feeding the irrigation dam, was considered a threat to Tanowong's livelihood and was regarded with immense anxiety and seriousness. Tanowong's old complacency dwindled and their vigilance mounted, for their drying fields had become a serious and alarming problem that touched everyone's heart.

As enormous as the Bwasao project was realized to be, the Tanowong people took to it with great enthusiasm. The meetings in the *dap-ay,* in which the underlying sentiment was "let us give it a try," must have strengthened their resolve, for there is no known case of anyone being fined or punished for outright refusal to cooperate on the project. Instead, there was eager support, and the work crews, composed of both sexes and of all ages, led by the old men, toiled hard and long. It is conceivable that this enthusiasm might have sagged later had not government funds been forthcoming. Government support was important in maintaining the morale of the people and in facilitating their continuing mobilization. The strong feeling of common purpose and solidarity generated among the Tanowong people by the Bwasao project still prevailed in 1972.

The Bwasao irrigation expansion heightened the awareness of the Tanowong people of the national government system as a source of financial, material, and technical aid. In addition, they discovered that the national government could exercise power on their behalf in the form of an irrevocable guarantee of their water rights in Bwasao and against commercial exploitation of their forests and watershed. This aid is an entirely different aspect of the government from the usual association the villagers have had with it. Earlier experience with the government was often negative, in that the government demanded taxes and free labor for roads and other kinds of construction, and demonstrated its power through incarceration of individuals in prisons. For a relatively isolated cultural minority group, this positive development is of critical importance in furthering Tanowong's identification with and integration into the national political system.

Finally, all of the formal and informal negotiations which the Tanowong undertook with the opposing village of Agawa, with her close neighbor of Pedlisan, and with the various government officials and agencies of the national government have been fertile fields of social and political experience for the villagers. This was especially true for the community leaders, who learned to deal with a much wider universe than the village and who progressively built skills and confidence in coping with internal and external challenges in their delicate and often painful role as intermediaries between their communities and outside forces. Through all the problems and processes that the Bwasao irrigation project led to, the Tanowong gained social and political skills, particularly in the art of negotiation,[12] which are of benefit to them in their increasing extravillage dealings in the contemporary Philippines.

The expansion of the Tanowong irrigation system is an example of the flexibility of traditional Bontoc social organization in incorporating greatly enlarged communal tasks without a change in its structural form. At the same time, the successful accomplishment of the Bwasao irrigation works projected the small Tanowong community into new external relationships at the local, regional, and national levels. In the Philippines, with increasing demand for water and other limited natural resources occasioned largely by a growing population, the Tanowong example may represent a more general pattern of change, whereby ecological necessity brings the once relatively isolated agricultural village into closer relations with the national forms of economic and political organization.

References

JENKS, A. E. 1905. *The Bontoc Igorot.* Manila.

KEESING, F. M. 1949. "Some Notes on Bontok Social Organization." *American Anthropologist* 51: 578–601.

12. To my knowledge the Bwasao problem is the first instance in which the Tanowong people had to negotiate over a protracted period a settlement with another village. It is also the first case in which binding agreements between Tanowong and another group were put in writing.

9

Traditional Customs and Irrigation Development in Sri Lanka

Michael Roberts
UNIVERSITY OF ADELAIDE

While water was a crying need for many cultivators in Ceylon, in the first half century of British rule this does not seem to have struck administrators so forcibly as to lead to any worthwhile activity in maintaining and restoring irrigation works. North and Maitland showed some interest in the subject and investigations were undertaken by men like Johnston and Schneider; others spoke of furthering the prosperity of the peasantry in grandiose terms; according to their lights, government sought to help peasant agriculture on other fronts by introducing freehold tenure and seeking to make the paddy tax uniform and fixed. But in the vital sphere of irrigation their achievements were negligible, lamentably and disastrously so. Decline continued apace where it had already begun. In regions such as those which neighbored the Patipola-aar in Batticaloa District and the Urubokka and Kirama dams in the Southern Province, decline set in where relative stability is said to have prevailed in Dutch times.[1] This was partly the result of government's preoccupation with military and strategic interests and, in the 1830's and 1840's, with the plantation enterprise. It was partly due to limited finances in an age when deficit budgeting was taboo. It was partly because laissez-faire notions were very strong in the second quarter of the century; North and Maitland might have accepted government's responsibility in maintaining irrigation works without question, but Earl Grey in the Colonial Office and officials in Ceylon used up several pages in the 1840's and 1850's justifying—to them-

Formerly published as "The Paddy Lands Irrigation Ordinances and the Revival of Traditional Irrigation Customs, 1856–1871," *Ceylon Journal of Historical and Social Studies* 10 (1967): 114–30. Reprinted by permission.

1. CO 54/328, Ward-Labouchere, no. 31, 27 February 1857.

selves and the world—state intervention in this sphere.[2] It was partly due to ignorance of conditions, an ignorance that was furthered by difficulties of communication and terrain. Even as late as the 1840's and 1850's, the tone of writing indicates that officials were bringing something little known to light when they described the remains of the ancient irrigation works in the dry zone.

Ignorance is also reflected in the patent optimism with which such men as Robert Wilmot Horton spoke of restoring peasant prosperity, and with which Tennent believed that the restoration of the Giant's Tank in Mannar would provide 134,000 bags of rice annually.[3] In a sense, it was just as well that the 1848 rebellion and the financial depression of the late 1840's forced government to shelve plans for restoring irrigation works which had been drawn up in 1848. When actual restoration was begun in Ward's time, it was on a relatively realistic note, though not altogether free from optimism.

Irrigation projects, then, commenced only with the advent of an active and practical governor, Sir Henry Ward, in 1855. He was aided by an expanding revenue but his interest in irrigation knew no bounds. Largely suppressed in earlier regimes, the enthusiasm of some of the A.G.A's and G.A's now came to the fore and a spirit of youthful activity in improving irrigation facilities pervaded the scene. They were not wholly blind to the difficulties, however, and a serious problem was soon placed before government by some of the younger officers. The village scene was being rent by disputes over the use of water as well as land disputes. The spirit and practice of mutual obligation which was so important a feature of peasant agriculture was being undermined by the forces of individualism, litigiousness, and apathy. The implications were serious. Any increase in irrigation facilities would be nullified and confusion worse confounded unless a remedy was provided.[4]

Ironically, the problem was partly, if not largely, that of the British rulers' own making. In abolishing forced services (*rajakariya*) in 1832 and in creating minor courts in 1843, they had deprived *vel vidanes* (irrigation headmen),

2. CO 54/252, Torrington–Earl Grey, no. 202, 13 November 1848, Minute by Grey, 18 January 1849 and draft reply, Earl Grey–Torrington, no. 345, 24 January 1849, CO 54/316, Ward–Lord John Russell, no. 53, 11 July 1855 and draft reply [by William Strachey], Molesworth–Ward, no. 19, 27 August 1855.

3. Philalethes [Sir R. W. Horton]. *Letters containing Observations on Colonial Policy originally printed in Ceylon, in 1832* (London, 1839), *passim; Reports on the Finance and Commerce of the Island of Ceylon* (London, 1848), Report on the Finance and Commerce of the Island of Ceylon by Sir James Emerson Tennent, 22 October 1846. This is abbreviated hereafter to *R.F.C.*, Tennent's Report.

4. CO 54/328, Ward–Labouchere, no. 31, 27 February 1857, encl. [Report on the Patipola-aar] by J. W. W. Birch, Acting A.G.A., Batticaloa, 16 December 1856 *Volume of Speeches and Minutes by Sir Henry Ward* [hereafter cited as VSMW], Report on Irrigation by John Bailey, A.G.A., Badulla, 19 December 1855, pp. 99–105, and the letter which conveyed this report, R. Power [G.A., Central Province]–Col. Sec., 21 February 1856. A. O. Brodie (another A.G.A.) is said to have presented similar views earlier [A.M. and J. Ferguson, *Taxation in Ceylon* (Observer Press, Colombo, 1890) Appendix, p. cxxx].

headmen, and *gansabhawas* (village councils) of the only effective means of compelling obedience to village agricultural customs. One of these practices had been the marshaling of village labor. One consequence, therefore, was a decline in the condition of village tanks; another, an increase in village squabbles. As oft stated, one lazy or litigious man could nullify the agricultural activities of a whole village. Such disruption was deepened and even encouraged by the fact that the British judges were quite at sea amidst the unfamiliar tenurial and agricultural practices and gave decisions that were generally faulty and inequitable.[5]

Prejudicial effects originating from the abolition of *rajakariya* and a decline in the cooperative spirit were referred to by officials at least by 1848 and 1851, but it was only with Ward that the Secretariat in Colombo took active notice. The credit for bringing the problem to light in a graphic and comprehensive manner goes to John Bailey, who had also achieved the status of Ward's son-in-law within a year of Ward's arrival.[6]

The malaise was serious. The prescription, "the Paddy Lands Irrigation Ordinance (No. 9) of 1856," was drawn up with infinite care. This ordinance was the work of many hands and did not spring full-blown from Bailey's brow. Nowhere in his memorandum does Bailey allude to the *gansabhawa* as a useful mode of enforcing village customs. But someone, somewhere, must have done; for, in the months immediately preceding Ordinance No. 9, we find Ward stating that the revival of the ancient customs and the *gansabhawas* were "indispensable preliminaries to any attempt at improvement."[7]

The elements and ingredients which went into the formation of Ordinance No. 9 merit detailed study. The first problem was that of defining the local customs. Bailey had suggested a scheme which implied legislation from the center in defining customs on an island-wide scale. Ward eschewed this idea and opted for the more sensible plan of legislation which empowered G.A's (or A.G.A's) and assemblies of proprietors to make their own "rules."[8] It followed a line of action suggested by Henry Selby, the Queen's Advocate, in his scheme for the commutation of the paddy tax.[9] Far more flexible, this plan solved the problem of codification and provided for varying local practices. It was also in keeping with a principle which Earl Grey had advocated in 1849 in

5. *VSMW*, Report on Irrigation by John Bailey, 19 December 1855, pp. 99–105. In Uganda, too, British courts could not be relied on to give correct decisions on tenurial questions. See C. K. Meek, *Land Law and Custom in the Colonies* (London, 1949), p. 285.

6. While touring certain provinces, Ward visited Uva in January 1856 and would certainly have discussed the problem with Bailey, who was A.G.A. there.

7. CO 54/321. Ward–Labouchere, no. 128, 22 July 1856.

8. Idem and its enclosure, Ordinance no. 9 of 1856, clauses 3–9. Note the democratic manner in which the committees empowered with the task of defining the customs—or "rules," as they were called—were chosen.

9. *Sessional Paper XXX of 1876*, Paddy Commutation, Draft of an Ordinance laid before the Executive Council, 1856, pp. 35–36; *VSMW*, Minute by Ward, [May] 1856, p. 90.

his instructions on irrigation policy, namely, that government activity should be in conjunction with communal peasant aid. Grey's despatch showed enlightened self-interest in suggesting such a principle as far cheaper and more effective, while stressing that such measures should be without compulsion and adapted to local "feelings, habits, and interests" and that government should not attempt too much at once.[10]

The second major problem was the question of choosing a person or persons who chould be vested with the power of settling disputes on matters pertaining to cultivation. There were conflicting opinions on this point among officials of the time. Sir Charles MacCarthy (Colonial Secretary) and Rawdon Power (G.A., Central Province) were adamant that a "native" body could not be safely entrusted with such power. Others, including Ward, were reluctant to rely solely on the district officers because this would vest them with great personal power. Eventually, the working compromise suggested by MacCarthy and Power of a "judicious intermingling of European with Native Agency" was decided on by the subcommittee of the Legislative Council which finalized the scheme.[11] *Gansabhawas* under the chairmanship of G.A's or G.A's nominees were entrusted with the task. They were to enforce their decisions through fines. They would not be bodies which were permanently in existence but would be called into action only when a "rule" was broken. In effect, the *gansabhawa* had been resuscitated in partial form. This decision also reflects the principle of cooperation with the people. Significantly, Grey had suggested that they should follow the Indian practice in utilizing any local bodies that existed to aid government's irrigation activities and to act as liaison between government and people; and that such bodies should be created if none existed.

The revived *gansabhawa* was to be simple in form and summary in its action. All the officials were agreed on this point and were particularly insistent on excluding legal procedure, lawyers, and proctors from *gansabhawa* proceedings. The emphasis was to be on amicable settlement and substantial justice. Ward was inclined to think that appeals against *gansabhawa* decisions should be permitted but MacCarthy dissented and the subcommittee of the Legislative agreed with the latter.

Another feature was the permissive character of the Ordinance. The remedies suggested by Bailey had implied an element of compulsion. This was deliberately avoided. In keeping with Grey's ideas, the operation of the ordinance was made permissive, though government was given room to impose it

10. CO 54/252. Torrington–Earl Grey, no. 202, 13 November 1848; draft reply, Earl Grey–Torrington, no. 345, 24 January 1849.

11. *VSMW*, Minute by Ward, [May] 1856, pp. 99–90; "Remarks on Plan" by R. Power, 6 June 1856, p. 91; Minute by MacCarthy, 17 June 1856, p. 92; Report of the Subcommittee of the Legislative Council appointed to report on [Ordinance No. 9 of 1856], 27 September 1856, pp. 113–25.

if necessary. This indicates a healthy respect for local prejudices. The truth was that government moved with circumspection and in a conciliatory manner. The ordinance was considered experimental because their fund of knowledge was limited and the reaction of the peasantry considered uncertain—hence its limitation to five years, hence the tentativeness, hence the desire to make the first experiments in the field successful ones.[12]

Such were the main features of the ordinance. Of good conception, its principles and details were well worked out and its machinery was simple and flexible. It was as perfect a piece of legislation as one could expect at an initial stage. A traditional institution had been refashioned in new circumstances and newfangled innovationism avoided: so Murdoch and Rogers (Emigration Commissioners) in London applauded the fact that government had ''decided very wisely rather to endeavour to adapt the ancient regulations to existing circumstances, extending and revising them where necessary, than to introduce an entirely new system to which the feeling of the natives might be opposed''.[13] While the ordinance aimed at the settlement of irrigation disputes and the maintenance of minor irrigation works, it was also considered a means of encouraging self-help. So Rawdon Power enthused that it would have ''a marked tendency to throw more of the management of native communities into the hands of their members and teach the people both to think and to act for themselves.''[14]

Under Ward the operation of the ordinance was allied to another of Earl Grey's suggestions which he had adopted, that of irrigation ''grants-in-aid'' where government voted half the estimated expense of a village irrigation work if the villagers concerned provided the other half in labor or money. Ward stipulated that those localities which adopted Ordinance No. 9 of 1856 would be given preference in ''grants-in-aid.''[15] It would appear that £ 8899-18-1 was spent by the district officers on smaller irrigation works, including, ''grants-in-aid,'' between 1855 and 1860.[16]

Though they were highly enthusiastic and hopeful about the potentialities of this ordinance, its architects were far from being confident about its results. They did not expect it to be free from defects and looked to experience to cure them. Ward was equally aware that its usefulness hinged on its proper im-

12. Ibid., Col. Sec.-G.A's, Circular, 7 November 1856, p. 117.

13. CO 54/323, Murdoch and Rogers–Merivale [Permanent Under Sec. of State for the Colonies], 22 January 1857.

14. *VSMW*, ''Remarks on Plan'' by R. Power, 6 June 1856, p. 92.

15. CO 54/329, Ward–Labouchere, no. 195, 23 October 1856.

16. 1859 *Blue Book Reports*, Ward–Duke of Newcastle, no. 120, 15 June 1860, Appendix 12, An Epitome of the Principal Works Executed and in Progress 1855–1860, p. 143. There is some uncertainty *re* the statistics on this subject. £ 100,413–5–4No/No:¼ was voted for ''grants-in-aid'' in Ward's time, of which £ 10,187–16–9No/No:¼ was spent by December 1862, according to CO 54/378, MacCarthy–Duke of Newcastle, no. 141, 20 August 1863, Encl. 7; Return shewing the Progress of Expenditure under the Surplus Funds Ordinances to the year 1862 and the estimated expenditure in 1863, 1 June 1863.

plementation and on the active cooperation and drive of individual G.A's and A.G.A's.[17] The working of the Ordinance was carefully watched. The Executive Council scrutinized each and every rule agreed to at the proprietors' assemblies and refused assent to a few.[18] Their supervision was directed toward freeing the proceedings from any "semblance of official minuteness" and preventing rules which gave headmen room to exercise arbitrary or illegal powers. In this period there was a distinct unwillingness to clothe the headmen with more authority and to depute to them any of the powers which the revived *gansabhawas* possessed.[19] Nor did they trust the working of village councils "without the wholesome control of the Agents' presence."[20] In mid-1859 Ward made it a point to investigate the working of this scheme by calling for returns which were reviewed by Bailey, who had been promoted to Assistant Colonial Secretary. Bailey's verdict was that "while in some [districts] . . . its provisions [had] been carried out in perfect conformity with the spirit in which they were enacted, in others there [had] been apparently, a mis-apprehension as to [government's] intentions . . . as expressed in 1856."[21] The principal complaint was that, as a rule, the district officers had not presided over the proprietors' assemblies or the *gansabhawas,* particularly in the Southern and Western Provinces. A circular was eventually sent, sharply reprimanding the G.A's for thus deviating from the spirit of the ordinance. It added that government was well aware that the G.A's themselves were hard-worked men but felt that there was "no better school" for their assistants and for the newly arrived cadets attached to the *kachcheries* (District headquarters) than the implementation of this ordinance.[22] Whatever the merits of these ideas, government showed a refreshing tendency to keep the district officers on their toes.

Ward was not in any hurry to apply the ordinance on an island-wide scale.[23] The first experiments were in Badulla, Matale, and Colombo Districts. By 1859 it had been brought into operation in many parts of the Seven Korales, Nuwara-kalawiya and Batticaloa Districts, as well as the Southern Province. But it was not implemented on a comprehensive island-wide scale. Even by 1867 such districts as Yatinuwara, Udunuwara, Bintenne, and Nuwara Eliya had not come under its operation.[24]

17. *VSMW,* Col. Sec.-G.A's, Circular, 7 November 1856, p. 117.

18. CO 57/24, Executive Council Minutes, 15 July and 17 October 1857.

19. Ibid., 17 October 1857; CO 57/25, Executive Council Minutes, 4 May 1859.

20. CO 57/29, Executive Council Minutes, 18 February 1861, Printed Encl., *Col. Sec.-G.A's, Circular,* 28 February 1861.

21. Idem. This circular was "in the sense" of Bailey's report.

22. Idem.

23. CO 54/329, Ward–Labouchere, no. 63, 9 April 1857.

24. CO 54/343, Ward–Sir E. B. Lytton, no. 8, 8 January 1859; *VSMW,* pp. 304, 351, 354, 356; CO 54/432, Robinson–Duke of Buckingham and Chandes, no. 7, 12 January 1868, Encl. 3, Morgan [Queen's Advocate]-Col. Sec., no. 343,3 September 1867. Hereafter abbreviated to Morgan's Report, 3 September 1867.

What of the success and influence of the irrigation ordinance in its testing time under Ward? All the official accounts report that the villagers were highly pleased with the ordinance and that there was nothing which surpassed its popularity (where applied), especially in the Kandyan districts.[25] Officials, of course, were flushed with enthusiasm for a project they had pioneered and one must consider the possibility of exaggeration. But the addresses presented to Ward by the local peoples during his tours round the island and those presented to A.G.A's and G.A's when they were transferred from their districts indicate that the operation of this ordinance was highly valued. Ward himself was greatly pleased by the character of the irrigation assemblies and the lively interest shown by the people. He felt that their discussions were of "a very sober and practical character and shew [ed] a perfect consciousness of the possession of equal rights."[26]

In conjunction with the irrigation "grants-in-aid" the revival of *gan-sabhawas* certainly brought concrete benefits to several districts, not only improving the irrigation facilities but also raising the value of land to previously unknown heights.[27] The improvements were the most marked in Batticaloa and Upper Uva Districts and to lesser extent in Colombo District.[28] In Upper Uva John Bailey had recommended the restoration of certain village irrigation works "with a desire to improve the country and to induce the people to appreciate the readiness with which Government [was] prepared to advance its interests" and without any intention of securing financial profits for government. Nevertheless, the monetary returns turned out to be fruitful. What pleased and surprised Bailey most, however, was the chain reaction: "[v]illage vied with village in satisfactorily completing the work allotted to each, and a feeling of pride that they were setting an example to the whole District had sprung up among the people, producing a healthy spirit of rivalry which had been invaluable in its results."[29]

In its primary intention of settling disputes connected with the customs relating to cultivation, success seems to have been marked. The people of

25. CO 54/334, Ward-Labouchere, no. 48, 20 March 1858.

26. CO 54/329, Ward-Labouchere, no. 63, 9 April 1857.

27. CO 54/334, Ward-Labouchere, no. 48, 20 March 1858, and Encl., John Bailey-G.A., Central Province, 25 March 1858; CO 54/361, MacCarthy-Duke of Newcastle, no. 125, 2 July 1861, Encl., Statement showing Profit and Loss on Irrigation Works in the Central Province; *Governor's Addresses,* Vol. 1, Ward, 19 November 1857, p. 395.

28. *Sessional Paper IV of 1867,* Report of the Committee appointed by the Legislative Council to inquire into and report upon Irrigation Works and Rice Cultivation in the island of Ceylon, 18 September 1867 [abbreviated hereafter to RIWRC], Appendix, Part II, Annual Report for 1858, J. Morphew, A.G.A., Batticaloa, n.d., p. 172; 1862 *Blue Book Reports,* Encl. [1862 Administration Report, Western Province and Colombo District], C. P. Layard, G.A., 14 April 1863, p. 164; W. C. Ondatjie, "Notes on the district of Badulla and its natural products," *JRASCB* (1860), p. 386.

29. CO 54/334, Ward-Labouchere, no. 48, 20 March 1858, Encl., Bailey-G.A., Central Province, 25 March 1858.

Badulla reported that it had "not only worked with great advantage, but [had] even rescued paddy growers from expensive and tedious law proceedings."[30] The A.G.A., Nuwarakalawiya, made a similar remark.[31]

In its object of drawing forth popular cooperation and instilling the principle of self-help there appears to have been some improvement. C. P. Layard, the G.A., Western Province, considered that both the irrigation ordinance and the extension of communications had stimulated industry and influenced the character of the people unmistakeably.[32] Victorians were prone to place much emphasis on character, and it is not surprising that Ward spoke of the scheme as having effected a "moral and social change."[33] Such factors, however, are hard to measure. It is equally hard to trace them to a particular cause. In any event, they have to be enduring in order to be of any value, so too much regard must not be paid to such views.

It is not surprising that government renewed the Paddy Lands Irrigation Ordinance in 1861 and instructed G.A's to extend its provisions as far as practicable throughout their districts.[34] Like that of 1856, Ordinance No. 21 of 1861 was flexible. A modification made it even more elastic. With regard to the machinery for settling disputes and organizing the distribution of water, the proprietors of each irrigation division were provided with several options: that of *gansabhawas* only, that of village headmen only, or that of *gansabhawas* in combination with headmen. In the result one found that the last choice was the more popular one in most provinces, though in the Eastern Province the preference was for *gansabhawas* only and in the Norther Province for headmen untrammeled by councils.[35]

On the whole, the modifications inserted in the Paddy Lands Irrigation Ordinance of 1861 were minor. The only other changes of some consequence were the provisions enabling headmen to prevent offenses, to repair damages promptly, and, once the proprietors agreed to such a rule, to cultivate neglected lands. This was a distinct trend away from policy in Ward's time when government was reluctant to trust headmen with such powers and desired district officials to participate in all the deliberations of the *gansabhawas*. Experience in the 1860's forced government to modify these ideas further. Insistence on supervision by district officers only led to arrears in dealing with breaches of irrigation rules, while the reluctance to trust headmen limited the effectiveness of the ordinance. Evidence collected in 1866 showed little

30. Ibid., Encl., Address to the Governor by the inhabitants of Badulla, n.d. [1858].

31. C.G.A., Lot 41/174, R.W.T. Morris–G.A., Northern Province, no. 18, 13 April 1861.

32. 1862 *Blue Book Reports*, Encl.

33. *Governors Addresses*, vol. I, Ward, 19 June 1860, p. 511; Newcastle *MSS*, folio 10988, Ward–Duke of Newcastle, 26 April 1860.

34. CO 54/367, MacCarthy–Duke of Newcastle, no. 5, 9 January 1862; CO 57/29, Executive Council Minutes, 18 February 1861, Printed Encl., *Col. Sec.–G.A's, Circular*, 28 February 1861.

35. Morgan's Report, 3 September 1867, Question 11.

ground for the fears that *gansabhawas* would abuse their powers and headmen exceed their authority.[36] With this rise in confidence, greater independence was permitted to the *gansabhawas* and greater power was vested in the headmen. Richard Morgan, who conducted this investigation, considered that although the system was open to abuse, it was a lesser evil than the dilatory and blundering process of the district courts.

While the irrigation ordinance was renewed in Governor MacCarthy's time, money was refused for irrigation "grants-in-aid."[37] His period of office was characterized by a niggardly policy in many and vital spheres, including that of irrigation works. The crimping accountancy of his colonial secretary, W. C. Gibson, reigned supreme. The maintenance, repair, and extension of minor irrigation work was thrown on the narrow shoulders of local labor and local contributions. The enthusiasm and activity of district officals were stifled. No wonder, then, that such an irrigation enthusiast as J. W. W. Birch raised a cry against this policy as soon as a new governor arrived.[38] The parsimony with regard to minor irrigation works was rank blindness and was utterly inexcusable. Ward had not been far wrong in telling the Colonial Office that they would be "pleased and surprised at the smallness of [the expenditure] in comparison with the good effected."[39] £2,000–3,000 a year on such projects would not have brought government down.

The direction and emphasis of policy at the center was only one of several factors which influenced the extent to which the irrigation ordinance was applied as well as the quality of its performance. As indicated earlier, indifferent results arose in part from misunderstandings or inadequate attention on the part of some G.A's and A.G.A's. To note specific instances, Thomas Power not only devoted his time to private financial ventures but actually "disapproved" of the irrigation ordinance and "would not give it his support and countenance."[40] Philip Braybrooke, G.A. Central Province, was an industrious officer but was so overwhelmed with work that he could not apply the ordinance in such localities as Yatinuwara and Udunuwara in Kandy District.[41]

36. Idem. Morgan was the Queen's Advocate. A Burgher, he was native to Ceylon. His report was a comprehensive review and a precis of the conclusions drawn from answers to a questionnaire circulated among G.A's and A.G.A's.

37. CO 54/415, Robinson–Earl of Carnarvon, no. 232, 15 October 1866, Encl. 7, Return shewing the Progress of Expenditure under the Surplus Funds Ordinances to the year 1865 and the Estimated Expenditure in 1866.

38. 1864 *Blue Book Reports,* Encl. in Encl. 7 [1864 Administration Report, Sabaragamuwa], J. W. W. Birch, A.G.A., 30 September 1865, pp. 143–44.

39. CO 54/334, Ward–Labouchere, no. 48, 20 March 1858.

40. 1864 *Blue Book Reports,* Encl. in Encl. 7, p. 143. His private ventures included that of gemming. He was reduced in rank when this was brought to light. See CO 54/379, MacCarthy–Duke of Newcastle, no. 156, 14 September 1863.

41. CO 54/432, Robinson–Duke of Buckingham and Chandos, no. 10, 14 January 1868, Encl. 4, *Irrigation and Rice Cultivation in Ceylon,* printed extract from debates in the Legislative Council as reported in Ceylon Observer, December 1867. See M. Coomaraswamy's speech.

Partiality of application and of success also arose from shortcomings among the irrigation headmen. Such appointments were not specified in Ordinance No. 9 of 1856 but had been envisaged in the preceding discussion and in practice *vel vidances* had been appointed, following elections by the landowners in each irrigation division[42]—a practice which was regulated for in Ordinance No. 21 of 1861. Some of these *vel vidanes* were overworked. Some neglected their duties.[43] In this field, it was a pity that the British raj sought to administer on the cheap. Neither village headmen nor *vel vidanes* were properly remunerated. It was the decided opinion of a select committee of the Legislative Council who studied the subject of irrigation in 1866–67 that adequate remuneration was essential. The efficacy of the irrigation ordinance would also have been maximized if they had used surveyors in its implementation, wherever possible.[44]

The irrigation ordinance also encountered difficulties posed by climatic and demographic factors. In many parts of the dry zone, the tank country *par excellence,* Ordinance No. 9 of 1856 was not very practicable because the population was so scanty and so widely scattered that the peasants lacked identical interests.[45] It was in order to meet this difficulty that the ordinance was modified in 1861 so as to enable people and government to rely solely on the agency of headmen rather than that of *gansabhawas*. In Lower Uva, which lay within the dry zone, the ordinance was not applied because there appear to have been few streams and irrigation works besides a sparse population. There were heavier concentrations of population in Kegalle and Galle Districts in the wet zone but they had few irrigation works because of the very abundance of rain. Here too the ordinance was adopted to a very limited extent.[46]

Since other localities in the wet zone with very similar conditions adopted the ordinance, one cannot help feeling that the predilections of the peasantry of Galle and Kegalle contributed to this state of affairs. This brings us to another factor. It is of some significance that in many districts the *vel vidanes* had earned such uncomplimentary epithets as *vel moodia* (field cork) and *vel panuwa* (field worm).[47] Whatever the good they effected, their interference

42. CO 54/529, Ward-Labouchere, no. 195, 23 October 1856.

43. *RIWRC*, Appendix II [1866 Administration Report, Matara], Liesching, A.G.A., n.d. [1867], p. 177, and Notes on Irrigation by Captain Fyers, Surveyor-General, n.d. [1867], p. 31, and Replies from C. P. Layard, G.A., Western Province, p. 43; *Ceylon Observer,* Supplement, 16 September 1867, Report on a meeting of the Committee of the Ceylon Agricultural Society, 4 September 1867.

44. CO 54/404, Robinson-Cardwell, no. 134, 16 September 1865, Encl. in Encl., J. W. W. Birch [A.G.A. Sabaragamuwa]-C. P. Layard, G.A., Western Province, no. 409, 6 September 1864.

45. *RIWRC,* Appendix 11, Replies from W. C. Twynam [G.A., Northern Province], 13 June 1867, pp. 62–63.

46. 1864 *Blue Book Reports,* Encl. in Encl. 7 [1864 Administration Report, Kegalle], F. R. Saunders, A.G.A., 28 February 1865, p. 134; Morgan's Report, 3 September 1867, paras 2, 6, 8.

47. Ibid, para 34.

would appear to have incurred some unpopularity. Indifference on the part of the people was certainly responsible for lack of uniformity in the application and success of the irrigation ordinance.[48] Though it was accepted eagerly in many parts of Ceylon, there were several exceptions to the rule. In some of these instances climatic and physical-health factors contributed to this indifference. As W. C. Macready noted of Puttalam District, the fact that the peasants were "poorer and less energetic" influenced the success or failure of the ordinance.[49] Malaria and *parangi* (yaws) were endemic to most parts of the dry zone, including Puttalam, and malnutrition would have been a natural and universal result. That there was apathy is hardly matter for surprise. In three specific divisions of the Western Province, on the other hand, the ordinance was consistently rejected by the cultivators on grounds which officials regarded as reasonable, though unsupported by evidence: namely, a fear that the *gansabhawas* would be dominated by cliques and would abuse their powers.[50]

Several factors, then, influenced the success or failure of government's policy in reviving the traditional irrigation customs. It is not surprising that both in its range and performance, the working of the irrigation ordinance was patchy. In investigating the general state of irrigation works and rice cultivation in Ceylon, the Select Committee of the Legislative Council reached the following conclusion:

> although [the ordinance] has been represented in the replies as working satisfactorily in several districts, there are scarcely any Irrigation works in the Island in a stage of efficient repair with the exception of those undertaken by Government. There appears to be little doubt that this Ordinance has not been introduced as generally or as completely as might have been the case and that when introduced, it has not been worked in a satisfactory manner.[51]

It is clear that the state of village tanks was the yardstick by which the success of the ordinance was measured. It is perhaps of significance that in his own review, Morgan, who was a member of this committee, felt that the papers "furnish[ed] very little accurate information on the point involved," but concluded that the sum spent on irrigation works "appear[ed] to have been small and the works themselves of no importance."[52] In effect, they were asking more of the ordinance than its architects had in mind. This was particularly true of unofficials in the 1860's. Their criticisms were coloured by highly optimistic and unrealistic expectations: *The Ceylon Times,* for in-

48. *RIWRC,* 18 September 1867, p. 9.

49. 1867 *Administration Reports,* Puttalam, W. C. Macready, n.d. [1868], pp. 62–23.

50. Morgan's Report, 3 September 1867.

51. *RIWRC,* 18 September 1867, p. 9. The members of this Committee were Richard Morgan, A. Wise, and E. J. Dehigame (both unofficials in the Legislative Council), Captain Fyers (Surveyor-General), John Capper (editor of *The Ceylon Times*) and John Parsons (Acting Collector of Customs and a senior G.A.).

52. Morgan's Report, 3 September 1867, para 68.

stance, complained that the ordinance had "not sufficed to restore native agriculture to its old position after such long neglect."[53] Their intentions were good but their expectations were too high.

Even on this basis it would be a mistake to consider the ordinance a total failure. The description of the good done in Upper Uva and Batticaloa in Ward's time shows that much was done in these districts. In Matale District the ordinance seems to have been of commendable value in maintaining minor irrigation works, while in Colombo District "numerous works [were] executed with and without the aid of Government which could not have been undertaken before the Ordinance."[54]

It is a mistake, however, to assess the ordinance purely from the state of irrigation works. It is significant that after studying answers to a questionnaire that had been circulated in connection with the projected Irrigation Ordinance No. 21 of 1867, a subcommittee of the Legislative Council concluded that it was "beyond all doubt that wherever the Ordinance [had] received a fair trial, it [had] worked successfully."[55] This largely contradicted the conclusion drawn by the select committee earlier in the year, though several individuals were members of both committees. In terms of the motives which had led to Ordinance No. 9 of 1856, theirs was the substantially fairer verdict. Their angle of vision was a wider and more general one: had the ordinance been of value in settling disputes, in organizing agricultural activities, and in stimulating cooperation, enthusiasm, and industry among the cultivators?

Cooperation was far from perfect and lack of it affected the maintenance of tanks, but unofficial and official evidence indicates that the ordinance assisted the settlement of disputes in the 1860's just as much as in Ward's time.[56] Macready was convinced that it also induced "more systematic and industrious habits," the district officers in general that it was greatly appreciated by the people, particularly those in the Kandyan provinces.[57] Voicing ideas that were clearly borrowed from the British, the *mudaliyar* of

53. *Ceylon Times,* 29 October 1867. The editor was a member of the select committee. Also see *Ceylon Times,* 5 November 1867 and CO 54/432. Robinson–Duke of Buckingham and Chandos, no. 10, 14 January 1868, Encl. 4, M. Coomaraswamy's speech.

54. *RIWRC,* 18 September 1867, p. 11; 1867 *Administration Reports,* Matale, R. C. Pole, A.G.A., 5 May 1868, p. 11 1862 *Blue Book Reports,* Encl. 7 [1862 Administration Report, Western Province and Colombo District, C. P. Layard, 14 April 1863, p. 164.

55. *Sessional Paper XVI of 1867,* Report of the Subcommittee of the Legislative Council appointed to report on the bill, "An Ordinance to promote the maintenance and extension of Paddy Cultivation in the island," 21 December 1867. Its members were Morgan, Wise, Dehigame, M. Coomaraswamy (another unofficial), and C. P. Layard and P. W. Braybrooke (G.A's). Answers were sent by headmen as well as staff officers, and also by some unofficials: e.g., Capper, a Mr. Wickremesekera.

56. *Ceylon Hansard,* 1871–72, 10 November 1871, p. 21, Morgan quoting a report from H. S. O. Russell, G.A., Central Province; *RIWRC,* Appendix 11, Memo on Rice Cultivation in the Seven Korales by J. Capper, 12 February 1867.

57. Ibid., Replies from W. C. Macready, A.G.A., Puttalam [1867], p. 129. Morgan's Report, 3 September 1867, para 6.

Raigam Korale said that it was one of the most useful ordinances enacted for many years past and would "serve as the germ of other future institutions of a popular character."[58] Earl Grey had been thinking in terms of the training in self-government which such local bodies would provide when he advocated the use of local committees in constructing irrigation works and roads, though that was not his sole object. Though the irrigation ordinance was conceived largely in terms of utilitarian and nonpolitical objectives, this idea had not been altogether lost sight of by officials in Ceylon. It was a line of thinking which eventually led them to enact the Village Communities Ordinance of 1871 (No. 26) and extend the powers of *gansabhawas* to that of general administration and petty judicial cases. The very fact that officials and headmen advocated such a step[59] is a measure of the useful role of the irrigation ordinance. No wonder that a reading of the 1867 memoranda led Fredric Rogers in the Colonial Office to write that it was "a real pleasure to see a really important matter move and consolidate itself as this appear[ed] to be doing."[60]

The irrigation ordinance had brought improvements. There was, equally, room for more improvement.

Ordinance No. 21 of 1867 sought to effect these improvements. This ordinance was of a more general scope and had wider objects than the previous efforts and was therefore entitled "An Ordinance to promote the maintenance and extension of Paddy Cultivation in this Island." It had two aspects. One pertained to the continuation and improvement of the previous measures concerning the revival of ancient customs, the other to the principles on which government should consider and undertake irrigation works. The former alone concerns us in this study.

It was not considered practicable or desirable to consolidate the rules agreed to in the irrigation divisions that existed. As before, each locality was allowed to formulate its own rules. To answer the problem of localities with a dispersed population, it was decreed that government agents could make rules on their own authority without convening assemblies of proprietors. Headmen could be fined up to a maximum of £5 for acts of commission or omission. Cultivators who resisted headmen in the course of their duties were liable to a similar fine. Headmen were appointed for three years and it was left open for the proprietors (or G.A's where no assemblies and councils were utilized) to remunerate them in kind or in cash. There was also a fresh emphasis. Whereas the previous ordinances had been aimed at the settling of disputes and the

58. Ibid., para 35, Morgan's words conveying the views of the *mudaliyar*.

59. Ibid., para S. 5, 17.

60. CO 54/432, Robinson-Duke of Buckingham and Chandos, no. 7, 12 January 1868, Minute by Rogers, 12 March 1868. Rogers was Permanent Under-Secretary of State for the Colonies from 1860 to 1871.

maintenance of existing irrigation works, Ordinance No. 21 of 1867 wished the *gansabhawas* to be utilized for the extension of irrigation facilities as well. Even more positive activity was being demanded.

In Robinson's time the ordinance was extended to several localties which had not adopted it previously. In the course of the nineteenth century it was applied in all parts of Batticaloa, Trincomalee, Uva, Matale, Sabaragamuwa, and Matara Districts, while a further 69 irrigation districts were proclaimed in the Western Province, 59 in the Northwestern Province, 14 in Nuwarakalawiya, and 2 in Kegalle District.[61] The Northwestern Province and Kegalle and Galle Districts seem to have been the most backward in adopting the ordinance. District officers continued to report that the ordinance was very useful in settling parochial disputes and maintaining village tanks.[62] The *vidane muhandiram* of Gampaha certainly found it useful and felt that it had played a large part in increasing the cultivation of paddy in his district.[63] Writing in 1911 and speaking of the nineteenth century, a retired civil servant named Edward Elliott referred to the irrigation ordinance as the "Magna Charta of the paddy cultivator." He was convinced that it had given a great impetus to paddy culture and "secured . . . the rescue of an important agricultural interest probably from extinction."[64] Elliott had been in the civil service from 1863 to 1896, shown untiring interest in paddy cultivation, and ventured to open a paddy farm in Ceylon on retirement. His views, therefore, deserve weight; but it is possible that he let enthusiasm sway judgment so they must not be accepted wholly or too readily.

From the nature of things the improvements effected in 1867 did not remove the difficulties posed by climatic and demographic factors or those arising from poverty and apathy among the peasantry. Nor did it alter the basic reliance on G.A's, A.G.A's, and headmen to keep things moving. The improvements themselves were not altogether satisfactory. Though district officers were given the power to make rules on their own authority in regions in which *gansabhawas* could not conveniently be formed, they had to depend on the dilatory and ineffective process of civil courts to enforce these rules.[65] Punishment through fines was not always effective among people with little

61. E. Elliott, "Paddy Cultivation in Ceylon during the Nineteenth Century," *Tropical Agriculturist,* vol. 37 (December 1911) p. 503.

62. 1870 *Administration Reports,* Northwestern Province and Kurunegala District, J. Parsons, G.A., 3 February 1871, pp. 176–77; *Ceylon Hansard,* 1871–72, 10 November 1871, p. 21, Morgan quoting Russell, G.S., Central Province.

63. *Sessional paper XVL of 1877* [Report of the Grain Tax Commission], 30 October 1877, Appendix, no. 69. Translation of Answers sent by C. De Cumat, n.d. [1877]. p. xcviii.

64. E. Elliott, "Paddy Cultivation in Ceylon," pp. 227, 503, 504. Elliott was Dr. Christopher Elliott's son and was so enthusiastic about paddy cultivation that he spent his time and money looking for suitable machinery and studying improved methods of culture.

65. 1872 *Administration Reports,* Mullaittivu, G. H. Withers, Acting A.G.A., 28 February 1873, p. 138.

money and sometimes with no property. Some of the irrigation headmen had far too much work and served too many masters.[66] Favoration marred efficiency. A popular rhyme in Batticaloa ran thus: "Water, water, where will you go? I will go first to the Wanniah's [*sic*] land, then to the Udaiyar's, then to the Vattai Vidahn's [*sic*]; after that to anybody who pays more."[67] In practice, moreover, *vel vidanes* remained unpaid;[68] the remedy provided under Ordinance No. 21 of 1867 was a mere half-measure which proved unsatisfactory; the cultivators in most districts were reluctant or unable to remunerate *vel vidanes*. Since officials had been unanimous in demanding that these headmen should be paid, one would have thought that government could have borne this burden and ensured efficiency through the incentive of bonus payments or commissions.

Perhaps the greatest problem bedeviling the work of the irrigation ordinance was the inability of so many villages to sustain their corporate activities without the compulsive fiat of government. However much local influence the headmen were supposed to have, they were not altogether successful in enforcing obedience to irrigation rules and in supervising irrigation works. The tale presented by Elliott in 1871 is worth quotation:

> Each year shows that incessant personal attention on the part of the Assistant Agent is necessary to carry out irrigation works by villages; to simply order the Mudaliyar or Headmen to carry out any work may sound very fine, but, practically, the results are small, unless the Headmen be encouraged and supported by the Assistant Agent taking an active interest in their efforts; if the villagers see this and know that once they agree to any undertaking, everyone must contribute and that no shirking is allowed, all will combine cheerfully to carry out the work. But endless watching and numerous inspections are necessary and many difficulties arise to contend with, of which the natural procrastination and dilatoriness of the people are by no means the least. Now, as all this takes much time, I need scarcely say that with a heavy office and a large population (143,500) it is exceedingly difficult to attend satisfactorily to what may be termed the outdoor part of one's duties [which includes the supervision of minor roads]. It is when the time comes for recording what has been actually accomplished during the past year that one feels how little has been really done, although the whole year has been fully and more than fully occupied by the daily routine—"yesterday's face twin image of today"—little of it worthy of record, but all more or less important to the people immediately concerned.[69]

Not all the G.A's and A.G.A's took this degree of interest in irrigation works, so one can well imagine that even less was achieved under some officers.

66. 1869 *Administration Reports,* Matara, A. R. Dawson, A.G.A., 16 May 1870, p. 87.

67. *Sessional Paper XVI of 1877*. [Report of the Grain Tax Commission], 30 October 1877, Appendix, no. 2, Answers from J. Crowther Esq., Batticaloa, 15 March 1877. p. iv. Crowther was a resident in Batticaloa, occupation unknown.

68. Ibid, No. 3, Answers from T. B. Panabokke, Esq., Amunugame Walauwe, 19 March 1877, p. iv. 1872 *Administration Reports,* Nuwara Eliya, B. F. Hartsborne, A.G.A., 29 March 1873, p. 68.

69. 1871 *Administration Reports,* Matara, E. Elliott, A.G.A., 12 June 1872, p. 161.

Woolf's experience in Hambantota District in the early twentieth century indicates that he had to be constantly pushing the villagers and village headmen in order to achieve anything.[70] There is no reason to believe that this problem was not an island-wide one in both the nineteenth and twentieth centuries. The work of the villagers, moreover, was generally of poor quality. Woolf found that the repairs effected on the majority of village tanks were very unsatisfactory though a few were done well.[71] Part of the problem was technical. Village tanks were generally crude affairs without spill-weirs or sluices.[72] The practice of cutting the bund annually in order to supply the fields had a lethal effect in the long run.

Part of the problem arose from indiscipline and procrastination among the people, rising almost to the proportion of national failings. Such factors cannot easily be weighed, but they need emphasis. It could be said that indiscipline increased as the years passed and as the eroding force of individualism increased and spread among the villagers. This was the result of the changes and policies generated under the British raj. The British had sought to change the ethos and were suceeding in doing so, with results good and bad. It would be safe to generalize that with the spread of civilization in the nineteenth and twentieth centuries, the greater the sophistication of a village, the less communal cooperation there was and the less effective *vel vidanes* and irrigation customs were. One of the problems that arose was "the stratification of village society into rival interests which profited from each other's misfortunes."[73] This arose in part from the rise of a class of noncultivating landowners in many localities. These men sharecropped their land and drew rents and interest payments while disassociating themselves from the technical requirements of agricultural production.[74] In pre-British times village agriculture had been based on a community of economic action and purpose built around mudland held on an individualistic basis. As this corporate structure was undermined by the new forces of trade, individualism, and indiscipline, and as the long endemic virus of litigation prospered in this more congenial habitat, the continuation of traditional agricultural customs grew the more difficult. A mark of these new influences is the fact that some *gansabhawas* had begun to assume "the ways of a Court of Law" as early as 1870,[75] though

70. Leonard Woolf, *Diaries in Ceylon,* 1908–1911: *Records of a Colonial Administrator* (Hogarth Press, London, 1963), pp. 19–20, 63, 79–80, 85, 90–91.

71. Ibid., pp. 6–8, 16, 85, 93.

72. *RIWRC,* 18 September 1867, p. 7 and Appendix I, Notes on Irrigation by Captain Fyers, Surveyor-General, and Answers from J. W. W. Birch, Acting A.G.A., Hambantota, p. 78; A.O. Brodie, "Statistical Account of the District of Chilaw and Puttalam," *J.R.A.S. CB* (1853), p. 37. This remains largely true today. See E. R. Leach, "Hydraulic Society in Ceylon," *Past and Present,* no. 15, April 1959, p. 8.

73. S. B. D. De Silva, "Investment and Economic Growth in Ceylon," Ph.D. thesis, economics (London, 1961), pp. 280–81.

74. Idem.

75. 1869 *Administration Reports,* Matale, G. S. Williams, A.G.A., 15 June 1870, p. 54.

they had been expressly designed toward informality and simplicity. This was a feature that was to mar the working of the Village Communities Ordinance of 1871 as well. However much government tried to provide the people with simpler forms and institutions, participants and recipients molded them into other forms or sought the ordinary civil tribunals, Ephraim was being wedded to new idols.

If, then, the revival of traditional irrigation customs had some pleasing results in the period 1856–71, one would be hasty to assume that this level of success was consistently maintained. These traditional methods were encountering a new environment and new strains and stresses.

10

Management Themes in Community Irrigation Systems

E. Walter Coward, Jr.
CORNELL UNIVERSITY

Introduction

As many have noted, while the problems of irrigation management are not new, they have gained in perceived importance with the availability of new plant materials which, on the one hand, hold the promise of increased yields while, on the other hand, requiring improved water control to allow the complementary use of chemical fertilizers, appropriate weed control through water management, and control of excess levels of water because of the dwarf features of the new cereal varieties.

Attempts to deal with these problems have generally fallen into three categories (Wade 1976). First, and perhaps foremost, we have attempted to improve the irrigation technology and engineering structures used in new and existing systems. There has been a great deal of attention given to the need for terminal unit facilities (Small 1974, Thavaraj 1973). (Terminal units refer to the smallest section of contiguous farm fields served by a single outlet, or turnout, of the irrigation system. They are the most elementary irrigation blocks defined by the physical layout of the system and constitute an important management unit. In many modern irrigation systems the facilities to manage water within these terminal units are absent. The most important of these facilities are farm ditches, or field channels, but they might also include

Previously published as "Irrigation Management Alternatives: Themes from Indigenous Irrigation Systems," *Agricultural Administration* 4 (1977): 223-37. Reprinted by permission.

An earlier version of this paper was presented at the Workshop on Choices in Irrigation Management organized by the Overseas Development Institute and held at the University of Kent in September 1976.

measuring devices and checks, field weirs, and field drains (see Thavaraj 1973). Second, some attention has been given to the creation of economic incentives through water pricing as a means of improving water management. Third, there is currently a great deal of interest in the organizational dimension of irrigation management, particularly the organization of water users at the terminal unit level.

This paper dicusses examples of irrigation management alternatives based on the numerous cases of locally constructed and operated irrigation systems that exist around the world. Within this set of experiences are a large number of "silent successes": traditional, usually small-scale systems, many of which have operated for decades and centuries.

These traditional irrigation systems offer important insights regarding the solution of organizational problems, particularly at the level of the terminal unit; not that they can simply be duplicated in other situations but that they suggest important principles of organization which can be applied in other specific settings.

The paper is organized in two parts; first, a discussion of illustrative organizational themes from the study of indigenous irrigation organizations and, second, a discussion of the ways in which such organizational themes might be used to achieve irrigation development.

Illustrative Organizational Themes From Indigenous Systems[1]

As suggested above, the world is rich with a variety of local organizational arrangements for managing irrigation water. These systems have been constructed, maintained, and operated by local people, occasionally with outside assistance of some type. The local group has sometimes been a single community, or subset of the community, but often it is a multicommunity enterprise. While the material on these systems is rather extensive (see two recent bibliographies: Coward 1976a and International Rice Research Institute and Agricultural Development Council 1976, of particular interest are materials from the following countries: Ecuador (Anderson 1973), India (Chambers 1974 and Harriss 1974), Indonesia (Geertz 1967 and Satya Wacana University 1975), Laos (Coward 1971 and Taillard 1972), Mexico (Downing 1974), Peru (Mitchell 1976), the Philippines (Coward 1979 and Lewis 1971), Sri Lanka (Leach 1961), Tanzania (Gray 1963) and Thailand (Bruneau 1968, Frutchey 1969, Moerman 1968, and Wijeyewardene 1965, 1973).

These studies, while providing a great deal of detail about various organizational arrangements and the principles of organization that define them,

1. Many of the ideas in this section draw on material from two previously prepared papers (Coward 1976c, 1976d).

often fail to include critical information regarding basic elements of the situation such as amount of rainfall, types of soil to be irrigated, cropping patterns being utilized, amount of water diverted into the system, and other items that would give us a more complete picture of the environment in which the irrigation organization operates. Nevertheless, it is possible to see in these various cases some common themes and principles that are relevant to current efforts of irrigation development.

The Theme of Accountable Leadership

A major problem with developing organization for the terminal unit is identifying and maintaining adequate leaders. In many modern irrigation systems such "leaders" may actually be employees of the irrigation agency, be selected by the agency or, if neither selected nor paid by the agency, nevertheless very much subject to its interests. Often, local leaders of this type fail to achieve the important linkage required between the water users and the water authorities.

In contrast to these agency-directed local leaders are the various irrigation leaders found in traditional systems. In my view, the traditional form of irrigation leadership can be characterized as an *accountability model* (Coward 1976b). Notwithstanding the novel and idiosyncratic characteristics of irrigation organization in various local contexts, several basic features, which act to make the leader responsible to his irrigation group, are common to this role in numerous settings.

The actual responsibilities of traditional irrigation leaders vary, in part, according to the size and physical aspects of the irrigation system and the location of ultimate control of the system. Moerman (1968) has described a case in northern Thailand where the irrigation system is a community-constructed,-maintained, and -operated entity. Here we see that the major duty of the irrigation leader is the mobilization of labor, and materials for the annual repair of the diversion structure. In contrast, Nash (1965) describes a local role which is organized around the function of canal maintenance. The village studied obtains its water from a large, government-built irrigation system and thus has no responsibility with regard to maintaining the diversion structure(s).

Maintenance and repair of either the diversion structure or the canals is, of course, a basic activity for which all irrigation groups must arrange. However, there are instances in which the water headman's responsibilities have included considerably more complex and controversial activities. One example is the *klian subak* described by Geertz (1967). In the organization of this irrigation society (*subak*) the irrigation headman and his assistants are also involved in the distribution of water through the assignment of planting times for the various subsections of the *subak*. While it is not clear what determines this more complex leadership role, one might hypothesize that the higher population density in Bali requires greater coordination of water use among various groups obtaining water from the same stream.

While this range of activities exists, several common denominators persist. Three are of importance for this discussion. First, traditional irrigation leaders serve relatively small groups of water users, not necessarily from the same village, and small areas of irrigated land surface. Second, they are selected, in some manner, by the members of the local group which they serve. While leaders often serve for long periods of time, perhaps a lifetime, they are nevertheless subject to review and replacement, as is clearly illustrated in Wijeyewardene's report (1965). Third, they receive compensation directly from the group members whom they serve: sometimes in the form of exemption from labour (Moerman), sometimes in the form of land for cultivation (Jay 1969, Coward 1979), sometimes based on a proportion of the crop harvested (Harriss 1974), or a fixed amount of the crop (Coward 1971).

These three elements—small scale, local selection, and direct compensation—are the basis of the accountability model of irrigation leadership. Leadership roles based on these organizational principles pattern the incumbent's actions in directions acceptable to the irrigation group.

It may be hypothesized that successful terminal unit organization in modern irrigation systems will be dependent, in part, upon the establishment of local leaders who are accountable to the water users whom they serve and not exclusively to the system-wide irrigation agency (more on this point in the subsequent discussion on utilizing indigenous organizations).

The Theme of Mini-unit Organization

As already noted, most traditional irrigation systems are small; for example, in one province in the northern Philippines (Ilocos Norte) the National Irrigation Administration reports more than 600 traditional irrigation systems (*zangjeras*) that range in size from a few hectares to more than 1,000 hectares (see also Lewis 1971 for an indication of sizes in this region). It may be somewhat surprising, therefore, to learn that these small systems are almost always further divided into smaller subunits. These subunits usually have a set of local leaders to operate the system at this mini-unit level.

Geertz (1967) has provided some rich detail on the organization of an irrigation system in Bali that illustrates this theme. He discusses one *subak* which serves a total area of 159 hectares but which is further divided into seven subunits (locally referred to as *tempek*), each of which has its own locally selected leader. Thus, the system is divided into seven mini-units which, on average, cover an area of about 23 hectares and serve about 65 water users.

Another example of this theme derives from field research that I have been doing in the northern Philippines. Observations have been made in an irrigation system (*zangjera*) in the Ilocos region which serves approximately 1,500 hectares of paddy land in the wet season. Here, the entire service area is divided into 32 mini-units (locally referred to as *sitios*) comprised of 15–70 hectares and serving about the same range in number of water users.

There are several leadership roles within each of these mini-units. Water users within the mini-unit select one individual to serve as the unit leader (the *panglakayen*). He is assisted by another individual selected in the same manner (the *segundo*) and sometimes by a third individual who serves as the unit's secretary (the *escribiente*). These leaders are responsible for the distribution and allocation of water and the settlement of disputes and enforcement of rules within their mini-unit.

While there are a number of other fascinating details about this system not presented here, it should be mentioned that these mini-units are also used as the basis for mobilizing labor for routine repair and maintenance tasks in the system. Briefly, work groups are organized to carry out these activities by having each mini-unit contribute to the work group a number of men in approximate proportion to the amount of paddy land served in that mini-unit.

The creation of these mini-unites and the establishment of various leadership roles for each unit results in a pattern of organization that might be characterized as "labor intensive." For example, in the *zangjera* just described there are 32 mini-unit leaders and three canal leaders, as well as an overall president and secretary for managing a service area of about 1,500 hectares. By way of contrast, an irrigation system of comparable size (1,849 hectares served in the wet season) operated by the National Irrigation Administration in the southern Luzon region of the Philippines was operated by 12 ditch tenders directed by a supervising watermaster (Coward 1973). These differences in management intensity are not uncommon when traditional systems are compared with government-managed systems.

It is important to note that these mini-units are not merely organizational units. Typically, they are also discreet physical units within the larger systems. They may be arranged so that each unit is served directly from the main canal or a lateral and is not dependent on a water supply that passes over the territory of another mini-unit (although it may pass through neighboring units). The importance of the congruence between the physical and the organizational components of mini-units is shown in the discussion of terminal facilities by Thavaraj (1973). He notes the following: "A terminal network also helps to cut down a large irrigation compartment or block into smaller independent units. These small units would be self-contained and any irrigation problem could be easily isolated and overcome within such small units. . . . The conversion of large compartments into smaller units will also improve farmer cooperation and simplify irrigation extension." Interesting experiments with mini-unit organization are going on in the MUDA project with the division of farmer associations into small agricultural units, which are in turn being divided into working groups of 7–15 members.) Thus an important point to remember is that these mini-units are established so as to make the boundaries of physical units coterminous with the boundaries of the social groups.

The presence of these mini-units within already small irrigation systems is important in facilitating successful terminal unit organization. As Taillard (1972) had suggested with regard to Laos: "Here one discovers the elements of village sociology which explain the small size of the irrigation association. Lao society is founded on reciprocal solidarity bonds connecting the members of a group; in order for these bonds to function satisfactorily the group must not have more than 70 or 80 members."

What Taillard suggests for Laos can be applied more broadly to the Asian region; perhaps more appropriately to irrigation organization everywhere. Given some degree of water scarcity, variations in soil and plant conditions between farms and the differential behavior of individual water users, problems of coordination, conflict, and cooperation are ubiquitous in irrigation operation. Small groups (especially if differences of social status and social class are relatively minor) are able to employ special mechanisms of reciprocity to achieve relative order and conformity. While this can also be achieved with large-scale organizations such as bureaucracies, these organizational arrangements are often dependent upon technologies and infrastructures (for example, roads and telephones to facilitate mobility and communication) not available in developing countries.

The Theme of Canal-Based Organizations

Rural society in Asia is usually viewed as a village-based society; it is somewhat surprising, therefore, to discover how infrequently traditional irrigation systems in the region are coterminous with village boundaries. One needs to carefully avoid the problem of assuming that traditional irrigation systems are village systems; they might better be thought of as *villager* systems.

This point is important because of the frequent propensity of irrigation developers to want to establish irrigation associations along village lines. For purposes of irrigation organization the critical unit is the "irrigation community," composed of field neighbors, and not the village community, composed of residential neighbors, although in some instances the two groups may be the same. The indigenous systems provide some interesting examples of organizing irrigation communities that are not also village communities.

Three approaches to establishing canal-based irrigation organization can be noted: the establishment of nonvillage irrigation associations, the establishment of pan-village associations, and the creation of differential membership categories.

The classic illustration of the establishment of nonvillage irrigation organization is the Balinese *subak*. In this approach village membership is muted in the context of irrigation since one is simultaneously both a "citizen" of a village and a "citizen" of a *subak*, although not all citizens of the one village are citizens of the same *subak*, and vice versa (Geertz 1967). While several members of a *subak* may come from the same village, there is no indication that these individuals form a subunit with the irrigation system unless, of course, their fields are so located as to place them in the same mini-unit of the system, as described above.

While it may be erroneous to assume that there is no overlap between the leaders of the village and the leaders of the *subak* (see Birkelbach 1973) or that this overlap is undesirable, the distinguishing feature of the *subak* is that it represents a task-oriented coalition of individuals whose membership is defined by field locality.

Nonvillage irrigation associations have been reported in the northern Philippines also. The *zangjeras* of Ilocos Norte often deliver water to the lands of more than one village and to water users who come from many villages (Lewis 1971). In both cases the pattern of nonvillage irrigation associations seems related to the presence of water users from various villages, not only at the system level but also at the level of the terminal unit. When village membership is diverse within the terminal units it precludes the possibility of effectively organizing irrigation activities along village lines. Several examples of traditional irrigation organization illustrate the ability of water users to coalesce in organizational patterns other than village or kinship units.

In other situations the pattern of land use is such that field neighbors in individual terminal units are also village neighbors, although the water users from the entire system are members of two or more villages. This pattern frequently occurs, for example, in northern Thailand (for a useful survey of irrigation in this area, see Agricultural Development Council 1974 and Sektheera and Thodey 1975). In these systems, referred to as pan-village organizations, there is an overall irrigation leader (often referred to as the *ke muang* or *houa na*) assisted by a number of associate leaders from the various terminal units of the system. Since the terminal units tend to be composed of water users from the same village, organization of the terminal unit is along village lines and the village leader often serves as the associate irrigation leader representing his terminal group. In this pattern of organization there are obvious advantages with regard to communication between water users and the water authorities, and the congruence of village and terminal unit may enhance the ability informally to enforce rules and obtain compliance with group procedures. On the other hand, difficulties can arise when conflicts between villages are transferred to the arena of the irrigation system and impair the smooth distribution of water between terminal units.

Irrigation organization that builds on village leaders also increases the risk that village leaders will use the power and influence gained through their authority roles unfairly to influence the distribution of water and other management procedures in the irrigation system. As recently reported for selected systems in Indonesia (Satya Wacana University 1975), such patterns may be especially vulnerable as village headmen are increasingly oriented toward the external bureaucratic and commercial interests of their region. Again, there is important evidence to demonstrate that local people have had considerable experience and success with creating organizational formats that build on village units while, at the same time, incorporating them into a larger, supravillage pattern. Since the command areas of irrigation systems nearly always exceed the territory of a single village, these traditional patterns provide important examples for potential organizational arrangements to be considered in our efforts toward irrigation development.

A third arrangement for dealing with multivillage membership at the terminal level is to create two (or perhaps more) categories of association membership. This pattern seems to occur in situations where a large portion, but not all, of the water users in a terminal unit or an entire system live in the village whose land is served while others come from villages outside this immediate area. When this occurs two situations arise: (1) the irrigation leaders find it difficult to communicate messages to those living outside the terminal village (often this is information relative to mobilizing labor; setting dates for maintenance work) and (2) those outside the terminal area find it costly to spend the time required for traveling to the terminal unit and providing labor for operation and maintenance activities in addition to the time required to travel and perform their farming tasks. In the northern Philippines these mutual dissatisfactions have led to a system of differentiated membership classes in the terminal unit. One, the *residential* member, participates in the system and meets his obligations by providing labor as required to operate and maintain the system. Since he is close at hand, the water authorities can easily communicate with him, and he expends little time in traveling to the actual work area. The other category is the *external* member. This member is allowed to meet his obligations to the terminal unit by providing a cash payment to the group in lieu of his labor contribution. Thus the water authorities are freed of the need for frequent communication with the external member and at the same time are able to accumulate a small cash supply to purchase a variety of items—from a sack of cement to food and drink—to be used at the time of communal work activities.

Utilizing Indigenous Irrigation Organizations

Having illustrated something of the rich diversity of indigenous irrigation organizations, some general ideas regarding the use of such organizational arrangements to facilitate contemporary irrigation development will be discussed. This utilization is not a simple matter and some important cautions need to be kept in mind.

As Wade (1976) has noted, the utility of these indigenous arrangements for the solution of current difficulties must be viewed dubiously, particularly in those cases where the irrigation system is the "property" of government. As I have commented elsewhere (Coward 1976c), in most instances irrigation development will be the responsibility of national governments and will be implemented by some national or regional agency. Here the critical question becomes whether or not the introduction of a major new element in the environs of the traditional organization, the irrigation bureaucracy, will preclude its successful functioning.

However, not all irrigation development takes the form of large-scale, bureaucratically managed systems. Several countries have extensive portions of their irrigated land served by traditional systems and are interested in improving the facilities and operations of these systems already in place (for example, the Philippines and Thailand). Other countries are investing a considerable amount of funds in the development of new, small-scale irrigation systems (for example, the small-system development program of the Farm Systems Development Corporation in the Philippines and the *sederhana,* or simple, irrigation program of Indonesia). A third situation of relevance would be in situations where large-scale systems are designed so as to create smaller subunits relatively independent of the total system. This approach has been suggested by Downs and Mountstephens (1974) and can be observed in systems in China using the so-called melons-on-a-vine design (Nickum 1977). (The melons-on-a-vine design is one in which a main canal system is used to supply a series of small reservoirs, or ponds, which in turn serve a limited command area. Thus, while part of a larger system, each pond group has some independence of action regarding water allocation.) In each of these three alternative strategies there are significant opportunities to utilize either existing indigenous organizations or specific organizational principles from them. However, the utilization of these organizational arrangements must be a conscious policy used to guide choices between technological alternatives, strategies for providing material and informational assistance, and means for linking the local systems with the national bureaucracies. Some examples will illustrate the potential problems.

Assisting Indigenous Systems

The Philippine government currently has an active program of technical and financial assistance to communal systems throughout the country. In many cases this assistance will provide significant improvements to critical elements of the physical infrastructure: the replacement of a bamboo diversion structure or the installation of a permanent headgate to reduce the risk of flood damage to the canal system. Assistance of this type contributes to a total program of irrigation development by improving the effectiveness of existing small-scale systems which remain the operational responsibility of local groups. Thus, it represents a strategy in which traditional systems are further mobilized to contribute to national goals.

But we should also recognize that this strategy of irrigation development represents a real challenge: how can assistance be provided (or received) while still maintaining a viable degree of local control over the irrigation system? For example, the introduction of sophisticated irrigation structures into traditional systems may serve to make them more dependent upon the outside bureaucracy. In contrast to the bamboo and stone weir of the traditional system, the concrete weir may require financial resources, masonry skills, and other prerequisites not available within the local community.

The Philippine case illustrates another dimension of the problem; the attempt to apply standard procedures to situations and groups that are very diverse. In the Philippine context material assistance is being given to the indigenous systems on a loan basis. The government is, correctly, concerned with ensuring the repayment of this investment. Thus, each indigenous system receiving government assistance is required to organize formally in accordance with standardized rules laid down by the Securities and Exchange Commission. These rules include detailed statements about the officers required, committees to be organized, and other procedural rules. While this approach may be advantageous in situations where the local irrigation association has fallen into disorganization, it seems unnecessary in situations such as Ilocos Norte, where a *zangjera* may have been operating successfully for several decades, or even a century. Hopefully, the resilience of these local associations will enable them to adapt to this latest "environmental" challenge.

These points are important not because indigenous systems should be left untouched or because concrete weirs are considered undesirable. Obviously, for many of the individual water users who annually spend a great deal of effort repairing their locally constructed weir, it may be considered a highly desirable form of assistance from the government. The need is to implement and provide the assistance in as unobtrusive a manner as possible.

A variation of the above approach is the situation in which irrigation development takes the form of constructing a large diversion structure which is able to command an area previously served by numerous small, individual traditional systems. Two cases illustrate this situation: the Laoag-Vintar Irrigation System in Ilocos Norte (Ongkingco 1973) and the Mae Taeng Irrigation System in Chieng Mai, Thailand (Sektheera and Thodey 1975). In both cases the new structures provide water to rice fields which were previously irrigated by numerous indigenous systems, each with its own leadership structure, rules for behavior, and patterns of action. In effect, each of these existing systems became terminal units in the single encompassing system, although not always remaining intact as a unit.

Ideally, in a situation such as this the physical layout and administrative structure of the new system would be designed so as to make maximum use of these already existing units. That is, the new system might be designed to deliver water not only through many of the existing terminal-level ditches, but also through the existing terminal-level leaders and patterns of organization. In neither case, ditches or social organization, would this be entirely possible, but the procedure would be to maximize the use of these existing "systems" in so far as possible. Of course, these local "systems" can be used only if they are known and understood. While it is relatively easy to know about the existing physical systems, it is easy to be unaware of the organizational arrangements that are in place.

As with the strategy of providing assistance to indigenous systems, the strategy of incorporating them into a larger government-managed system has the attached risk that their very incorporation will undermine processes and procedures associated with the internal dynamics of the systems and thus prevent their continued viability. As discussed above, this may occur if the new irrigation structures require skills and resources beyond the resources of the local unit or if bureaucratic rules and requirements inhibit any meaningful responsibilities for local irrigation leaders. Again, it is a strategy that may have considerable benefit if undertaken with sensitivity and concern for maintaining the viability of the local groups and facilitating their contribution to the effective functioning of the new system.

Building Small-Scale Systems

As noted above, two interesting examples of small-scale irrigation development policy are found in the cases of the Philippines and Indonesia. In the Philippine case, the so-called BISA program (Barrio Irrigation Service Association) is largely oriented toward the installation of pump systems to service small command areas (usually several hundred hectares). A major aim of this program is to organize irrigation associations that will build, own, manage, and maintain the irrigation system. While this program has given a commendable degree of attention to the organizational as well as the technological aspects of the irrigation system, it is interesting to note that the model of oraganization is largely an urban-western form and apparently not at all based on the various indigenous experiences of irrigation that exist in the Philippines.

Indonesia's *sederhana* program is a second interesting approach for making small-scale irrigation a part of the total irrigation development policy. (According to Bratamidjaja [1976] more than half of the new irrigation development during the second five-year plan will be simple irrigation: 0.55 million of a total of 0.95 million hectares.) This program is aimed at developing command areas of 2,000 hectares or less with the use of physical structures that are simple to design and construct. It is anticipated that much of the construction will be with labor-intensive techniques. There has been little discussion of the organizational forms that will be used to operate these simple systems other than the suggestion that water-user associations will be organized.

The rich variety of traditional experiences with irrigation organization in Indonesia suggests that there may be little need to "import" Taiwanese irrigation associations or other exogenous organizational forms. Obviously, these traditional forms can be considered only if policy makers are aware of their existence and persuaded to test their adaptability under these new circumstances.

Small Units in Large-Scale Systems

A third interesting situation in which principles of indigenous organization may be used for irrigation development are situations in which the physical layout of the large system creates smaller, somewhat discrete service units. At the level of these small-service or terminal-level units the sociological patterns of interaction among water users—and perhaps between water users and the water authorities—can be managed with indigenous organizational forms. One example of this approach is based on observations I have made in western Laos (Coward 1976b).

This modern irrigation system, commanding about 2,000 hectares in the wet season, is divided into 21 small zones, or mini-units, each composed of water users whose fields are contiguous and (in most cases) who receive their water from a common watercourse in the system. Each of these small groups has a rather elaborate leadership structure selected by members of the small group. Included in this leadership group is a water headman.

The water headman is an adaptation of a leadership role found in this area of Laos prior to the present irrigation development activities. In villages that constructed one large diversion dam to serve nearly all farmers in the village, specialized roles for administering irrigation activities were present. These villages selected an individual to act as water headman (*nai nam*), a role separate from that of village headman (*nai ban*). In some instances the water headman was assisted by a group who acted as a water committee or council.

Villagers reported that previously the water headman provided the leadership for the annual reconstruction of the diversion structure and the canals and also served in settling disputes regarding water allocation. He was paid for his services by each water user who annually contributed ten kilos of unmilled rice. In this pre-project organization the water headmen required little or no articulation with external authorities. The village systems which they administered were self-contained local systems not dependent upon or linked with entities outside the village. The traditional water headman primarily focused on the mobilization of intravillage action and typically did not involve the articulation of village interests with those of some external organization. This aspect of the water headman's role, of course, has changed fundamentally in the context of the modern system.

In the modern context the basic responsibility of the water headman is to see that water is actually delivered to the members of his small group during their scheduled period. The responsibility has shifted from a concern with maintenance to a concern with water distribution. Furthermore, to perform this new role satisfactorily the water headman must act as a link between his water-user group and the external project administration. He must also act as a guard to protect the group's delivery from other water users who are proximate to the supply channel. For their activities and responsibilities the water headmen are paid a small fee by each of the irrigators, reported to be about 16 kilos of unmilled rice for each hectare of land served. In addition to this fee, each water user also pays a water fee to the project administration at the rate of 80 kilos of unmilled rice per hectare.

The role of water headman in this situation, while significantly different from the traditional role in some aspects (particularly regarding duties and relationships with government agents), is highly analogous in others (particularly in the manner in which incumbents are selected, rewarded, and judged). It is important to note that the methods of recruitment and reward and the expectations of the water headman held by the water users, even in the modern context, make the role incumbent accountable to the water users whom he serves. First, they are involved in his selection and have the ability to review his performance to consider his continuance or dismissal. Second, the incumbent is dependent upon the water users and not the bureaucracy for the payment of his fees. Reluctance to pay, or the actual nonpayment of irrigation fees to the water headman, provide a very direct form of job evaluation. Both of these procedures reinforce the expectation that the water headman will act to meet the needs of the immediate water-user group that he serves.

Having continued elements of the traditional role that maintain the incumbent's local accountability, the water headman's role may be altered with regard to its basic activities and expanded to include relations with project administrators. These alterations make the role compatible with selected administrative and management needs, especially the need to facilitate the distribution of water within the area of each small group while also achieving the equitable distribution of water among these groups.

Discussion

The current indigenous models of irrigation organization that exist throughout the world represent an important pool of experiences and examples for our current efforts at irrigation development. Clearly, these indigenous models are not equally applicable to all the irrigation projects being funded by the World Bank and other national and international agencies.

Nevertheless, we can see in them locally derived solutions to recurring problems in our modern systems. The applicability of these local solutions to new settings obviously depends on the degree of similarity that exists between the old and the new settings. An important point to note is that it may be possible in the design of large-scale, modern irrigation systems to create some of the contextual properties that characterize many indigenous systems; one interesting example is the melons-on-a-vine design being used in some large systems in China.

Finally, it has been noted that some nations have, as part of their total irrigation development strategy, a focus on small-scale irrigation systems. In these cases, the utilization of indigenous principles of organization may be even more applicable.

Planners and policy makers may have some erroneous ideas about indigenous irrigation systems, if they are aware of them at all. These myths can be dispelled only through the accumulation of more detailed information about a wide array of local systems operating in various natural and socioeconomic conditions.

References

Agricultural Development Council. 1974. "Irrigated Agriculture in Northern Thailand." Bangkok, Agricultural Development Council, National Seminar Report no. 5.

Anderson, D. 1973. "Irrigation Water Management in Ecuador." M. S. thesis, Department of Political Science, Utah State University.

Birkelbach, A. W., Jr. 1973. "The Subak Association." *Indonesia* 16:153–69.

Bratamidjaja, Otje S. R. 1976. "Water Management on Rice Fields in Indonesia." Paper presented at Asian Productivity Organization Symposium on Farm Water Management, Tokyo.

Bruneau, M. M. 1968. "Irrigation tranditionelle dans le Nord de la Thailande: L'exemple du bassin de Chiengmai." *Bulletin de l'Association de Géographes Français,* pp. 362–63 and 155–65.

Chambers, R. 1974. "On Substituting Political and Administrative Will for Foreign Exchange: The Potential for Water Management in the Dry Zone of Sri Lanka." Paper presented at Conference on Agriculture in the Economic Development of Sri Lanka, Peradeniya, Sri Lanka.

Coward, E. W., Jr. 1971. "Agrarian Modernization and Village Leadership: Irrigation Leaders in Laos." *Asian Forum* 3:158–63.

———. 1973. "Institutional and Social Organizational Factors Affecting Irrigation: Their Application to a Specific Case." In *Water Management in Philippine Irrigation Systems: Research and Operations,* pp. 207–18. International Rice Research Institute, Los Baños, Philippines.

———. 1976a. "Irrigation Institutions and Organizations: An International Bibliography." Cornell University International Agriculture Mimeograph Bulletin no. 49.
———. 1976b. "Indigenous Organization, Bureaucracy, and Development: The Case of Irrigation." *Journal of Development Studies* 13:92–105. [Chapter 15 below.]
———. 1976c. "Peasants and the Dilemma of Irrigation Development: Bureaucracy and Local Organization." Paper prepared for Fourth World Congress of Rural Sociology, Torun, Poland.
———. 1976d. "Indigenous Irrigation Institutions and Irrigation Development in Southeast Asia: Current Knowledge and Needed Research." Paper presented at Asian Productivity Organization Symposium on Farm Water Management, Tokyo.
———. 1979. "Principles of Social Organization in an Indigenous Irrigation System." *Human Organization* 38, no 1:28–36.
DOWNING, T. E. 1974. "Irrigation and Moisture-Sensitive Periods: A Zapotec Case." In *Irrigation's Impact on Society*. Ed. Theodore E. Downing and Gibson McGuire. Tucson: University of Arizona Press. Pp. 113 and 122.
DOWNS, J. B., and MOUNTSTEPHENS, N. 1974. "Farmer Participation in Irrigation Schemes: Northern Thailand." Paper presented at Second International Seminar on Change in Agriculture, Reading, England.
FRUTCHEY, R. H. 1969. "Socioeconomic Observation Study of Existing Irrigation Projects in Thailand." Paper prepared for the U.S. Bureau of Reclamation, Department of Interior.
GEERTZ, C. 1967. "Tihingan: A Balinese Village." In *Villages in Indonesia*. Ed. Koentjaraningrat. Ithaca: Cornell University Press. [Chapter 4 above.]
GRAY, R. F. 1963. *The Sonjo of Tanganyika: An Anthropological Study of an Irrigation-Based Society*. London: Oxford University Press.
HARRISS, J. C. 1974. "Problems of Water Management in Relation to Social Organization in Hambantota District." Paper presented at Seminar on Project on Agrarian Change in Rice-Growing Areas of Tamil Nadu and Sri Lanka, St. John's College, Cambridge, December.
International Rice Research Institute and Agricultural Development Council. 1976. *Bibliography on Socio-economic Aspects of Asian Irrigation*. Laguna, Philippines.
JAY, R. R. 1969. *Javanese Villagers: Social Relations in Rural Modjokuto*. Cambridge: M.I.T. Press.
LEACH, E. R. 1961. *Pul Eliya, a Village in Ceylon: A Study of Land Tenure and Kinship*. Cambridge: The University Press. [Chapter 5 above.]
LEWIS, H. T. 1971. *Ilocano Rice Farmers: A Comparative Study of Two Philippine Barrios*. Honolulu: University of Hawaii Press. [Chapter 7 above.]
MITCHELL, W. P. 1976. "Irrigation Farming in the Andes: Evolutionary Implications." In *Studies in Peasant Livelihood*. Ed. Rhoda Halperin and James Dow. New York: St. Martin's Press.
MOERMAN, M. 1968. *Agricultural Change and Peasant Choice in a Thai Village*. Berkeley: University of California Press.
NASH, M. 1965. *The Golden Road to Modernity: Village Life in Contemporary Burma*. New York: John Wiley.
NICKUM, J. 1977. "Local Irrigation Management Organizations in the People's Republic of China." *China Geographer* 5:1–12. [Chapter 13 below.]

ONGKINGCO, P. S. 1973. "Case Studies of Laoag-Vintar and Nazareno-Gamutan Irrigation Systems." *Philippine Agriculturist* 56:374–89.

Sataya Wacana University, Research Institute in Social Studies. 1975. "A Research on Water Management at the Farm Level: An Indonesian Case Study." Paper presented at Workshop on Technical and Social Progress, Penang, Malaysia, December 16–19.

SEKTHEERA, R., and THODEY, A. R. 1975. "Irrigation Systems in the Chieng Mai Valley: Organization and Management." *Chieng Mai University Journal* 11, no. 2.

SMALL, L. E. 1974. "Water Control and Development in the Central Plain of Thailand." *Southeast Asia* 3:679–97.

TAILLARD, C. 1972. "L'irrigation dans le nord du Laos: L'example du bassin de la Nam Song a Vang Vieng." In *Etudes de geographie tropicale offertes a Pierre Gourou,* pp. 241–56. Paris: Mouton.

THAVARAJ, S. H. 1973. "The Necessity of Terminal Facilities for Water Management at the Terminal Level." Paper presented at the National Seminar on Water Management at the Farm Level, Alor Setar, Keoah, Malaysia.

WADE, R. 1976. "Comments on 'Canal Irrigation and Local Social Organization.'" *Current Anthropology* 17:404–5.

WIJEYEWARDENE, G. 1965. "A Note on Irrigation and Agriculture in a North Thai Village." In *Siam Society, Felicitation Volumes of Southeast Asian Studies,* vol. 2, pp. 255–59. Bangkok: Siam Society.

———. 1973. "Hydraulic Society in Contemporary Thailand?" In *Studies of Contemporary Thailand.* Ed. Robert Ho and E. C. Chapman. Australian National University, Department of Human Geography, Canberra. Pp. 89–110.

PART III

Bureaucratically Operated Irrigation Systems

Introduction

In contrast to the irrigation systems that are designed, built, and managed by local communities, in numerous systems some or all of these functions are handled by agencies of the state. Each of the chapters in this section deals with irrigation organization in the context of bureaucratically managed systems.

A distinctive organizational feature of agency-managed systems is the separation of responsibilities between the water users and the water authorities. In general, the individuals who staff the irrigation agency are not farmers in the system; in fact, their understanding of irrigated farming may be quite limited. On the other hand, those individuals who use the water may be intimately familiar with the water requirements associated with their modes of production but uninformed regarding the technical requirements of the engineered infrastructures on which they depend.

Thus a critical organizational process in any bureaucratically managed irrigation system is articulation between water users and water authorities. The nature and degree of interaction between these two groups is an important factor in determining the effective management of these systems.

The search for effective means of articulating water users and water authorities is fundamental to irrigation development. Attempts to improve articulation may be placed in one of three categories:

- *Technical improvements,* particularly the provision of engineered structures at the field and small-zone level to improve the ability of both farmers and agency to control water predictably.
- *Economic incentives*. Of special interest has been the institution of water pricing. It is assumed that attaching an economic cost to the

volume of water used will positively influence the water-management behavior of the farmer (though no assumptions are made about its effect on the agency's management behavior). Thus farmer and agency will be linked through the device of the market.

- *Organizational modifications*. Considerable attention has been paid to the need to organize groups of water users into formal associations as a means of enhancing the farmer–agency linkage.

The papers selected for inclusion in this section have two characteristics: (1) each chapter reports on an irrigation situation in which an extralocal agency plays an important role in system management, and (2) each provides information on the patterns of relationship between water users and water authorities in the system.

Canute VanderMeer (Chapter 11) discusses the organizational changes that followed two major innovations in a small gravity system in Taiwan. These two innovations, the introduction of rotational water distribution and the structural improvement of canals and headgates, were related to significant changes in the relationships among water users and between water users and the management agency. Change was in the direction of a new suprasystem irrigation association (the Neng-kao Irrigation Association) and the formalization of farmer groups within the irrigation system (the rotation groups and their subunits, the working teams). These farmer groups, acting through their elected chairmen, provide a regularized structure of communication between the management staff and the farmer irrigators.

In Chapter 12 Richard B. Reidinger discusses the pattern of water allocation that he observed in a large gravity system in northern India. In this setting the water users and water authorities are rigidly linked through the device of administratively determined water turns. Each water user in the system is assigned a specific time for receiving an irrigation supply. While his turn is assured, however, the water user has no assurance that water actually will be available in the channels at the time of his turn. Furthermore, since the irrigator is limited in his ability to negotiate with either the bureaucracy or other irrigators, the fixed-turn procedure results in great uncertainty regarding the actual timing and quantity of water that will be available. In this situation, many farmers continue heavy reliance on dry-land crops in their farming patterns. Reidinger notes that the uncertainty that this form of farmer–agency linkage creates greatly constrains the users' ability to adopt modern varieties that require more precise water management. He concludes that solutions to this problem may be found in new agency procedures (particularly better coordination of schedules for rotation between canals and among water users on a common watercourse) and in the creation of water-user associations established for each watercourse.

Relatively little information is available concerning the operation of irrigation systems in the People's Republic of China. An important exception is the work of James E. Nickum, as reported in Chapter 13. This chapter provides the reader with useful detail regarding the organization of the Meich'uan Reservoir Irrigation District, which serves an area of approximately 8,000 hectares in Hupeh Province.

Here, as elsewhere, problems arise in articulating the irrigation staff and the water-user groups at the commune, brigade, and production-team levels. As Nickum reports, several procedures and arrangements are designed to deal with this issue: a biannual conference of irrigation district representatives, involvement of the party committee in the irrigation district's affairs through representation on the management committee, and other points of contact.

Two other facts are of special interest. First, irrigation agencies are urged to minimize the cost of water to water users by themselves engaging in activities either to produce their own subsistence needs directly or to obtain income through the sale of products. In Meich'uan the staff has planted grains, cotton, and oil crops and has raised both hogs and fish.

Second, in some irrigation districts the irrigation system includes, in addition to one or more main reservoirs, various small tanks and ponds that may have been constructed previously by individual production teams. These small reservoirs are included in the overall water management plan for the district. Though we presently lack sufficient detail as to how these local ponds are coordinated with the larger system, they suggest an important means for disaggregating large systems into smaller units for local management.

The rigid form of articulation between the irrigation bureaucracy and water users in parts of India is reported in Chapter 12 and is an important feature of the system analyzed by Edward J. Vander Velde in Chapter 14. As Vander Velde points out, the administrative procedure of a predetermined and fixed water allowance serves to accentuate the importance of soils, relief, and field location. Since water users are unable to receive adjustments in their supply of canal water, they have concentrated on adjusting their landholdings so as to acquire fields with good soils and suitable relief in locations proximate to water sources. Obviously, these rearrangements were more feasible for some than for others. As a result, access to the benefits of irrigation development are highly uneven.

Vander Velde's research demonstrates the important interactions that occur among system design, management procedures, and the socioeconomic conditions existing in the command area. Low supplies of water inflexibly allocated to a stratified group of water users resulted in accomodations among the users that greatly benefited some to the detriment of others.

In Chapter 15, I discuss a case in which the procedures of the irrigation agency and the actions of the water users are joined through use of an indigenous irrigation leadership role. In western Laos, small-scale indigenous irrigation systems have existed for a long time. In many, direction and coordination are provided through a locally selected irrigation headman.

When a modern irrigation project was built in this area, the irrigation agency successfully incorporated this indigenous role and made it part of its staffing pattern. The agency divided the water users into twenty-one small groups with contiguous fields, and had each group select an irrigation leader. These group leaders serve as links between the users and the water authorities while remaining primarily accountable to the group by which they were selected. Thus, one can suggest that when bureaucratically operated systems are established in settings with indigenous irrigation experiences, the suitability of existing irrigation roles for project administration should be considered carefully.

In the final chapter, Robert Wade discusses one of the major efforts to improve the effectiveness of irrigation performance now in progress: India's Command Area Development (CAD) Programme. This program focuses on two levels of activity.

The first level deals with the administrative coordination of various development and state agencies at the system level—coordinating the activities of departments such as irrigation, agriculture, extension, credit, and others. The second level of activities deals with changes at the outlet level—the small water-user units within the system.

Here three specific actions are being implemented: the consolidation of fragmented landholdings, the implementation of rotational irrigation, and the organization of water-user associations. Any of these activities individually would be significant; collectively they suggest profound changes in the allocation and distribution of water at the local level.

A major point in Wade's discussion is that while these enormous changes are now being considered and implemented, little social science research is being done to guide implementation procedures or to assess actual results.

Many such "experiments" are occurring in the arena of irrigation development, though not always on the magnitude of the Indian efforts. Wade's concern may be equally applicable to other irrigation development projects. Thus an increasing number of social scientists is needed to work on issues of irrigation organization and administration.

11

Changing Local Patterns in a Taiwanese Irrigation System

Canute VanderMeer
UNIVERSITY OF VERMONT

The recent past has seen significant changes in water control methods in many irrigation systems on Taiwan. The most important have occurred largely as a result of research on water needs of rice plants. Whereas farmers had long felt that it was sometimes desirable to send a gentle flow of water through rice fields and that water had to stand on fields during almost the entire growing period in order to maintain high yields, research conducted on Taiwan prior to 1953 indicated that the plant could tolerate periods without water on its fields. The discovery meant that larger areas could be irrigated without increasing the water supply. This fact has allowed farmers of irrigation systems providing water to over 60 percent of the rice-crop area on Taiwan to adopt a new water-control method called rotation irrigation.[1] Water-control methods in

Previously published as "Changing Water Control in a Taiwanese Rice-field Irrigation System," *Anuals of the Association of American Geographers* 58 (1968): 720–748. Reprinted by permission. Footnotes have been renumbered.

The writer is indebted to the National Science Foundation for support which permitted twelve months of field work on Taiwan in 1965–1966; to J. S. Liu, H. S. Yang, and L. J. Wen of the Irrigation and Engineering Division of the Joint Commission on Rural Reconstruction in Taipei and to Chung-hsing Wu, Mu-hou Chang, Hsi-ming Hsieh, Wan-te Tsai, and Kuei-hsiang Cheng of the Neng-kao Irrigation Assocation in Pu-li for considerable assistance, advice, and information; to the many farmers in the Nan-hung irrigation system area who gave freely of their time and knowledge to provide most of the information upon which this study is based; to James J. Flannery, David A. Schoen, Matthias U. Igbozurike, and Abhaya Attanayake for preparation of the final maps; and to Barbara Zakrzewska for her valuable suggestions on the paper in draft form.

1. L. Chow, "Development of Rotational Irrigation in Taiwan," *Journal of the Irrigation and Drainage Division: Proceedings of the American Society of Civil Engineers* 86, no. IR3 (1960): 1–12; Taiwan Provincial Water Conservancy Bureau, *Report on the 1964 Irrigation Land Survey of Irrigation Associations in Taiwan, Republic of China* (Taipei: Taiwan Provincial Water Conservancy Bureau, 1965), p. 36.

some systems have been further altered as a result of the construction of reservoirs, modern water gates, wells, and better water intake structures in river beds, by the installation of pumps, and by lining canals with concrete.

An irrigation system is an arrangement by which water is conveyed from a source to an area needing water to facilitate the production of desired crops. As such, a system involves four elements: (1) one or more sources of water; (2) fields; (3) physical structures such as canals and ditches which can carry water from its source to the fields; and (4) a functioning set of principles and techniques adopted by humans to create a water-flow pattern within the physical structures related to the amount of water available from the source, the characteristics and locations of the physical structures, and the varying water needs of fields. The qualities of the source of water, the physical structures, and the fields provide the framework within which humans must devise methods of water control based upon acceptable principles and techniques to effect the water-flow pattern which best meets their needs. If the basic qualities of any one element are changed, the water-control methods may also be changed. The water needs of many fields and the physical structures of many irrigation systems on Taiwan have changed in recent years. How did the farmers within those systems alter their water-control methods in response to these changes? What effect have the changes had upon the pattern of water availability from place to place within the systems? The answers to these questions may have relevance for rice farmers in other Asian countries where irrigation systems must be improved to support growing populations.

The objective of the present study is to describe and analyze the changing water-control methods and water-availability patterns between 1936 and 1966 during the first rice-crop season in one small gravity irrigation system, the Nan-hung system near the city of Pu-li in central Taiwan.[2] Since 1955 this

2. The years from about 1936 to 1966 comprise the period for which reliable information on water-control methods was obtained through interviews with farmers. The term "water-control method" as used here refers to an association of techniques whereby water is directed by human effort from a natural source onto individual fields. Although many studies of East Asian irrigation systems exist, none describe the detailed techniques actually used by farmers to course water from a river through a canal into ditches and onto fields. See, for example, J. D. Eyre, "Water Controls in a Japanese Irrigation System," *Geographical Review* 45 (1955): 197–216; N. Kaiso, "The Old and Present Customary Law on Water Use in Taiwan," *Agricultural Economics Research* 2 (1926): 218–61 (in Japanese); Y. H. Djang, *The Chia Nan Irrigation System and Its Problems,* Irrigation and Engineering Series no. 4 (Taipei: Chinese-American Joint Commission on Rural Reconstruction, 1953); E. G. Hipol, Irrigation of Lowland Rice in the Philippines," *Proceedings of the Far East Regional Irrigation Seminar: Taipei, 1961* (Washington, D.C.: International Cooperation Administration, n.d.), pp. 287–93. The Nan-hung system was selected for study for a number of reasons. (1) It is a fairly small system so that the writer could more easily discover and grasp the intricacies of its operations. (2) Water was not everywhere equally available within the system area prior to 1955. Some land regularly produced two crops of irrigated rice annually, some could produce one rice crop with certainty and a second only if rainfall was sufficient, and some yielded only one rice crop a year. Since water-control methods varied somewhat within each of these types of land, the new methods adopted after 1955 and the new physical structures constructed in 1959 had a variety of effects. Within one small system, then, some rather complex changes occurred. (3) The Nan-hung system, although small, was the

system has seen two major innovations which have enlarged the irrigated area, increased rice yields on some fields, reduced the amount of irrigation labor needed, and improved relations among the farmers who control the fields. In 1955 Nan-hung farmers adopted the rotation method of distributing water. In doing so, those operating land near the head of the canal lost the ability to take water generally at times of their choice during the first crop season so as to allow farmers whose land lay near the end of the canal a reasonably sure water supply. In 1959, portions of the canal were lined with concrete and modern water gates were constructed along the canal where important irrigation ditches received water. The structural improvements conserved water and facilitated control of the water-flow pattern from the canal to fields. Although the economic benefits wrought by these innovations were considerable, the primary concern of this study is an analysis of the ways by which farmers who receive water from the Nan-hung canal altered their water-control methods in response to the innovations, and of the resultant changes in the pattern of water availability from place to place within the area during the first rice-crop season.

The many irrigation systems on Taiwan which have experienced changing water-control methods vary in the qualities of their physical structures, irrigable fields, water-availability patterns, and human leadership. Therefore, no single system is wholly representative of all. However, since the government of Taiwan has strongly encouraged adoption of rotation irrigation wherever it would result in expansion of the rice area and has enacted a number of regulations governing the general procedure for accomplishing a change in water-control methods, the process of change was similar in its major characteristics among most irrigation systems. Thus, the changes which occurred within the Nan-hung system are probably representative of the changes elsewhere.

The Setting

The Nan-hung irrigation system today distributes water to almost 600 hectares of fields in the Pu-li basin on Taiwan. The basin floor lies at an elevation of about 450 meters and is surrounded by mountains reaching to 1,500 meters. The area is drained by the Mei River, which flows west to the western plains of the island. A subtropical climate with temperatures that rarely drop to the freezing point and rainfall averaging over 200 centimeters annually, combined with good alluvial soils which dominate the basin floor, permit intensive

most important system controlled by the irrigation cooperative in the Pu-li area and also the system nearest the cooperative headquarters. Therefore, officials of the cooperative had considerable personal knowedge of its operations. (4) The new physical structures had been constructed recently so that farmers could recall the details of water-control methods utilized prior to the construction. (5) Finally, officials of the cooperative sincerely welcomed an examination of the changing water-control methods.

agricultural activity. As a result, the seventy square kilometers of nearly level to gently sloping land in the basin carries a population of 70,000 and a farming population of almost 700 persons per square kilometer of cultivated land, half of which is irrigable.

Construction on the canal of the Nan-hung system began about 140 years ago and was completed to its present length and in its present location in 1907. The major steps in the extension and relocation of the canal are depicted in Figure 11.1. The original canal, later called the Lower Nan-hung Canal, began to take form in 1826. At first it provided irrigation water for about twenty-five hectares. Little by little, however, the irrigated area was extended downslope toward the north. In 1887 the canal length was more than doubled, but much or all of this extension was destroyed ten years later by floods. Also in 1887, another, shorter canal was dug taking water from a point higher on the river. This Upper Nan-hung Canal irrigated ten hectares lying above the Lower Nan-hung Canal. By the turn of the century population pressures forced the people in the area to consider making major improvements on their two canals. Therefore, a new canal was planned in 1906 and constructed a year later.[3] The new canal, called the Nan-hung Canal, utilized the water intakes of both former canals and the bed of the old upper canal. The remainder of the new canal was located almost two hundred meters upslope from the former lower canal.

The main water intake of the Nan-hung Canal from 1907 to 1963 was that lettered *B* on the Nan-hung River in Figure 12.1. Since the river occasionally carries violent floodwaters which have destroyed intake structures, several such structures were located at site *B*. In 1963, a new intake structure was constructed at site *A*. The intake lettered *C* has been a supplementary intake since 1907, receiving water from the river only during the second rice-crop season, when the water level in the river is sufficiently high to reach the structure. Intakes *D* and *E* served a similar function for some years prior to 1945, supplying additional water to the lower portion of a long ditch.

Lowland or wet rice is the most popular crop raised in the Nan-hung area for two basic reasons. The first derives from the fact that farmers consume rice as the staple in their diet. By producing as much rice as they can, they know their families will less likely suffer hunger. The second reason is that the market price of rice on Taiwan is about the most stable of all the desirable crops which could be raised on irrigable fields. Although sugarcane and bananas have yielded larger profits in some recent years, their market values fluctuate. In addition, the banana plant faces hazards of excessive moisture in many fields and of low temperatures. Thus, most farmers feel that they must devote the largest parts of their fields to rice.

3. Pu-li and Hok-ka Canals Irrigation Association, "A Collection of Historical Information on Dikes and Canals," bound collection of unpaginated manuscripts dated from 1923 to 1942, on file, Neng-kao Irrigation Association, Pu-li (in Chinese).

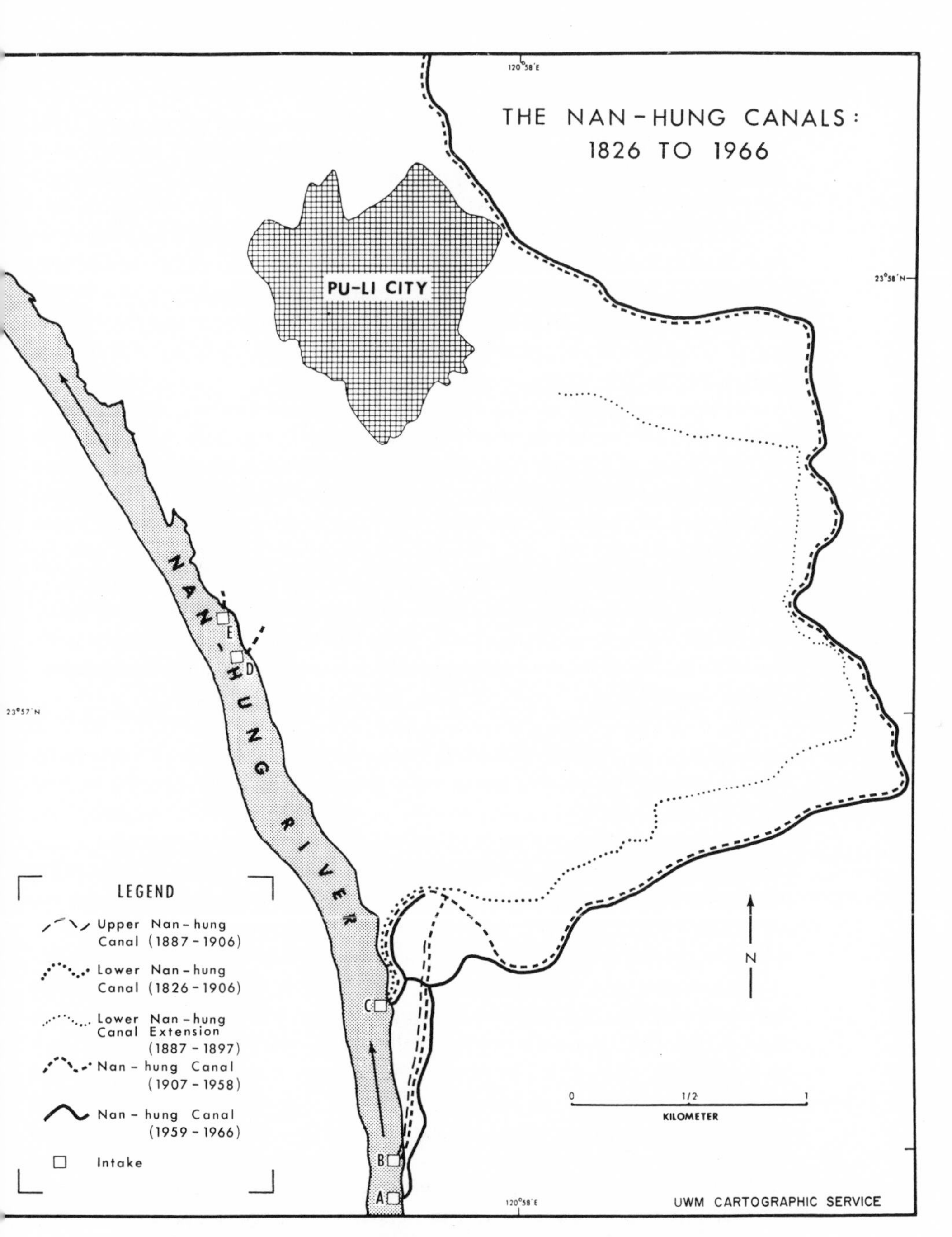

Figure 11.1. The development of the Nan-hung Canal. Sources: Pu-li and Hok-ka Canals Irrigation Association, as cited in footnote 3; map of the Nan-hung irrigation system, late 1950s, scale of 1:4,800, in the files of the Neng-kao Irrigation Association, Pu-li; interviews with knowledgeable farmers and officials of the Neng-kao Irrigation Association in 1966.

Temperatures in the Pu-li basin allow farmers to produce two crops of wet rice a year, generally between early January and late October. However, wet rice has rather stringent water requirements if it is to yield well. In the basin, rice requires more water than will fall as rain on a field. Rice also demands standing water on its field during a number of specific periods in its growth cycle and during some of the remaining time as it grows to maturity. Because the volume and timing of water applications are so significant, rice needs irrigation. Whether a sufficient volume of water is available to a field at the proper times determines whether or not the field may produce one or two crops a year.

Nearly all irrigable fields within the reach of water from the Nan-hung system obtain water only from rainfall and the Nan-hung Canal.[4] The seasonal rainfall pattern in the area evidences a distinct drier period from October into February, when each month receives an average of less than 75 millimeters of precipitation. By contrast, the months of May through August each have mean rainfall figures of over 300 millimeters. However, twenty-seven years of rainfall records reveal that monthly rainfall varies considerably from year to year. For example, all the drier months have experienced less than 2 millimeters, whereas May and July have had under 65 millimeters. The canal usually receives maximum water from the river in July and minimum water in April. Fourteen years of daily water records for the canal and estimates made by irrigation officials on the amount of water needed in the canal to produce two crops of rice on all irrigable land indicate that water in the canal is lowest in relation to the need for it between early February and mid-April. Water has generally been plentiful during the second rice-crop season, from late June through October. On the other hand, before 1955, when farmers applied water to fields with traditional techniques, serious shortages occurred frequently during the first crop season, from January to June, and many were unable to raise as much rice as they desired. Yet today all farmers can devote as much of their irrigable land to rice as they wish. Although the new physical structures on the canal and the new technique of applying less water to fields conserve water and thereby allow more fields to be irrigated, some of the expanded rice area during the first crop season is the result of changed water-control methods.

Water-Control Methods Before 1955

Prior to 1955, many farmers expended considerable effort in bringing water to their fields during the first rice-crop season and nearly all had to cooperate

4. Fifteen private irrigation wells exist in the area. However, fourteen are on land which does not receive water from the Nan-hung Canal during the first rice-crop season and so fall outside the scope of this study. Since the fifteenth was constructed by a farmer who wanted to be able to put water onto his fields at times of his choice and not because of a water shortage, it is also ignored.

with other farmers. Although water almost always flowed in a portion of the canal the individual farmer was responsible for coursing water from the canal through a ditch to his fields. When water was plentiful, all ditch headings could be left open and water would flow continually through the ditches.[5] Farmers could then take it as they desired. However, during much of the first rice-crop season water was scarce. Therefore, special water-control methods had to be devised for periods of low water to allow rice production on as much land as possible. The methods adopted evolved within certain natural and cultural limitations.

The Limiting Factors

The methods utilized by farmers to direct water from the canal onto their fields developed within bounds set by the amount of water which entered the intake of the canal as described above, the physical structures of the irrigation system, the customs concerning water rights of fields, and the limitations of individual human effort. The resultant methods suited many farmers and were accepted by nearly all.

The major physical structures of the Nan-hung irrigation system prior to 1955 are depicted in Figure 11.2. They consisted of four water intakes in the river bed, only the highest of which functioned during the first rice-crop season; the Nan-hung Canal, unlined except for stones located in places to reduce leakage; the irrigation ditches, also unlined, which carried water from the canal to fields; the ditch headings, which were holes in the canal bank through which ditches received water; and drainage ditches. Many portions of irrigation ditches also served as drainage ditches. There were probably about fifty ditch headings in the left bank of the canal, some supplying water to ditches which irrigated large areas and others to ditches only a few meters long. As such, the headings ranged from about 10 to 900 square centimeters. Most holes were rectangular and lined with wood, though some of the larger were lined with concrete tile and the smaller with iron pipes or bamboo tubes.[6] A mixture of dirt, grass, brush, and rocks was stuffed in a hole to prevent a flow of water into its ditch.

5. Irrigation engineers use the term "turnout" to denote an opening in a canal or ditch bank through which water passes into a ditch or field. The term reflects the outward direction of water flow which is the concern of engineers. In this study, a turnout on the canal is called a ditch heading. The heading in Taiwanese terminology is a dark water hole, but farmers call the spot on which the heading is located the head of the ditch. See a definition of heading in P. G. Afable and others, *Glossary of Irrigation Terms Recommended for Use in the Philippines,* Technical Bulletin 13 (College, Laguna: University of the Philippines, n.d.), p. 25.

6. Neng-kao Irrigation Association, "Collection of Statistics on the Irrigated Areas of Various Canals," bound collection of undated and unpaginated manuscripts, on file, Neng-kao Irrigation Association, Pu-li (in Chinese). This source provides data on the sizes and types of thirty ditch headings in 1934. Interviews with farmers and irrigation officials indicated that many more than thirty existed in the late 1930s.

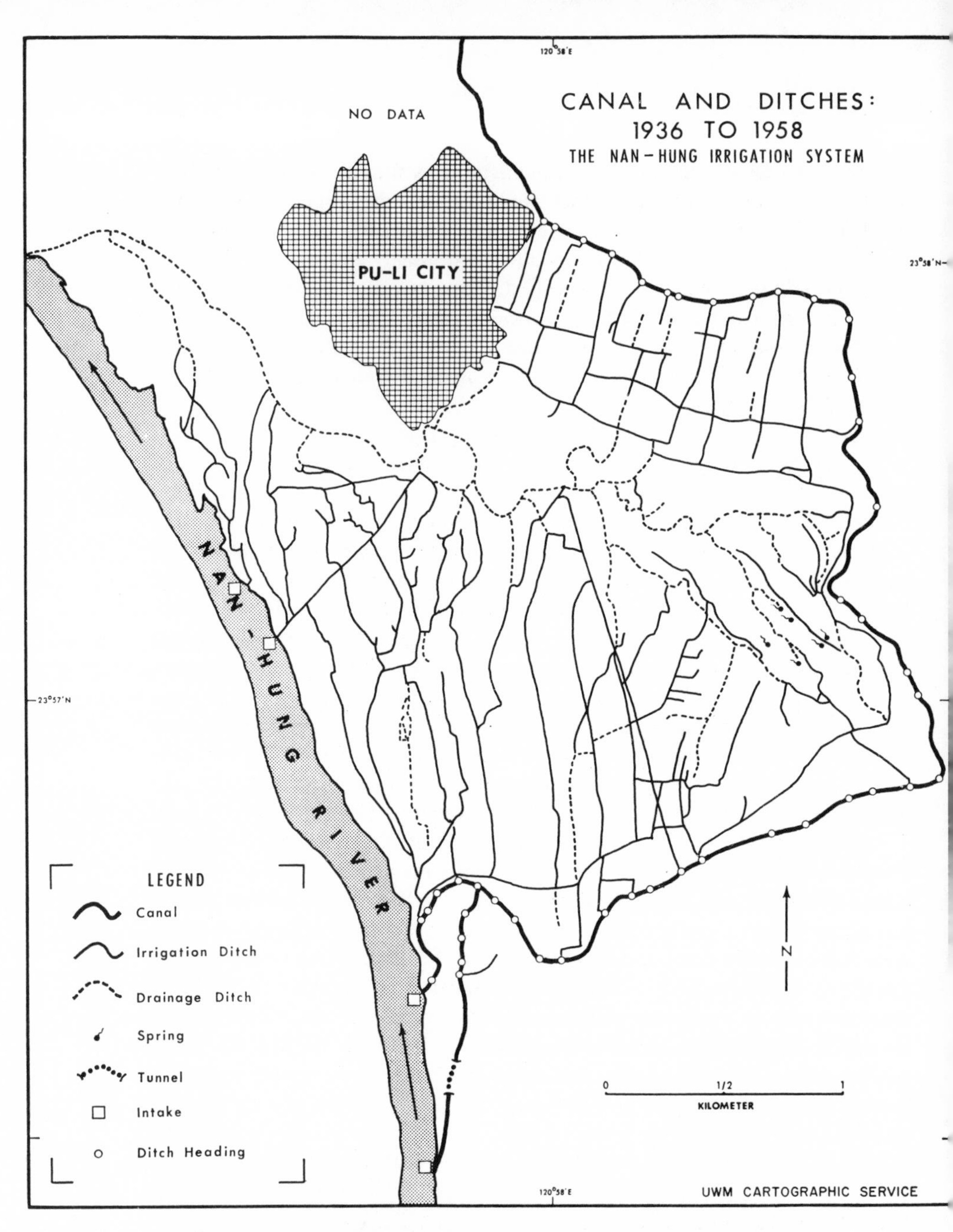

Figure 11.2. The physical structures of the Nan-hung irrigation system prior to 1958. The canal and ditches on this map are taken from the map listed as the first source below. The source map does not depict some short ditches and small ditch headings that certainly existed. Additional ditch headings were located with information provided by the other sources. Nevertheless, some small headings and short ditches that existed prior to 1958 are missing from this map. The springs yielded water only during the second rice crop season. Sources: Neng-kao Irrigation Committee, "Basic Investigation Map" of the Pu-li area in the late 1930s or early 1940s, undated, scale of 1:2,500, in the files of the Neng-kao Irrigation Association, Pu-li, sheets 6, 10, 11, 12, 18, 19, 20, 21; Neng-kao Irrigation Association, "Collection of Statistics on the Irrigated Areas of Various Canals," as cited in footnote 6; map of the Nan-hung irrigation system, late 1950s, scale of 1:4,800, in the files of the Neng-kao Irrigation Association, Pu-li; interviews with knowledgeable farmers and officials of the Neng-kao Irrigation Association in 1966.

The customary water rights of irrigable fields within the Nan-hung system concerned the amount of water which could be applied and the seasons during which they could receive water. All irrigable land was divided into three classes. Class *A* land had primary rights, assuring it of sufficient water to produce rice during the first and second rice-crop seasons. Class *B* land could take only that water not needed by Class *A* land. This meant that it could support rice during the second crop season every year and during the first in some years. Class *C* land received water from the irrigation system only during the second crop season. The approximate limits of these three land classes appear in Figure 11.3. This aspect of the water rights seems to have been based upon the dates on which the fields first received water from the canal and the amount of water which flows in the canal during the cropping seasons. In 1907, when the present canal was constructed, the fields served by the Upper and the Lower Nan-hung canals had been receiving irrigation water for twenty years or more. Since the upper portion of the new canal could satisfy the water needs of these fields during both rice-crop seasons, they automatically obtained primary water rights. In addition, some fields located near and taking water from the middle portion of the canal received primary rights. Figures 11.1. and 11.3. reveal that these fields were located upslope from the extension of the Lower Nan-hung Canal and so could not have been irrigated by Nan-hung River water prior to 1907. Therefore, they probably derived primary water rights by reason of their location near the middle portion of the canal, where they could receive water fairly easily.

Figures 11.2. and 11.3. indicate that Class *B* land obtained water from the lower portion of the canal or from an extremely long ditch. As this ditch and canal lay in large part on rather permeable soils, only a relatively large head of water could reach fields served by their middle and lower sections. When water was low, it was often impossible to build and maintain such a head in the ditch. Thus, most land served by the ditch was not assured of sufficient water to produce rice every year during the first crop season. In the second crop season, the lower portion of this ditch obtained supplemental water from rainy season springs. Although a fairly large head of water could always be created in the middle portion of the canal, during periods of abnormally low water it would not reach the lower portion. The Class *C* land, like some Class *B* land, was located along the end portion of a very long ditch. It was distinguished from the Class *A* and *B* land because it became a part of the Nan-hung system some years later.

A critical aspect of the water rights was that any field could receive only as much water as it needed at times of water shortage. The water needs of individual fields differed according to the crops they supported and the permeability of their soils. Since the precise timing of water applications was not as important to other crops, only fields of rice and sugarcane could receive water during low water periods. However, as sugarcane rarely needed irriga-

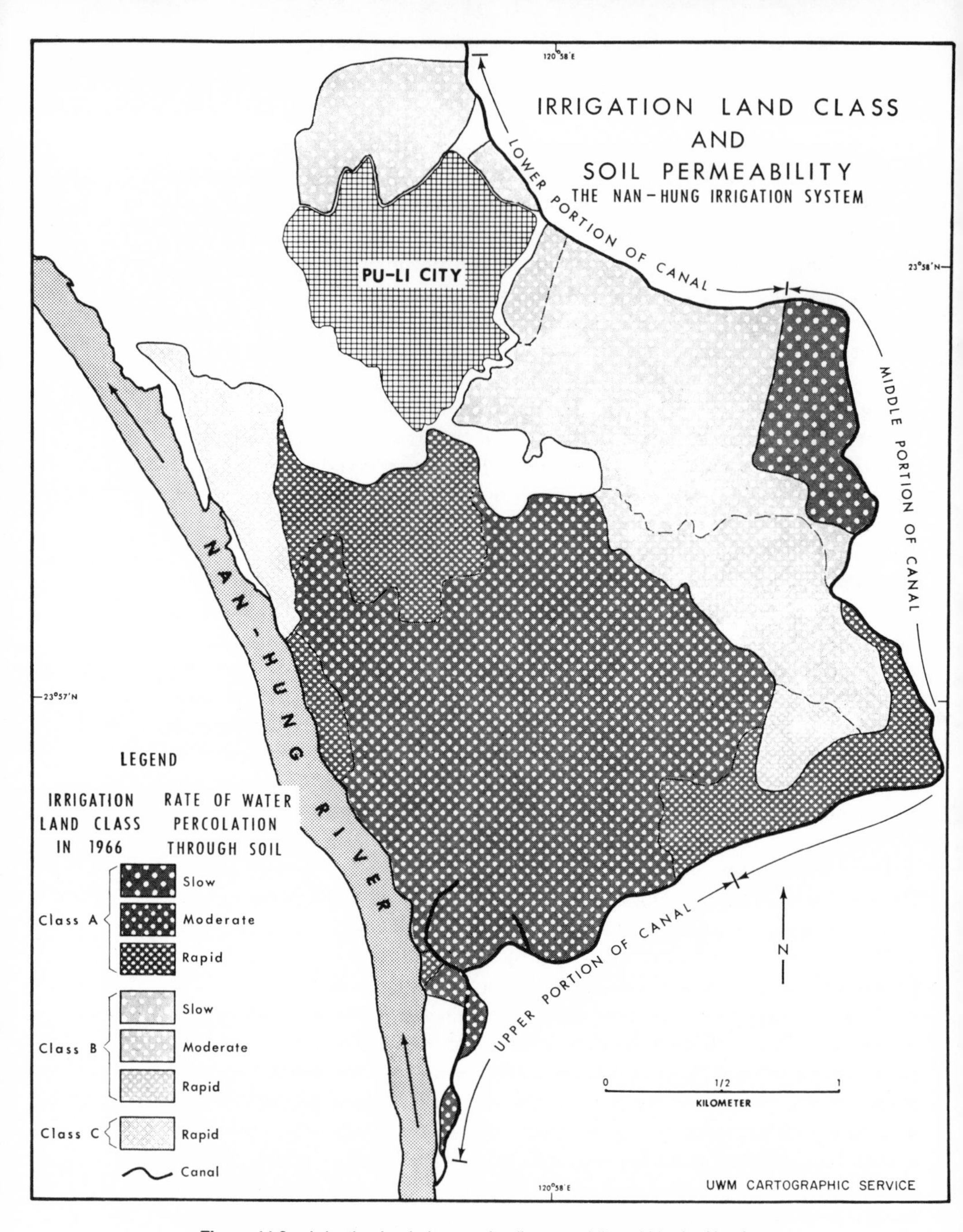

Figure 11.3. Irrigation land class and soil permeability within the Nan-hung system area. The irrigation land class information on this map is dated 1966. During the late 1930s and the 1940s the area of Class A land along the middle portion of the canal was smaller than in 1966. Sources: land property quality and ownership record cards in the files of the Neng-kao Irrigation Association, Pu-li; Taiwan Provincial Agricultural Research Institute, "Land Capability Map of Marginal Zones between Forest Land and Crop Land in Taiwan," scale of 1:20,000, published in 1956 by the Taiwan Provincial Agricultural Research Institute, Taipei, sheets J-15 and J-16; Neng-kao Irrigation Association, *Nan-hung Canal Rotation Irrigation Improvements: Engineering Plans and Cost Estimates Book* (Pu-li: Neng-kao Irrigation Association, 1958), map of soils scaled at 1:25,000; interviews with knowledgeable farmers and officials of the Neng-kao Irrigation Association in 1966.

tion water and its area was small, for all practial purposes any rice field had priority over a field carrying another crop. Before 1955, farmers felt that a rice plant had to stand in at least three and up to fifteen centimeters of water during nearly all of its growing period to yield well. The seedling plants required only a few centimeters of water, and all standing water was usually drained during days of weeding, fertilizing, and harvest. Otherwise, farmers generally believed that the more water standing on a field, the better the rice harvest would be. They also felt that water should at times be allowed to flow slowly and gently through the fields. This irrigation method is called continuous irrigation. The soils in the Nan-hung system may be classified according to their permeability into those through which water drains rapidly, those which drain at a moderate rate, and those which drain slowly. Their distribution is depicted in Figure 11.3. In general, a field which drained rapidly was allowed to receive somewhat more water per application than one which drained slowly.

Since it was physically impossible for a single farmer to put water on his fields by himself when water was low in the canal, cooperation among farmers occurred quite naturally. The farmers of neighboring fields taking water from the same ditch would work together to move water from the canal through the ditch to their fields. However, there was no need for farmers on some ditches to cooperate with farmers on other ditches, except when repairing the canal. On the other hand, it was necessary for farmers taking water from many ditches to have the cooperation of farmers on other ditches if they were to receive any water during low water periods. As a result, two types of cooperatives developed. One was created by law of the government of Taiwan, which recognized the desirability of cooperation among all farmers in an irrigation system in order to utilize as fully as possible the water provided by the system. The second type evolved to facilitate cooperation among groups of farmers whose fields had common water rights.

The Pu-li Irrigation Society was constituted in 1906 to supervise the operations of all irrigation systems in the Pu-li area. It enforced the water rights of irrigation systems and of the several classes of irrigable fields within each system, maintained the main canals, intakes, and ditch headings, and proposed and carried out improvements on the physical structures of the systems. In addition, it could grant emergency water to Class *B* land. Prior to 1945, the manager of the Society was a Japanese national appointed by the highest executive officer of the county in which Pu-li was located. Ultimate authority for policy within the Society was held, however, by a committee of sixteen irrigaion judges. Eight judges were appointed by the manager, usually from the landlord class, and eight were elected from persons who owned irrigable land by the farmers in the area. The administrative staff of the Society was divided into two sections, both headed by a Japanese national appointed by the manager, one engaged in water management and engineering affairs and the other in financial, personnel, and general administrative affairs. For the farmers, an important staff member was the canal traveler. Each major irrigation system had a canal traveler who reported to the manager on water and crop conditions in his system and informally attempted to mediate water disputes. The Society derived it finances from an irrigation water fee collected from owners of irrigable land. The owner of a Class *A* property paid twice the fee obtained from the owner of a Class *B* property of the same size.[7]

Two associations constituted the second type of cooperative. They were informally organized groups which envolved during the 1930's as a result of a need for cooperation in maintaining ditches and in the distribution of water among and from ditches at times of water shortage. They also appear to have had no formal ties with the Pu-li Irrigation Society. One association was composed of farmers who worked the Class *A* land obtaining water from the upper portion of the canal and the other of farmers who worked land taking water from the middle and lower portions. Each group elected a chairman, who informed the canal traveler of the wishes of farmers concerning specific issues and problems, organized farmrs for the repair of ditches, and encouraged cooperation in the distribution of water during low water periods. The chairman had a number of assistants representing farmers who received water from one or more major ditches.

Three levels of management thus influenced the distribution of water in the Nan-hung system during the first rice-crop season prior to 1955. The Pu-li Irrigation Society controlled the amount of water which entered the canal intake and regulated the flow of water from the canal. The associations in turn regulated the flow within and from the ditches. Finally, the individual farmer, who was a member of both the society and an association, had complete control over the flow of water among his fields. However, it was also the responsibility of the farmer to open or close all ditch headings and field water mouths necessary to allow water to flow from the canal to his fields.[8] Since for a great majority of the farmers this was no simple task, they devised water control methods to accomplish the task.

The Water-Control Methods

The amount of labor and cooperation necessary to get water onto a field at any time in the first rice-crop season depended largely upon the water level in the canal and the location of the field with respect to the intake of the canal and to the head of the ditch from which it received water.[9] In general, the

7. Interviews with Hsi-ming Hsieh and Wan-te Tsai and with a number of farmers, including Huo-sheng Wang. The irrigation fee was paid on all irrigable land, whether planted to rice or not.

8. The term "field water mouth" refers to a ditch turnout through which water flows onto a field. See footnote 5.

9. Interviews with thirty-five farmers provided all information in this section.

more distant a field was from its ditch head and the farther the ditch head was from the canal intake, the more labor and cooperation were needed. In addition, however, the lower the water level in the canal, the more difficult it was for almost all farmers to get water. Thus, the water-control methods employed by farmers varied not only with the locations of their fields but also as the water level in the canal changed from that sufficient to supply the needs of all rice fields to that sufficient for Class *A* land but insufficient for much of the Class *B* land, and to that generally sufficient for the Class *A* land only.[10] Yet certain features of the methods were uniform among all farmers, and farmers in all types of field locations utilized a similar method of water control during periods of plentiful water. Therefore, these common elements are described first.

The individual farmer had to be able to make decisions at any time concerning the date and amount of the next application of water onto his fields. Each decision was based in part upon the stage of growth of his rice plants and the drainage qualities of his soils, but also upon the amount of water which he expected would be flowing in the canal and the demands which other farmers would have for that water during the next few days. Thus, he needed continuous information about the level of water in the canal and whether it was tending to rise or drop. This information he obtained by walking to the canal or from conversations with other farmers or the canal traveler. By observing the activities of farmers whose fields had greater water rights than his and the stages of growth of their rice plants, he could estimate their water needs during the next few days. If he expected problems in getting water on the date his fields would need it, he could try to obtain water on an earlier date or take a chance that rainfall would obviate the problems.

Any farmer wanting irrigation water had either to take part himself in the process of coursing water onto his fields or to designate someone to act on his behalf. It was his responsibility to see that the water mouths of his fields were opened and that obstructions were placed in the ditch, if necessary, to direct water from the ditch through the mouths and onto the fields. As a low earth dike circumscribed each field and lay in part between the field and the ditch, the mouth was simply an opening in the dike twenty or thirty centimeters long created by removing the necessary earth and sod. The obstruction placed in the ditch was usually a temporary one composed either of the earth and sod taken from the dike or of rocks, brush, and sod kept in the immediate area. Few farmers had obstructions of wood boards or stone slabs located permanently in the ditch to act as a check, raising the ditch water level so that it would flow into the field. The field water mouth was normally located near

10. Average cms figures for each of these water-level conditions are not available because water-volume records for the canal were initiated in the early 1950s. Since the water needs of a field on any day depended in part upon the amount of rain which fell during the previous fifteen days or so, it may be assumed that the cms figures would vary greatly.

the point where a head of water flowing down the ditch would first come in contact with the field dike. In a few locations within the Nan-hung system, the ditch bed lay well below the level of nearby fields so that it was impossible to direct water onto a field through a mouth located on its dike. As such, the field mouth had to be placed higher on the ditch and a separate water channel paralleling the ditch joined the mouth with the field. Figure 11.4. depicts the types of field water mouths. The several contiguous fields operated by one farmer generally received water from a single field mouth. A mouth allowed water to flow onto the highest of his fields from whence it drained onto lower fields. However, if the area of fields was so extended that water could be directed onto the more distant fields only with difficulty, the fields had two or more mouths.

Water on fields operated by a farmer was completely under his control except that he could not drain it onto fields of other farmers without their permission and except during periods of emergency irrigation declared by the Pu-li Irrigation Society. The water level on fields was regulated by opening or closing field drains, which were cuts in the dikes. Excess water could be drained onto a lower field or into a drainage ditch. In addition, surplus water

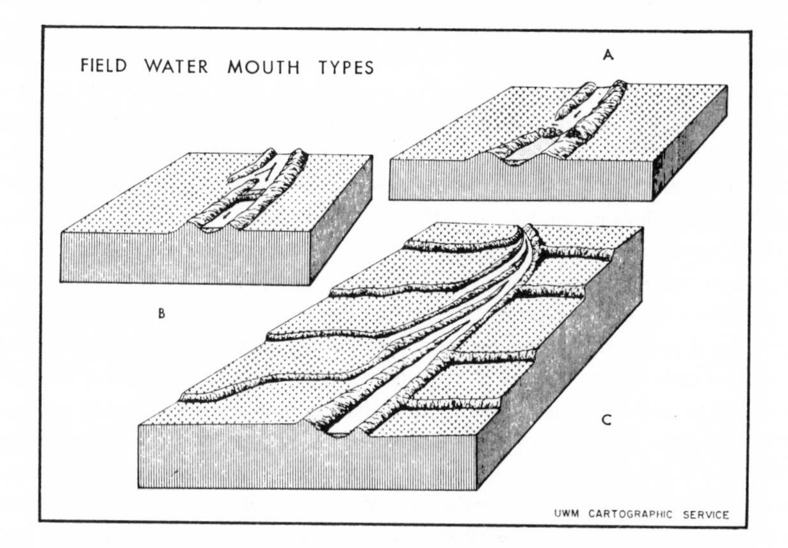

Figure 11.4. Diagrams illustrating the types of field water mouths in the Nan-hung system area. In diagram *A* the elevation of the ditch bed at the field water mouth is similar to or a few centimeters higher than the field, and a temporary obstruction lies in the ditch to direct water onto the field. Diagram *B* depicts the water mouth of a field that is a number of centimeters higher than the ditch bed. A permanent check in the ditch raises the water level. In diagram *C* the fields are thirty centimeters or more above the ditch at the spots their water mouths would normally be located. Thus, their mouths are placed higher on the ditch. The fields on the right-hand sides of the diagrams receive water from ditches not shown on the diagrams.

on many fields could be drained back into the irrigation ditches because the grades of many ditches were sufficient to allow such a water-conserving practice (Fig. 11.5).

During periods of plentiful water in the first rice-crop season, many or all ditch headings were left open so that water flowed continually in the ditches. Then almost any farmer wanting water had only to open his field mouths. If water happened not to be flowing in the ditch which fed his fields, he had also to open the ditch heading. This process, which constituted the normal procedure for obtaining water throughout the second rice-crop season as well, involved very little labor and no cooperation with others.

As soon as water became sufficient to supply the needs of all rice fields on Class *A* land but only some Class *B* fields, however, more labor and some cooperation were necessary to get water. Such conditions occurred frequently between early February and mid-April and commonly until the end of May. Since the water rights of *A* land farmers allowed them to satisfy their needs on days and to some extent at times of their choice, water often did not flow in the middle and lower portions of the canal if a large number of *A* land farmers were taking water.

Often when water was inadequate in the canal, it flowed in a ditch only if a number of farmers were trying to get water from the ditch. Thus, a farmer whose fields were on a ditch not carrying water and who wanted water had to open his field water mouths, close all open water mouths on the ditch above his mouths, open the ditch heading, and place an obstruction in the canal bed to direct water into his heading. If many farmers were taking water at higher ditch headings so that no water was in the canal at his heading, he had either to wait until they had received all they needed or to persuade them to allow some or more water to bypass their headings. Then when water began flowing in his ditch, it was necessary to guard his ditch heading, many or all other headings on the canal above, and all field water mouths on the ditch above his mouths in an effort to prevent other persons from taking water en route to his fields. On the canal he could prevent persons whose fields had the same or lesser water rights than his from closing his ditch heading and could try to persuade persons whose fields had the same or greater rights to delay opening their higher ditch headings until he had received the water he needed. Any unguarded ditch heading, however, might be opened or closed by other persons seeking water. According to custom, persons on the ditch above his field mouths could not take water unless they had participated in the effort to course water into the ditch. Nevertheless, the ditch had to be patrolled to prevent thievery. Obviously, some cooperation among farmers was imperative.

Groupings of farmers thus developed among those whose fields were near each other. Most farmers preferred to belong to temporary groups, though some groups existed for years. When a farmer decided his fields needed water, he would examine nearby fields which took water from the

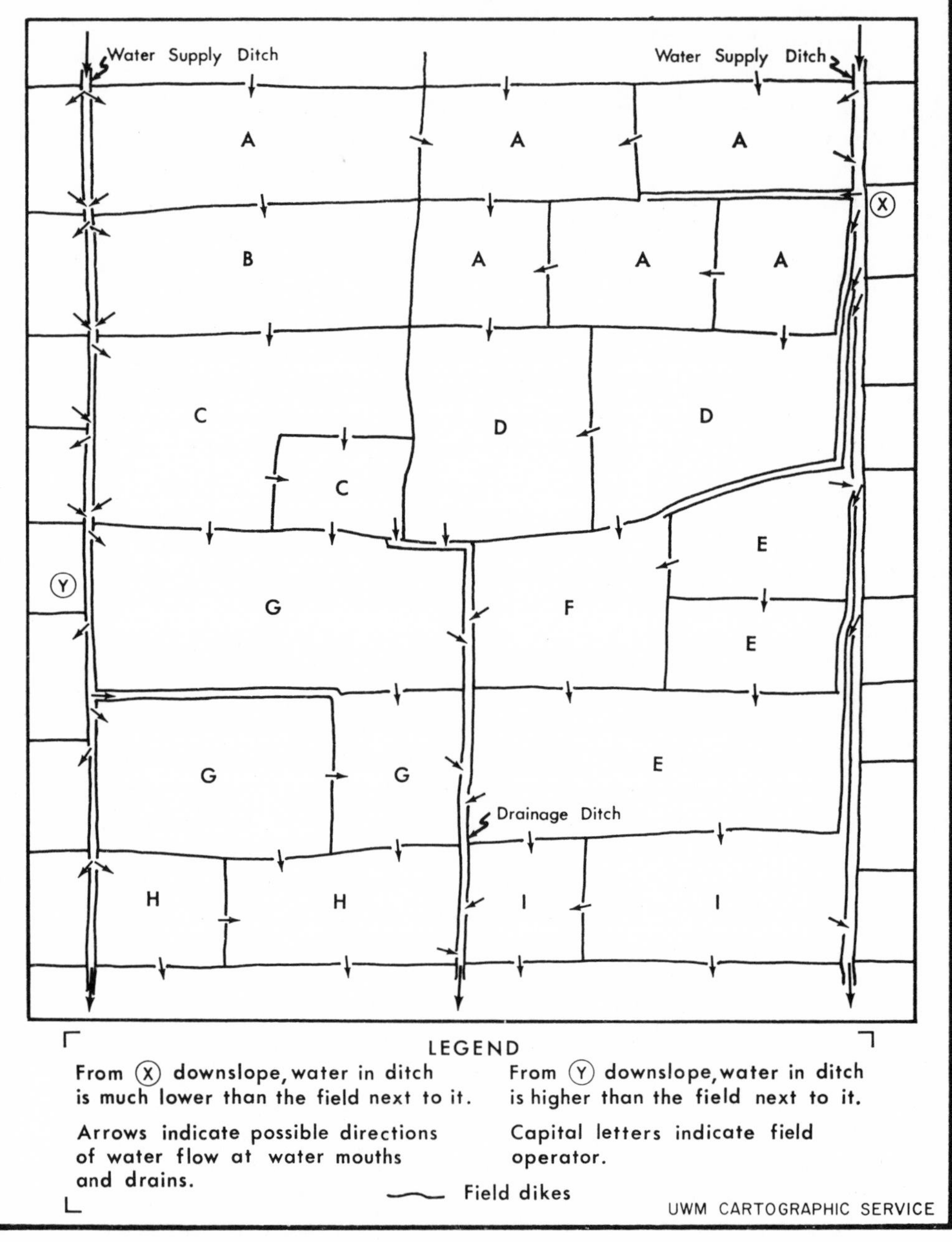

Figure 11.5. Diagram illustrating the possible directions of water flow within a hypothetical area of fields.

same ditch and contact the owners of those also needing water. Those farmers would then combine efforts to obtain water on a day and at a time which was mutually agreeable. A farmer operating widely scattered fields belonged to more than one group.

In general, a Class *A* land farmer with fields along the upper portion of a ditch receiving water from the upper portion of the canal required the help of one to three farmers during periods of water shortage. When three would work together, two covered the canal and one the ditch. If no water was in the canal at their ditch heading, the two would open their heading, put an obstruction in the canal, and walk up the canal to the heading which was receiving the last water. If no one was guarding the heading, they would close it. Then, if they felt that the water which flowed to their heading was sufficient to reach their fields, they needed only to watch the two headings. On the other hand, if the heading was guarded, they would attempt to induce the guard to allow some water to flow by. When the decision of the guard was affirmative and the two farmers believed the water was enough, one would move to watch their heading while the other remained at the higher heading to keep the guard honest. On the other hand, if the decision was negative or if the two felt that the water was insufficient to meet their needs, one or both would walk to a higher heading to try again. At these headings, however, they tried to persuade the guards to allow more water to flow past the headings than was already flowing past. Any water which they could obtain at higher headings was almost always allowed to flow past lower headings by lower guards.

In the meantime, the farmer covering the ditch closed all field mouths above those of fields belonging to members of his group and informed other farmers whose fields took water from the ditch and who happened by that his group was trying to get water. If the others needed water, they might be asked or allowed to join his group. Those not invited would work on their own. When water arrived in the ditch, it was his duty to direct as much onto the fields of members of his group as they needed and to thwart thievery by farmers higher on the ditch. If the flow was a good one, two or three field mouths might receive water at a time. Otherwise, one mouth would consume the whole flow. Since he could see anyone approaching the ditch above and since he knew the water needs of the higher fields and the identity of persons who might steal water, his efforts to protect the flow of water in the ditch consisted primarily of keeping his eyes open. On dark nights his son or wife would patrol the ditch above.

The water-control methods of Class *A* land farmers whose fields lay near the end of a long ditch were basically similar to those used by persons high on the ditch. However, the groups of cooperating farmers were larger and the time and effort required to obtain water was greater. Since all ditches in the Nan-hung system lost water through seepage, the greater the distance between a ditch heading and a field mouth, the larger had to be the head of water which flowed down the ditch to the mouth. Therefore, the energies of from three to eight or more men were required to direct water in the canal to the heading. In addition, two to four men worked the ditch, patrolling it and coursing water onto the fields.

Farmers of Class *B* land faced considerable difficulties during the first rice-crop season and had no assurance of receiving any water during low-water periods. Nevertheless, rice was the most desired crop and experience had shown that hard work would produce some harvest. Thus, most Class *B* land farmers tried to plant some fields, and especially those with less permeable soils, to rice. Within the Nan-hung system area, the ideal time to plant first rice-crop seedbeds was in late January. Fields could then be plowed, harrowed, and leveled in early February and transplanting could occur in middle or late February. This ideal schedule was adopted by Class *A* land farmers. As a result, water rarely flowed in the middle and lower portions of the canal during the first two weeks in February, when large quantities of water was needed to soak *A* land fields in preparation for plowing. So as to create for themselves a period of low water demand in early February, almost all farmers of Class *B* land and a few of *A* land planted their seedbeds in December or early January, after which they watched the water situation carefully. When sufficient water for soaking some of their fields became available, they plowed, harrowed, and leveled the fields and transplanted. Since the rice seedlings could be transplanted any time between twenty and sixty days after the seeds had been sown, farmers had some flexibility in choosing their field-soaking period.[11] If not enough water was available to soak all the Class *B* fields which farmers had planned to plant, they lost only the seed.

During all low-water periods, farmers of Class *B* land competed for water at unfavorable odds with the Class *A* land farmers, each of whom could take water from the canal before they did. This meant that although a group of Class *B* farmers might have obtained permission from a number of ditch-heading guards to course water past their headings, any Class *A* land farmer who subsequently appeared at the canal had the right to some or all of the water en route to the Class *B* land fields. The Class *B* farmers could then only try to persuade the Class *A* farmer to postpone his efforts. If the latter refused, as he often did, the former had no recourse but to wait and try again later. As the farmers with primary water rights preferred to irrigate in daylight, those with secondary rights usually attempted to obtain water during low-water periods at night. Even so, they frequently ran into competition with farmers of Class *A* ditch-end fields. Thus, it was commonly necessary for twenty to thirty-five Class *B* land farmers to band together for the purpose of coursing

11. The growth period of seedlings in a seedbed could be lengthened by reducing water and fertilizer applications.

water onto their fields. Nearly all these men would endeavor to close and guard each ditch heading above theirs on the canal. Not uncommonly, a group worked a full night or several consecutive nights for naught.

Although the Class *A* land farmers who took water from the middle portion of the canal had fewer water problems than the Class *B* land farmers, they were not as fortunate as other Class *A* farmers. Since the head of water required to create a flow to the middle portion of the canal had to be larger than one going to a heading in the upper portion, they spent more time working the canal and the groups of cooperating farmers averaged fifteen to twenty men.

On the average, farmers in the Nan-hung irrigation system area experienced a period of extremely low water supply two or three times a year. During such periods, the water in the canal was generally sufficient only for Class *A* fields, and little or no water flowed into the lower portion of the canal for many days at a time. Competition for water among Class *A* land farmers then became rather keen, although they knew the water would meet their needs. Class *B* land farmers, on the other hand, had no hope of receiving enough water. When their fields had not received water for ten to twelve days, the shortage could cause a significant reduction of rice yields. Under those circumstances, the farmers could request emergency irrigation, but since there was no assurance that it would be granted, they also redoubled their efforts to obtain water with normal water-control methods. Tensions then quite naturally rose along the canal between Class *A* and Class *B* land farmers, occasionally causing physical violence.

A request for emergency irrigation emanated from the chairman of the association of Class *B* land farmers and was directed to the canal traveler. The canal traveler, who had intimate knowledge of the water needs, passed the request to the manager of the Pu-li Irrigation Society along with his recommendations. If the manager felt the need was sufficient, he contacted the irrigation judges, who determined whether or not to declare a state of emergency irrigation. When the judges declared one, the head of the water management section of the Society created an emergency water-flow schedule in consultation with the canal traveler and the chairmen of the associations of Class A and Class B land farmers. The schedules were of two types, a rotation and a single-application schedule. Only the rotation type appears to have been used prior to 1945. This schedule gave all the water in the canal for a specified number of hours to each of the major localities of irrigable land within the system consecutively through a two-or three-day period. The schedule took effect on a date and at an hour determined by the manager and was repeated until rainfall alleviated the water shortage.[12]

12. Data from the looseleaf file of rotation and emergency irrigation schedules maintained by the Neng-kao Irrigation Association.

The water-flow pattern from the canal to fields was regulated by the associations. When water arrived at the highest heading serving a locality, the chairman of the association would decide whether all the water in the canal should be directed into one or more headings at a time. If the water was very low and the heading large, he might send it all into one heading. The water in the ditch was then coursed onto fields starting with the highest field mouths on the ditch and ending with the lowest. Usually the flow of water was sufficient to go into at least four mouths at a time. At the same time, it was required that drains be cut in the dikes of each field so as to allow any standing water to flow onto the next lower field regardless of its ownership. As such, the purpose of emergency irrigation was to give a field enough water to wet the soil only. If in the course of one rotation period all ditches in a locality did not receive water, the process continued at the beginning of the next rotation period where it ended during the previous period.

Although the rotation type of emergency irrigation was employed occasionally between 1945 and 1955, the single-application type was more common, largely because it was easily administered. This type of schedule designated on a one-time basis a number of hours of all the water in the canal to a locality needing it. Ordinarily the names of one, two, or three localities appeared on the schedules, but on a few, one locality was given a number of hours of all the water on several consecutive or alternate days. The date and time of the commencement of each period of water application was indicated on the schedules. Since the number of hours of water which a locality received was usually large, it appears that this type of schedule gave a locality more water than would be needed simply to wet the soil.

In summary, though the process of directing water from the canal to fields was basically similar everywhere within the Nan-hung irrigation system, the water-control methods utilized by farmers of Class *A* and of Class *B* land during periods of low water differed because of their different water rights and the differing water-flow distances involved. In order to make a realistic decision on when to try to obtain water for his fields, the Class *B* land farmer needed to be more closely aware of the water level in the canal and the water needs of fields with greater water rights. Since the distances he had to walk to obtain this information were great, he spent more time and effort at arriving at such decisions. When the decision was made, more cooperating farmers were needed to obtain water because a much larger head of water had to be created in the canal for the water to reach the ditch heading. In general a Class *B* land farmer expected his water-control methods to involve much more walking, talking, waiting, and cooperating than did his Class *A* land counterpart. Similar contrasts appeared on Class *A* land between the water-control methods of farmers with fields high on a ditch and those with fields low on the same ditch. Since the Class *A* land farmer with low fields could at any time preempt water en route to Class *B* land, however, his difficulties were much less severe than those of Class *B* land farmers. On Class *B* land, differences in the elevation of fields on a ditch usually had no effect upon water-control methods because the operators of lower fields either cooperated with higher farmers or did not plant rice during the first crop season.

Water-Control Methods After 1959

Between 1955 and 1959, two basic elements of the Nan-hung irrigation system were changed.[13] First the water needs of rice fields were reduced and later some physical structures were improved. As a result, it became possible to modify water-control methods for the benefit of a great majority of the farmers. Nevertheless, certain desirable modifications of the methods could be achieved only if accompanied by major changes in the organization of cooperative activity. Therefore, within the five-year period, the characteristics of two of the four factors which limited the development of water-control methods were altered.

Changes of the Limiting Factors

The factors limiting the form of water-control methods adopted by farmers after 1959 were basically the same factors effective in earlier years: the amount of water which entered the canal intake, the physical structures of the irrigation system, the customs concerning the water rights of fields, and the limitations of individual human effort. The amount of water received by the canal and the water rights principles remained the same as prior to 1955. Important features of the two other factors were modified, however, in part by powers beyond control of the farmers.

Experiments conducted on Taiwan between 1933 and 1943, the results of which were published in 1953, revealed that rice plants needed less water standing on their fields than had traditionally been thought desirable.[14] Yields could be maintained with periodic applications of relatively little water if their timing was coordinated with periods in the plant life cycle. Since adoption of this technique would result in water savings that could be applied to fields otherwise unable to produce rice, the government of Taiwan strongly encouraged farmers to devise water-control methods suitable for establishment of the technique in their own areas. The irrigation cooperatives in the Nan-hung system area assumed the task.

The essence of the new cultivation technique involved careful timing of water applications. The rice plants stood in water during certain periods of

13. Interviews with Mu-huo Chang, Hsi-ming Hsieh, Knei-hsiang Cheng, and Wan-te Tsai, and over eighty farmers, including Yun-peng Wen, Huo-sheng Wang, and Wang-kun Chang, provided all information in this section, except where otherwise noted.

14. Chow, "Development of Rotational Irrigation in Taiwan."

their growth and alternately in and out of water during other periods. In addition, the soil of the field could not become too dry. The technique could be utilized without any changes in water-control methods when water was plentiful in the canal and flowed continually in all ditches. However, much water would then be wasted as it was impossible to regulate the sizes of ditch headings, originally constructed to allow enough water flow to satisfy the needs of fields according to traditional techniques. With the new technique, smaller headings would suffice and also shorten the periods when water was too low in the canal to satisfy all fields.

If the new technique was to allow cultivation of rice on more land at times of low water, water would have to be applied to fields in various areas according to a rotating schedule granting it to any field every three to eight days. In order to facilitate execution of rotating water schedules, two type of modifications of the physical structures of the system were desirable and a reorganization of the irrigation cooperatives was neccessary. Most ditch headings needed adjustable and lockable gates. Adjustable gates would allow manipulation of the water-flow pattern to give each ditch sufficient but not surplus water when the canal water level was dropping toward a level too low to provide the needs of all fields. Lockable gates would obviate posting guards at headings. Similarly, the number of headings on the canal should be reduced so that small areas of fields would not require consideration in the rotation schedules. The cooperative required reorganization to facilitate greater cooperative activity and to enforce the rotation schedules. Since the government of Taiwan recognized these facts, it agreed to subsidize the construction of structural improvements on the canals of systems which would benefit from rotation irrigation and enacted a law reorganizing irrigation cooperatives.

In 1956 the Neng-kao Irrigation Association was created with administrative offices in Pu-li city. Its membership includes all farmers who benefit from irrigation systems in the Pu-li area. The members elect eleven representatives who formulate policy, approve budgets, and elect the Association chairman. The chairman administers Association affairs through five divisions headed by his appointees. The functions of the Association are similar to those of its forerunner, the Pu-li Irrigation Society. However, as the technique of rotating irrigation requires much closer regulation of the widely scattered irrigation systems than was necessary prior to 1955, the Association has three stations through which it deals with farmers. The Nan-hung system falls within the responsibility of the station located in Pu-li city, with a staff including a chief and canal travelers.

Under the station and within the structure of the Association are six rotation groups, which play the major role in determining the water-flow pattern in the Nan-hung system. All land within the system is divided into six rotation areas, each encompassing the fields irrigated from several consecutive ditch headings. All farmers who operate fields within a rotation area are members of the rotation group of the area. The members elect a group chairman to an unsalaried position of considerable authority. It is the responsibility of each group to manage the irrigation process within its area and to maintain ditches. In addition, the six chairmen act as a committee which determines when to institute rotation and emergency irrigation and creates the water-flow schedules. The rotation groups are subdivided into working teams of fifteen to twenty farmers who operate an area of about ten hectares of contiguous fields taking water from the same ditch.[15]

By early 1959, the Neng-kao Irrigation Association completed construction of the new physical structures desired for the Nan-hung system. In addition, portions of the canal which lost considerable water through seepage were lined with concrete. Figure 11.6 depicts the locations of the structures in 1966. As compared with earlier structures, the number of ditch headings decreased from about fifty to thirty-seven, twenty-three with modern gates. Since the ten hole headings near the end of the canal provide water to the lowest fields in the system and all receive water at the same time in a rotation schedule, they require neither gates nor guarding. The Association created the four hole headings on the upper part of the canal when it found that the ditches they supply did not receive sufficient water from gated headings. Certain alterations were also made in the pattern of ditches to provide more direct water routes from the canal to some areas of Class *B* land.

The Neng-kao Irrigation Association taught farmers the new cultivation technique and water-control methods. In general, the farmers of Class *B* land readily accepted the ideas not only because they benefited most from their acceptance but also because a germ of the ideas was already in their minds, for they had seen rice plants growing with no standing water on their fields for days at a time without much reduction of harvest. The most difficult problem was to convince some Class *A* land farmers that the new technique was desirable and would not lower yields. They did not object to receiving less water so that the Class *B* land farmers could have more. Their water-rights principles clearly indicated that at times of low water any field could receive only the water it needed. On the other hand, as they had been accustomed to taking water generally on days and at hours of their choosing, they objected to the idea of complying with a schedule. Their main question, however, concerned the effect of the new technique upon yields. Several years of experience in using the technique and of observing a number of demonstration fields were required to convince some farmers. By 1966 only a few still had doubts.

Since the new cultivation technique and water-control methods would in effect give Class *B* land some water which traditionally went to Class *A* land,

15. See also *Laws and Rules Concerning Rotational Irrigation* (Taipei: Rotational Irrigation Promotion Commission of Taiwan, n.d.), pp. 31–75; T. Y. Tsai, *Rotational Irrigation Practice* (Tainan: Chia-nan Irrigation Association, 1964), *passim*. In the Chinese literature the rotation group is called a small group and the team a section.

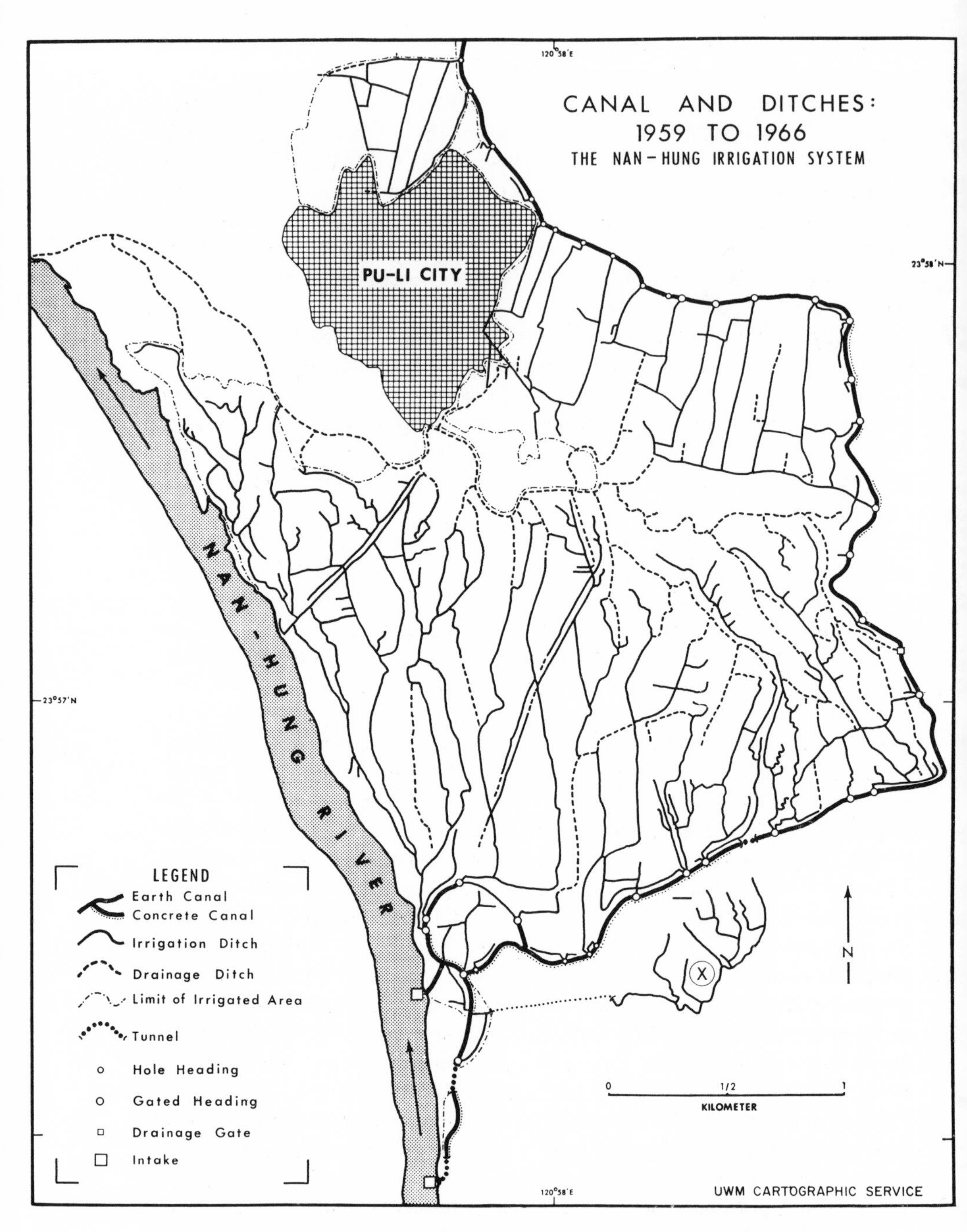

Figure 11.6. The physical structures of the Nan-hung irrigation system in 1966. Some unirrigable land is found within the boundary of the irrigated area. The area lettered *X* receives supplementary water from the Nan-hung system only during the second rice crop season. Source: map of the Nan-hung irrigation system, late 1950s, scale of 1:4,800, in the files of the Neng-kao Irrigation Association, Pu-li, as revised and updated by the writer through field mapping.

the representatives of the Association revised the irrigation fees. Whereas in the past, Class *B* land farmers paid half the fee collected from Class *A* land farmers per hectare of irrigable fields, they now pay three-fourths. In addition, the Association assessed each landowner a special fee to finance that portion of the new construction not paid for by the government.

Today nearly all farmers know enough about the growth cycle of rice to make valid decisions concerning applications of water to their fields. Rice-cultivation activities within the area irrigated by the Nan-hung system during the first cropping season proceed through five major stages: (1) planting and caring for seedbeds, (2) preparing fields for transplanting. (3) transplanting and caring for transplanted fields until the second weeding, (4) caring for fields between the second weeding and the start of rice flowering, and (5) caring for fields from flowering to harvest. Since the water demands of rice plants vary with these stages, it is useful here to describe the interrelations between the rice-cultivation process and its water needs within the stages. The description which follows is based entirely upon interviews with knowledgeable farmers. Therefore, the water needs described are those which the farmers feel and not necessarily the actual needs of rice plants.

The first rice-crop seedbeds are normally planted in January, but in the twelve years prior to 1966 planting started as early as December 1 and ended as late as February 10. The longest single planting period extended from December 1 to February 2 and the shortest from January 3 to January 24. The varying times of the commencement of planting seedbeds and the varying lengths of the planting period are determined largely by the Chinese agricultural calendar and by the availability of irrigation water. The calendar defines the limits of the period during which planting should take place. Under ideal water-availability conditions, farmers would prefer to plant late in the period, when temperatures are warmer and the seedlings grow more rapidly. More rapid growth means less time spent caring for seedlings, and especially checking and controlling the amount of water on the seedbed. Whether a farmer plants his seedbed early, midway, or late in the period depends primarily upon the locational and soil characteristics of his fields as related to the availability of irrigation water. While his seedlings are growing, the farmer prepares his fields so that the seedlings may be transplanted when about fifteen to twenty centimeters tall. Since three to five centimeters of water must stand on the field while it is plowed, harrowed, and leveled in preparation for transplanting and until transplanted, the seedbed planting date is chosen primarily according to his expectations regarding the availability of sufficient water for field preparation. As the canal never carries enough water in February to allow all farmers to prepare fields at one time, most farmers who receive water from the middle and lower portions of the canal still plant seedbeds early.[16]

This tendency to plant early is especially true among those with fields of more impervious clay soils. Since most rice-fields soils become dry in November and December, they need a soaking before being prepared for transplanting. The soaking requires much more water on clay soils than on more sandy soils. After being soaked, however, the clay soils require less water. Fortunately, relatively impervious soils dominate large parts of the area low in the Nan-hung system. Farmers who wish to plant rice on the more permeable soils must decide between planting seedbeds and preparing their fields early and taking a chance that the lower portion of the canal will have sufficient water to support their early transplanted fields on the one hand and planting late when there may be too little water for field preparation on the other.

Although the rice seedlings grow in the seedbed for thirty to forty days, the beds require very little water. Carefully selected, processed, and already germinated seeds are broadcast on the bed with the soil muddy but without standing water. After planting, farmers put two or three centimeters of water on the bed and then allow the soil to become a little dry over the next two or three days. A seedbed is dry when fine cracks begin to appear on the soil or when mud will not stick on a finger rubbed lightly over the soil. Then two or three centimenters of water is coursed onto the bed and the soil is allowed to dry again. This process of wetting and allowing the bed to become dry continues through the entire seedbed period.

While the seedlings are growing, the fields into which they will be transplanted are prepared. The preparation process makes the top twenty to thirty centimeters of soil into a loose wet mud and levels the field. The soaked field with three to five centimeters of standing water is first worked with a plow to turn over the soil which may carry stubble and a green manure. Then a knife tooth harrow breaks up the large clumps of soil. The clumps are further broken and the field leveling process begins with a comb harrow. A pulverizing roller then comminutes the surface soil and continues to level the field. Finally, the surface is smoothed with a tool composed of a comb harrow whose prongs are stuck through a three-meter-long bamboo pole ten or twelve centimenters in diameter.[17] The pole and harrow are pulled across the field until the water standing on the surface is everywhere the same depth.

The seedlings are transplanted with two to five centimeters of water on the field. The transplanting period within the Nan-hung system lasts about a month and falls on the average between mid-January and mid-February. Since 1954, however, it has begun as early as January 3 and ended as late as March 31. The date on which an individual farmer transplants his fields falls a few

17. For descriptions of these tools, see F. C. Ma, T. Takasaka, and C. W. Yang, *A Preliminary Study of Farm Implements Used in Taiwan Province*, Plant Industry Series no. 4 (Taipei: Chinese-American Joint Commission on Rural Reconstruction, 1958), pp. 259–87. Water buffaloes pull all these tools. No mechanized cultivators or tractors are used in the Nan-hung system area.

days after his fields are prepared. Water should stand on a transplanted field continually until the second weeding, forty to fifty days after transplanting, but may be drained on the day of the first weeding, normally thirty days after transplanting. During this period, since water is often scarce and the air cool, farmers will usually put water on their fields and not allow it to drain. An application of three centimeters of standing water is sufficient. However, farmers often take six centimeters in case there is not enough water when needed next. If three centimeters of water is put on the fields, the more permeable soils will need to receive water every two days or so while the less permeable soils may hold standing water for five days or more.

Following the second weeding, the field should be without standing water for two or three days until the soil is somewhat dry to encourage tillering. Sufficient dryness is indicated when the soil begins to crack or when a man stepping on the soil makes a foot imprint a centimeter or two deep. Thereafter, it should be alternately wet and somewhat dry until the plants reach the flowering stage. Water should be applied to a depth of about three centimeters and then the field should be allowed to dry without opening the field drain. Actually during this period, the field can go without receiving water for many days without much reduction of rice yield.

Just before the rice begins to flower, water must be applied to a depth of about three centimeters and kept at that depth for at least five or six days until all plants in the field have flowered. Some farmers then resume the alternating wet and somewhat dry water schedule. Others keep water standing on the field until the rice heads begin to bend, about twenty days after the start of flowering. While the heads are bending the field needs very little water. More than just a little water would weaken the base of rice stalks and increase chances of lodging. Twenty days after the heads begin bending, the rice is ready for harvest and any water on the field is drained off.

Nearly all farmers harvest their first rice crop within a two-week period in middle or late June. Although this fact indicates that rice plants everywhere in the Nan-hung system area ripen to maturity at about the same time, the plants everywhere do not pass through the various stages of their growth cycle at one time. Variations in the rates of growth of rice plants on various fields occur with variations in planting and transplanting dates, rice varieties, use of fertilizers and insecticides, and water availability. As a result, applications of water are not needed by all fields on any single day. In fact, with rotation irrigation, seldom do all fields in any three- or four-hectare area require water on the same day.

The Water-Control Methods

Nan-hung farmers now recognize three broad types of water distribution methods: continuous irrigation, rotation irrigation, and emergency irrigation. Continuous irrigation occurs when the canal carries sufficient water for a continuous flow in all ditches and farmers can take water as needed and wanted simply by opening their water mouths. Rotation irrigation begins when the water level falls below that sufficient for farmers to obtain water at times of their choice, although the needs of all fields can be met. When the water is too low to satisfy the needs of all fields, a state of emergency irrigation is declared. During the first rice-crop seasons of the years between 1955 and 1966, the Neng-kao Irrigation Association promulgated an average of two rotation or emergency irrigation schedules in March and one each in February, April, and May. January had an average of one such schedule every three years whereas June had none. Farmers expect continuous irrigation in June and during the entire second rice-crop season. The water-control methods utilized today with continuous irrigation are similar to those applied during periods of plentiful water prior to 1955. However, the methods employed with the lower levels of water availability have changed.

Rotation irrigation designates those methods of water control which permit periodic applications of water in amounts generally sufficient for the needs of fields. When continuous irrigation methods fail to satisfy all fields, the first to feel a water shortage are fields along ditches with the lowest headings on the canal and the lowest fields on long ditches. It then becomes necessary to control the flow of water from the canal so that only some ditches carry water at one time. As such, a rotation schedule must be compiled giving water to certain ditches at specified times. In addition, fields should be grouped into a few named areas, each encompassing all fields served by ditches taking water from several consecutive headings on the canal, so that the schedule may describe precisely the areas and the time periods during which they are to receive water. The Nan-hung system now has six of these rotation areas containing from 50 to 140 hectares of irrigable land. Although fields in a rotation area or along any single ditch do not all need water during a single time period, the restrictions of the schedule force some farmers to wait for water so that more fields need water than would be the case with continuous irrigation. Since few ditches are large enough to allow many field mouths to receive water at one time and since not all mouths can be sized so that each field receives only its needs of water during the entire period in which water flows in the ditch, the rotation areas are further divided into team areas. These then receive water by a schedule which divides the water designated for their rotation area. Figure 11.7 depicts the boundaries of rotation areas and the locations of team areas.

Today, any farmer realizing a water deficiency with continuous irrigation mentions the problem to his team chairman. The team chairman surveys the area. If the shortage seems widespread, he consults the chairman of his rotation group. If the group chairman feels it desirable to institute rotation irrigation, he calls a meeting of all group chairmen. These chairmen decide whether the situation demands rotation irrigation. A water shortage in a small

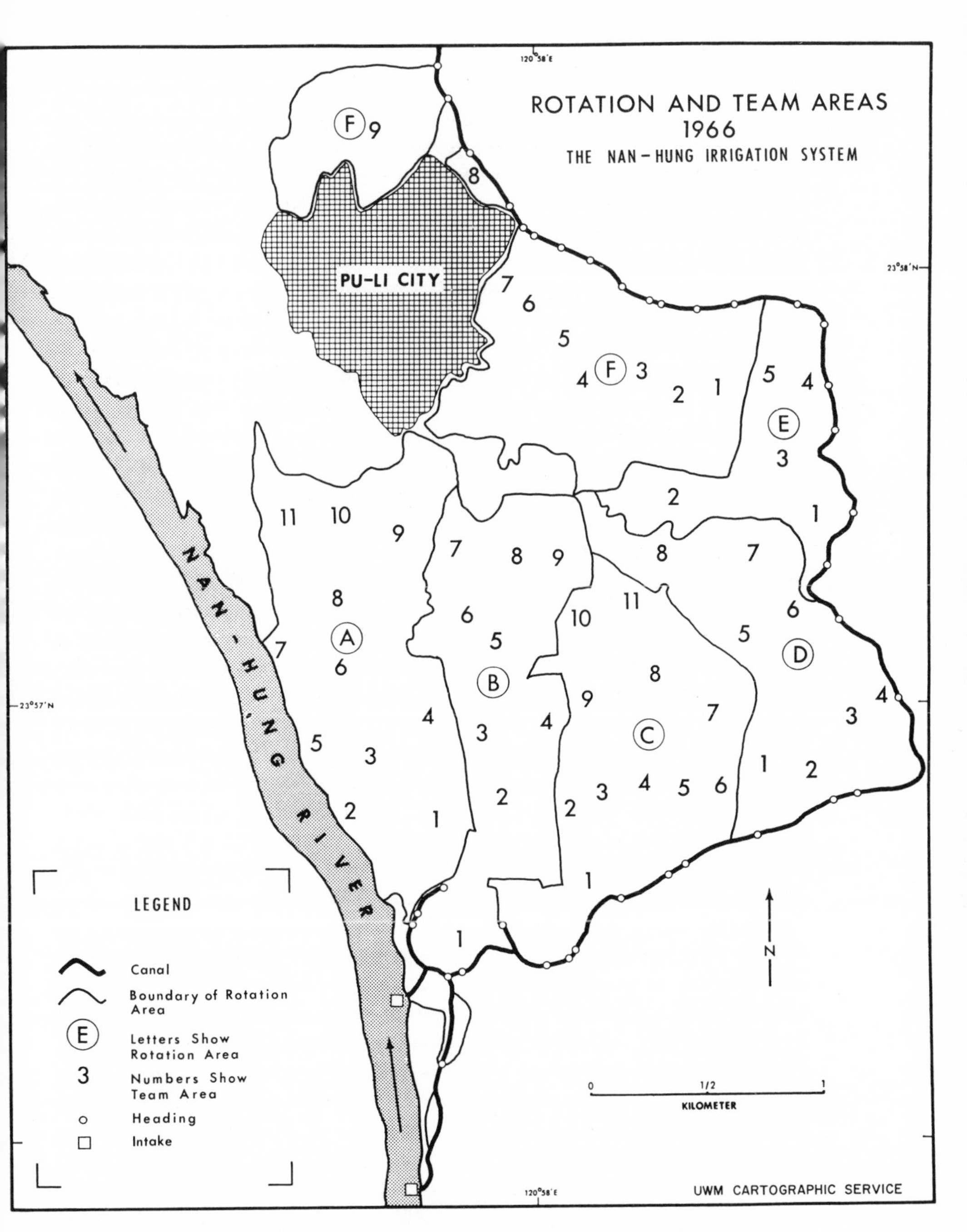

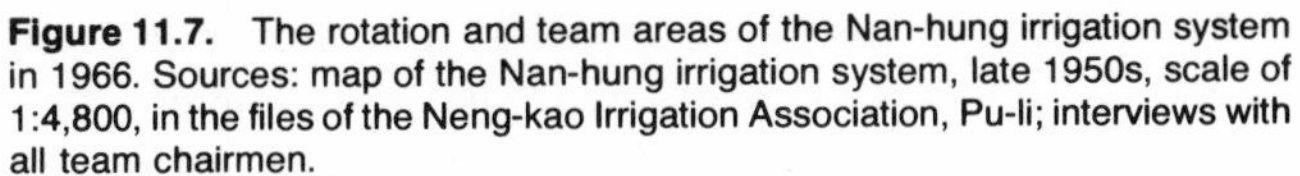

Figure 11.7. The rotation and team areas of the Nan-hung irrigation system in 1966. Sources: map of the Nan-hung irrigation system, late 1950s, scale of 1:4,800, in the files of the Neng-kao Irrigation Association, Pu-li; interviews with all team chairmen.

area may be alleviated by asking the canal traveler to reduce the sizes of the openings of some higher gated headings. If their decision favors rotation irrigation, they create a rotation irrigation schedule which names the rotation areas, indicates the number of hours of all the water in the canal which each area is to receive, and states the precise time and day when the schedule takes effect. The schedule is repeated until rainfall alleviates the situation. When the last area named on the schedule has received water, the schedule begins again with water going to the first area named.

In compiling the schedule, the chairmen have considerable latitude of choice. The total number of hours on the schedule, called the rotation period, may encompass any number of days: the rotation areas may receive water singly or as units of two or more areas and in any order; and any number of hours of water may be assigned to any area. Since each chairman has a veto power, however, the effort to compile a schedule may involve lengthy discussions. A number of factors guide the discussions. (1) The minimum water needs of all fields should be satisfied. As Class *B* land farmers pay a smaller irrigation fee, however, a greater portion of the water in excess of minimum needs should be distributed to Class *A* land then to Class *B* land. (2) Rotation periods should generally extend for forty-eight hours because some fields in the Nan-hung area have permeable soils requiring frequent water applications, especially prior to the second weeding of rice fields. On the other hand, if the needs of fields are fairly great and the canal contains relatively little water, a seventy-two-hour rotation period may be desirable. With the shorter rotation period, the time period during which each rotation area receives water may be too brief to satisfy the needs of all fields. With the longer period, the fields in the last area named on the schedule may suffer because they wait a day longer for their first water application. However, the chairmen can select as the last area one which has less permeable soils. (3) The names of all rotation areas should appear on the schedule with allocations of water time. (4) Some rotation areas should be grouped into a larger unit if the volume of water in the canal exceeds the carrying capacity of ditches in a roation area. (5) The time required for the head of water to flow from the headings of a higher rotation area to those of a lower area must be carefully estimated and not included in the water time allotted to the lower area. Similarly, if a lower rotation area is to receive water before a higher area, the time during which the water tail flows in the canal must be charged to the lower area.

Determination of the water-time allotment of each rotation area often involves considerable give and take on the part of the chairmen. Discussions start from a base of percent figures which indicate the portion of the entire water time in a rotation period which should be allotted to each rotation area under conditions of similar water shortage everywhere. These figures are derived from data on the areas of Class *A* and Class *B* land. The number of hectares of Class *A* land in each rotation area is added to three-fourths the number of Class *B* land hectares in the same area. These total are then divided by the total area of irrigable land in the system during the first rice-crop season to obtain the base percent figures. To or from these figures, the chairmen add or subtract points according to their knowledge of the water needs of fields in various areas. Areas which have more permeable soils, larger portions of their land in rice, or more young rice may get extra percentage points at the expense of other areas. The completed rotation schedule is distributed by each group chairman to his team chairmen on the day before the schedule takes effect. The chairmen then decide how to distribute water within their respective rotation areas. Since all fields needing water in a rotation area cannot take water at the same time, each team area receives a water-time allotment determined in the same manner as that of the rotation area.

With rotation irrigation, the chairman of each rotation area except area *F*, as shown in Figure 11.7, normally opens all ditch headings in his area during its entire watering period because of the relatively small carrying capacity of most ditches. Area *F* has so many headings that those with gates are opened for a part of its watering period and those without for the remainder of the period. Since each heading on the canal and each important bifurcation of a ditch is sized to receive water in approximate proportion to the area of fields it serves, each ditch receives its share of water. Then the team areas on a ditch receive their water-time allotments one after another, usually in order of their elevation on a ditch, the highest first. However, the order may be altered to suit special conditions.

When the water level in the canal is unusually low during rotation irrigation, and the chairmen realize that all farmers who want water will not be able to get it within one watering period, they have a choice between two methods of distributing water among ditches within a rotation area. They may continue to allow water to flow in all ditches or they may direct it into one or two ditches at a time. In the former instance, each team area will receive its normal water allotment so that all teams will obtain some water during the watering period. Not all farmers who want water within a team area, however, will be able to obtain it. Some will wait until a future rotation period, two or more days later. This method is utilized if fields needing water quickly are scattered throughout the rotation area. On the other hand, if the immediate water needs are concentrated in area, water may be sent into one or two ditches at a time. The group chairman then divides his rotation area into two parts of approximately equal size so that all farmers in one part receive water during one watering period. The areas with long ditches, such as areas *A* and *B* in Figure 11.7., tend to be divided into an upper and a lower part with the upper encompassing the higher teams along the ditches. The other areas are usually divided into parts encompassing fields served by several ditch headings. In both cases, that part with the greater need receives water during the first cycle of rotation irrigation.

The water distribution method within a team area is determined by the team chairman in consultation with members of the team. Three basic methods have evolved. The field-covering method allows a field to receive water until about three centimeters stands on its surface. The time-sharing method grants each rice-field property a certain number of minutes of water. The simultaneous-watering method has all field mouths open at one time.[18] The three methods have features in common. Any farmer desiring water must inform other team members of his wishes and participate in water-distribution activities or send a representative. While the team area receives water, one or two members patrol the ditch above to obviate stealing and another patrols the canal above the headings which supply their rotation area. The farmer working the canal stops any leaks of water from the water-control gate which prevents a flow to lower headings and from the higher headings which are not to receive water. Although the canal traveler is responsible for opening and closing these gates, some gates fit loosely within their frames and occasionally a farmer will try to cause leakage by placing a few stones at the bottom of the gate so it will not close completely. The farmer on patrol also observes whether the canal traveler opens and closes gates on time. If he does not, the fact is reported to the chief of the Pu-li city station of the Neng-kao Irrigation Association.

The most common water-control method utilized within team areas during rotation irrigation is the field-covering method. With this method, farmers direct the water in the ditch into as few as one or as many as five or six field water mouths at a time, with each owner or his representative opening his own mouths. The number of mouths opened is greater if the flow of water in the ditch is strong or if the rice plants are small so that those near the mouths would be damaged by a strong flow. When the fields obtaining water from a field mouth have enough, the next farmer to receive water usually closes the mouth. What constitutes enough for any field is determined by the team members who happen to be standing in the vicinity, except the field owner. Generally the mouth remains open until they feel sufficient water is on the fields it serves to permit three centimeters of standing water everywhere. This means, however, that the mouth must be closed before the far end of the most distant field it serves is even wet. Near the field mouth, then, the water should be four to six centimeters deep and it should have a head strong enough to flow to the far end of the fields. More than three centimeters of water may be granted to fields with permeable soils. The opening and closing of field mouths continues until the team watering period ends and a farmer from the next team scheduled for water closes those mouths which receive water last. If all fields in a team area do not obtain water in its allotted time, water distribution begins with the next rotation cycle two or three days later, where it

18. These methods have no special names in the Taiwanese language. In fact, some farmers who use the field-covering method are unaware of the other two methods.

previously ended. Rarely must a field wait for water until the third rotation cycle. Since the team chairman can predict fairly accurately on the basis of the water level in the canal whether some fields will be required to wait, he may grant water first to those fields having the greatest need. Otherwise, the highest fields along the ditch receive the first flow of water.

The time-sharing method of water control is practiced by five teams in rotation area A and one in area C (Fig. 11.7). Each farmer is alloted a percent of the total water time corresponding to the percent of the irrigable land in the team area occupied by his fields. The water mouths of fields operated by one person are opened at one time and water is directed into them for the predetermined number of minutes. The operator opens his own field mouths and the next person to obtain water closes them. As the fields may not receive enough water with this method if relatively little water flows in the ditch, the farmers must arrange to share their water allotments when necessary. For example, if one farmer has a half hectare of rice and two of his friends each have a quarter hectare, the one may accept half or all the water designated for the two, while the two either just wet their fields with the remaining half of their shares or receive no water at all. Then during the next rotation cycle, the two take half or all the water designated for the one. The disadvantage of this water-control method is that it requires the farmers to plan for sharing. The method persists, however, because one or more farmers consistently claim that their fields do not receive a fair share of the water whenever the field-covering method is tried. The time-sharing method thus has a reputation for being the method used by teams in which some farmers are uncooperative.

In most team areas in which each field mouth has a check in the ditch, farmers use the simultaneous-watering method of water control. These teams includes six in rotation area *D*, two in *B*, and one each in *C*, *E*, and *F* (Fig. 11.7). When water arrives in one of these team areas, all field mouths are open and sized so that each takes a share of water in approximate proportion to the share of irrigable fields it serves in the team area. The entire ditch within the team bounds carries water throughout the watering period and the ditch is blocked just below the lowest mouth to prevent loss of water to the next team. As it is impossible to size all field mouths with complete accuracy, they are closed one by one as the fields obtain enough water. Since this method may fail to provide water to the far ends of the most distant fields served by a mouth, the chairman often divides the team area into two parts when relatively little water flows in the ditch. Each part then receives water during alternate cycles of the rotation periods. Otherwise, farmers make water-sharing arrangements among themselves.

An analysis of the distribution of teams which utilize these three methods of water control during rotation irrigation reveals no patterns of relationships with factors such as the carrying capacity or length of the ditch, slope of the land, and distance to the ditch heading or canal intake. The six teams which practice the time-sharing method are all on Class *A* land but some are high on a ditch whereas others are low. In addition, four have relatively permeable soils. Nevertheless, they do not constitute a majority of the teams on Class *A* land or on the more permeable soils. Although the eleven teams using the simultaneous-watering method all have checks in their ditches at field water mouths, other teams also have checks. Only two of the eleven are on permeable soils, but the remainder comprise only a small part of the area of less permeable soils. Therefore, it appears that these water-control methods are randomly distributed over the Nan-hung system area except that the time-sharing teams are all on Class *A* land and receive water from the upper portion of the canal.[19]

When rotation irrigation methods fail to satisfy the water needs of any team area, the team chairman informs his rotation group chairman. If the shortage cannot be alleviated by adjustments of the water allotments within his own rotation area, the group chairman calls a meeting of all group chairmen. They may then revise the rotation irrigation schedule or declare a state of emergency irrigation. A decision for the latter requires an emergency irrigation water-flow schedule. In recent years, such a schedule has been promulgated about once a year. The emergency schedule is like a rotation schedule except that it almost always has a rotation period of seventy-two hours and rotation areas receive water individually beginning with area *A* and ending with area *F* (Fig. 11.7.). As with the rotation schedule, it is repeated until a substantial rainfall allows use of continuous or rotation irrigation water-control methods again.

The emergency irrigation water-control methods within a rotation area vary with the desires of the group chairman and his team chairmen and with the location of fields having the greatest water needs. If the fields needing water most are localized along one or two ditches, all water destined for the rotation area during the first cycle of the schedule may go into those ditch headings and the remaining headings wait until the second, the third, or even the fourth cycle. In this situation, team boundaries are usually ignored and field mouths along a ditch obtain water three or four at a time starting with the highest on the ditch. On the other hand, if the driest fields are scattered, each heading in the rotation area may receive a portion of the water time of each cycle so that most or all dry fields realize a watering during the first cycle. With widely scattered dry fields, the group chairman divides the water time of each ditch among all teams on the ditch. The teams then receive water in succession, giving to their neediest fields during the first cycle and to others in later cycles. With the driest fields located in a number of team areas, the

19. The time-sharing teams are those numbered 1,5,6,8, and 11 in rotation area *A* and number 6 in area *C* (Fig. 11.7). Those using the simultaneous-watering method are numbered 5 and 6 in area *B*, 9 in area *C*, 1,2,5,6,7, and 8 in area *D*, 2 in area *E*, and 6 in area *F*.

group chairman may grant water first to needy fields without a time allotment and after their soils are wet begin a team-by-team watering schedule.

Whether the water is distributed by team areas or by ditch areas during emergency irrigation, each field obtains enough only to wet its soil. Several field mouths are open at a time and the ditch is blocked just below the lowest open mouth. When the farmers standing in the vicinity feel that the fields have sufficient water, they direct that the mouths be closed. If a mouth is closed too early, the far end of the most distant field served by the mouth will remain dry. A late closing may deny water to another needy field during the cycle. Therefore, they decide carefully. Since water may not stand on a field, drains must be open so that excess water flows onto lower adjacent fields. Although in some team areas farmers open and close their own field mouths, generally these acts are performed by persons other than the operator and his immediate neighbors. While a team or ditch area receives water, several representatives from the area patrol the ditch and the canal above.

With both rotation and emergency irrigation, the opinions and decisions of group and team chairmen concerning water-control methods are respected. They have power to distribute water within their own areas as they see fit. To make the proper decisions, however, the group chairmen must be aware of water conditions everywhere within the Nan-hung system and each team chairman must be aware of conditions in his rotation area. In addition, they seek the advice and consent of members of their teams and groups. If farmers feel that their team or group chairman is unfair, they may complain to the group chairman or the Irrigation Association. The situation is then studied and resolved. Rarely do officials of the Irrigation Association hear such complaints. Similarly, very few chairmen complain about uncooperative farmers.

Conclusions

Farmers in the Nan-hung irrigation system area changed their water-control methods between 1955 and 1959 primarily to effect a more equitable pattern of water availability which would allow more wet rice cultivation during the first rice-crop season. This objective was accomplished with considerable success. In addition, they reaped other benefits from the new methods and from the improved water gates and concrete lining on the canal. None of the important advances would have been as significant as they are, however, without the highly effective organization of the irrigation cooperative and the remarkable cooperation of a great majority of the farmers.

During the late 1930s and the 1940s farmers who operated land now in rotation areas *A, B,* and *C* could have planted all their irrigable land to rice during the first crop season. However, as most farmers had no unirrigable land, they devoted an average of about 80 percent of their land to rice and divided the remainder between sugarcane for cash and sweet potatoes, vegetables, and other minor crops. The same basic crop pattern exists today in these areas except that rice has lost several percentage points of land to sugarcane because of its improved price. Prior to 1955, water shortages limited rice production in rotation areas *D, E,* and *F* to averages of about 65, 40, and 40 percent of the irrigable land, respectively, during the first crop season. Much land was left to rest while some was planted to sugarcane, sweet potatoes, peanuts, and vegetables. Water shortages also reduced rice yields in some years. Today, on the other hand, the crop pattern in rotation areas *D, E,* and *F* is similar in all respects to the patterns in areas *A, B*, and *C*.[20] The new water-control methods and the improvements on the canal have therefore eliminated the role of field location as a factor affecting the availability of water.

Most farmers also praise the labor-saving features of the new water-control methods. Farmers in all working teams of rotation areas *D, E,* and *F* now devote much less effort to the process of obtaining water, especially when considered in relation to the number of hectares irrigated or the rice produced. The members of thirteen teams operating most lower fields in areas *A, B,* and *C* likewise expend less labor today. Of the remaining sixteen teams, the members of ten scattered through areas *A, B,* and *C* experienced little or no change in labor input. The members of eight, however, find it more difficult to obtain water. In the four of those eight which operate land near the headings of ditches, farmers complain about the inconvenience of the new methods. In the past they could generally take water at their convenience. Now, because of rotation and emergency schedules, they must plan their water applications so that their fields will not face a shortage on days when water is unavailable. The other four team areas are among the five lowest in rotation area *C*. Three take water from a series of connected ditches, one of which, in its upper portions, has always served as a drainage ditch for fields now encompassed by teams 1 and 2 in rotation area *C* and team 2 in area *B*. Before the upper portion of the canal was lined with concrete, leakage from the canal was such that team area number 1 received too much water. Therefore, the drainage ditch carried water continually and farmers could take water as they pleased from the irrigation ditches it supplied. Today this drainage ditch carries much less water so that farmers in the lower portions of rotation area *C* have complaints similar to those of farmers whose land is near the ditch headings. Members of the fourth team low in area *C* feel that they now expend more effort in obtaining water because the heading of the ditch from which

20. Data on water availability and percent of irrigable land devoted to rice in each team area were obtained from interviews with each team chairman. The percent figures provided above are first rice-crop season averages for the rotation areas. Although the percent figures for team areas varied somewhat within rotation areas, the variations of the post-1955 figures were not caused by variations in water availability but rather in soil drainage qualities and farmer preferences as related largely to market prices. The figures for rotation area *F* above exclude team areas 8 and 9, which have traditionally emphasized sugarcane and vegetable production. Only about 10 percent of their land was put in rice during the late 1930s and 1940s and 40 percent since 1955.

they take water is, according to their calculations, too small to satisfy the fields it serves.

Although the new water-control methods have reduced the labor needed to put water onto a great majority of the fields in the Nan-hung system area, farmers who operate fields taking water from the lower ditch headings on the canal or from the lower portions of long ditches must still generally work more for their water than do other farmers. With rotation and emergency irrigation, however, water comes to team areas by a schedule. Since getting water to the field mouths now requires little or no labor, the minor variations in labor input from place to place are related almost entirely to the length of the ditch and canal which must be patrolled.[21]

Several other benefits have accrued as a result of the changed water-control methods and structural improvements on the canal. Many farmers feel that the innovations have promoted better relations among the farmers of various localities within the system. Water problems no longer cause numerous irritations between farmers with land low on a ditch or in rotation areas *D*, *E*, and *F* and farmers higher in the system. Class *B* land farmers need not beg for water. As water thievery has been reduced, the uncomfortable confrontations between farmers on patrol and culprits are fewer. In addition, emergency irrigation is now less common. The new water-control gates and concrete lining on the canal have also lessened the chance of flooding high in rotation areas *A*, *B*, and *C*. Whereas in the past, occasional heavy rains caused flood damage, now water flows more rapidly in the canal and so is less likely to overflow the banks. The concrete lining also reduces canal leakage so that the soils of fields in team area 1 of rotation area *C* now may dry out and dry field crops may be more easily raised if the farmers so desire. Finally, the value of land, and especially Class *B* land, has increased. During the late 1930s, Class *A* land was generally worth about one-third to one-fourth more than Class *B* land, all other factors being equal. Today the two classes are of approximately equal value.

Only those farmers who formerly obtained sufficient but not excessive amounts of water with ease express any dissatisfaction with the innovations. Besides the complaints mentioned above, their major lament concerns the fact that they, like all other owners of irrigable land in the Nan-hung system area, are assessed a special irrigation fee to pay for part of the cost of the structural improvements on the canal. Each owner of Class *A* land pays a fee seven-tenths the amount paid per hectare by owners of Class *B* land. They feel that this fee is probably fair for those Class *A* land farmers at the end of long

21. This generalization does not include the labor needed to maintain ditches. In theory, team members are responsible for ditch maintenance in their respective team areas. In fact, farmers who take water from the lower portions of ditches generally do more than their share of maintenance labor as they gain most from the lower seepage rate and more rapid flow of water in a well-maintained ditch.

ditches, who have benefited significantly from the new water-control methods, but not for them. In addition, a few farmers complain that they must now, as members of a team, cooperate with farmers not of their choice.

Notwithstanding these complaints, an overwelming majority of the farmers derived significant benefit from the innovations, in part because they have been able to achieve an extremely high level of cooperation in irrigation for a number of reasons. (1) Cooperation is not new to the farmers. Nearly all farmers had worked with other farmers in order to obtain water for their fields prior to 1955. Friends and neighbors have long shared labor at times of field preparation, rice transplanting, and harvest. (2) The Neng-kao Irrigation Association is an exceedingly effective instrument of cooperation. The fact that each member of the association is also a member of a working team of only fifteen or twenty persons helps him feel that his opinions are valued. He has some influence over the way water is distributed in his team area. He also sees many of his complaints about a water shortage produce action when the water-control methods are revised within his team area, his rotation area, or the entire Nan-hung system area in order to help meet his needs. It is a reflection of the high quality of the leadership in the Neng-kao Irrigation Association that his opinions are generally respected. The head of the water management division of the association headquarters in Pu-li city, the chief of the Pu-li city station, and the canal travelers are actively interested in his problems. His team and group chairmen participate in the water-distribution activities of rotation and emergency irrigation not only by drawing up the water-flow schedules but also by being present whenever possible in the field areas as they receive water, to observe or lead and to mediate disputes. Rare is the chairman who does not take his responsibilities seriously. (3) The amount of water available from the canal is almost always sufficient to satisfy the minimum needs of all fields. In addition, most farmers produce well above the subsistence level. Therefore, with proper planning, a farmer, a team, or a group has some latitude of operation as regards the distribution of water.

12

Water Management by Administrative Procedures in an Indian Irrigation System

Richard B. Reidinger
THE WORLD BANK

Irrigation has historically been especially important in India, with its monsoon climate. Along with the growing population pressure, the advent of the so-called green revolution has enhanced the importance of irrigation and heightened the need to use limited water resources with maximum effectiveness.[1] The system of rationing irrigation water directly affects the efficiency with which farmers can use water, and rationing used in large-scale canal irrigation is especially important because canal irrigation comprises a major component of the total Indian irrigation system.[2]

Previously published as "Institutional Rationing of Canal Water in Northern India: Conflict between Traditional Patterns and Modern Needs," *Economic Development and Cultural Change* 23: 79–104.

This paper presents some of the results from the author's Ph.D. thesis, "Canal Irrigation and Institutions in North India: Microstudy and Evaluation" (Durham, N.C.: Duke University, 1971), directed by Joseph J. Spengler and based on research in India under a Fulbright-Hays fellowship. The paper was written under a postdoctoral appointment in agricultural economics and water science and engineering at the University of California, Davis. Varden Fuller, Charles Moore, and Eric Gustafson provided helpful comments on earlier drafts of this paper. Responsibility for its shortcomings, however, lies with the author. The help and cooperation of the many active irrigation engineers interviewed in obtaining the largely unpublished data on the operation of the canal system are sincerely acknowledged, as are the contributions of the 168 farmers interviewed to obtain the necessary farm data.

1. The Fourth Five-Year Plan clearly recognizes this in stating that "the possibilities offered by the new seed varieties for increasing yields of cereal crops and intensifying cultivation are contingent on irrigation" (Planning Commission, *Fourth Five-Year Plan* [New Delhi: Government of India, 1969], p. 245).

2. At present, large-scale canals serve roughly 37 percent of the irrigated area in India. With 37.6 million hectares under irrigation in 1968, India has the largest amount of irrigated land in the world (with the possible exception of China) and more than double the irrigated area of the United States. The percentage of cultivated land receiving irrigation in India was roughly three times

This paper examines the rationing system used with the Bhakra canals, one of the major new canal systems in north India, and focuses on a study area in Hissar District.[3] Rationing occurs at three principal levels: first, at higher levels, with the allocation of water available in the reservoir; second, at the middle level, with the rotation among channels of the preference to receive water; and third, at the farm level, with the rotation of turns among farmers (*warabandi*) served by the same watercourse,[4] which is the ditch from the main channel to the farmers' fields.[5] Insitutional or administrative decision rules and schedules rather than market forces govern canal-water rationing.

The interaction of rationing at the various levels introduces a substantial degree of uncertainty in the farmers' water supply, in addition to a relative scarcity of water. This analysis will focus in detail on rationing and interaction at the two lower levels—the rotation among channels and the *warabandi*—that most directly affect farm water supply. Primarily consideration will be given to the winter season (*rabi*) when rainfall is negligible and irrigation most crucial. The effects of the rationing system may partially explain the often disappointing performance of many canal irrigation projects with respect to yields, profits, and utilization of irrigation potential.[6] These effects may also help explain the relatively sudden increase in private tube wells even in

higher than that in the United States. Although tube-well irrigation is currently receiving strong emphasis, the future importance of large-scale canal irrigation seems assured. During the first three plans, canal-irrigation potential created or anticipated from projects under construction increased by 17.2 million hectares, compared to 7.3 million hectares for tube-well irrigation, the second largest category (see Directorate of Economics and Statistics, Ministry of Food, Agriculture, Community Development, and Cooperation, *India Agriculture in Brief,* 9th ed. [New Delhi: Government of India, 1968], p. 47; and K. K. Framji and I. K. Mahajan, *Irrigation and Drainage in the World: A Global Review,* 2d ed. [New Delhi: International Commission on Irrigation and Drainage, 1969], p. 420).

3. Hissar District is the largest district in the state of Haryana. The case-study area is approximately 100 miles northwest of Delhi. Rainfall is about 16 inches yearly, and the area depends heavily on irrigation, especially in the dry winter season. Bhakra canals come several hundred miles to serve the area. These canals are fed from the 8-million-acre-foot reservoir of the Bhakra Dam. For comparison, Shasta Dam, the keystone of California's Central Valley Project, has only about 4.5 million acre-feet of storage.

4. In effect, the outlets from the main canals to the watercourses serving the cultivators' fields also ration the available water supply. These outlets have no gates for regulation. By their design, they automatically draw relatively less water from the main channel as its water level decreases, thus passing downstream a relatively larger proportion of the available water supply. This allows the tail reaches of the system to receive their share of the water even during low levels of flow in the main canal. In doing this, however, the outlets magnify the effect of canal-supply fluctuations on the water supply received by the cultivators.

5. The channels of the system vary in size from the largest mainline canals to the smaller branches, subbranches, distributaries, and minors of very small capacity. A watercourse serving many individual farms may receive its water directly from any of these except the largest.

6. As Nasim Ansari states: "Even in a region where irrigation has existed for long, agricultural development may remain quite restricted, hardly going beyond increased intensity of cropping and some changes in cropping patterns" (*The Economies of Irrigation Rates: A Study in Punjab and Uttar Pradesh* [New Delhi: Asia Publishing House, 1968], p. 170).

canal-irrigated areas, if cultivators find that their canal-water supply severely limits their ability to participate in the green revolution.[7]

In general, the rationing system will be evaluated in terms of its capability to supply water according to the needs of the individual cultivator, and in particular, the predictability, certainty, and controllability of the timing and quantity of water supply. Some data will be presented to indicate the canal-flow characteristics resulting from the operation of the system, but the evaluation is primarily qualitative due to the lack of necessary field date. For example, water supplied to individual farmers is not metered volumetrically. Nevertheless, a qualitative evaluation should help identify those elements of the system that most directly interfere with the predictability and usefulness of the cultivator's water supply, and thus those aspects most in need of alteration and additional research. Finally, several possible alterations will be suggested to improve predictability, certainty, and control of the cultivator's water supply with minimal change or disruption of the system.

Predictability, certainty, and control are relevant criteria for evaluating the cultivator's water supply. "There are important, fundamental agronomic reasons for delivering known amounts of water to fields. . . . "[8] The timing and quantity of water applied largely determine how much of that water will be captured for the consumptive use of the crop, rather than wasted. Control over timing and quantity of his water supply would enable the farmer to equate marginal revenues and costs from water applied, and to eliminate one source of risk in planning his farm operations. Even without water control, a highly predicatable and certain water supply would permit him to approach profit maximization with precise farm planning, and actual outcomes should closely match expectations. However, as the certainty of his water supply decreases, the farmer must reduce the level of all inputs, and thus output and profits, in order to decrease the chance of loss. Fertilizer, for example, is relatively costly, and the return from its use depends at least partly on the timing and quantity of water received by the crop. To ensure that the marginal returns from fertilizer at least equal its marginal cost, the farmer with an uncertain

7. A recent government report on the high-yielding varieties states that "a relatively larger proportion of plots under canals reported inadequate irrigation compared to plots under other sources," especially private tube wells (Programme Evaluation Organisation, Planning Commission, *Evaluation Study of the High Yielding Varieties Programme: Report for Rabi 1968–69—Wheat, Paddy, and Jowar* [New Delhi: Government of India, November 1969]; p. xi). The response has been a surge of tube-well construction since the high-yielding varieties were first introduced. Since 1965–66 the number of tube wells has tripled, though the total number (400,000) is still relatively small (F. A. Ahmed, "New Dimensions of the Green Revolution," *India News* [Washington, D.C., Indian Embassy], January 22, 1971, p. 1).

8. Chester Evans et al., *Need of and Plan for Research on Water Use and Soil Management toward Meeting India's Food Shortages,* report issued by the International Agricultural Development Service, U.S. Department of Agriculture; Agency for International Development, U.S. Department of State; and the Indian Council of Agricultural Research (New Delhi: Government of India, 1967), p. 20.

water supply should apply fertilizer at a relatively low level. Such risk adjustments sacrifice per acre output and profit for a higher probability of positive net returns.

A similar risk adjustment can also occur in water management, resulting in a water-application rate well below the economically optimum level in order to assure that marginal returns from water applied at least equal marginal cost, and certainly remain nonnegative.[9] Agronomic, climatic, and soil factors can, however, severely limit the possible reduction in water-application rates. The farmer may therefore diversify his crop pattern as an additional risk adjustment and include drought-resistant but low-valued crops, such as gram.[10] Wheat is generally more profitable than gram. Gram, however, is highly drought-resistant, and is grown extensively in the area as a hedge against the uncertainty of the canal-water supply.[11] If available water exceeds

9. Because of low water cost, the profit-maximizing point will likely occur very close to the point where marginal product becomes negative on the farmer's production function for water. The marginal cost of water is primarily the labor for application, because the water levy is on a per acre rather than volumetric basis; that is, the farmer pays according to the number of acres he irrigates rather than the amount of water he uses. While the marginal cost of water is very low, it nevertheless has a high opportunity cost because of its scarcity. The farmer has insufficient water to irrigate all his land and cannot afford to waste the water, and he certainly does not want to apply so much that marginal product becomes negative. Figure 12.1 shows a production function for water and the reduction in output and input due to the risk of an uncertain water supply. It indicates the economically feasible region—stage 2—between the point of maximum average (AP_{max}) product and the point where marginal product becomes zero ($MP = 0$). Assume prices and returns such that marginal returns equal marginal costs at W_1 units of water applied, with a corresponding yield Y_1. To maximize profit, the farmer should plan his farm so that each irrigated acre receives this amount. From past experience, however, he knows that he is unlikely to actually receive W_1 units of water. If he receives less than W_1, marginal returns still exceed marginal cost. But with more than W_1, profits decrease, and he could easily go past $MP = 0$. Therefore, he will probably plan to apply somewhat less than W_1 units of water, perhaps W_2 units with a corresponding yield Y_2. The yield difference, $Y_1 - Y_2$, represents the yield sacrifice in the risk adjustment to avoid losses. Depending on the accuracy of the farmer's estimate of expected water supply W_1, his ability to absorb temporary financial losses, and his degree of risk aversion, the farmer will consistently underirrigate, relative to the economic optimum. Evidence of this phenomenon has occurred in West Pakistan, which has an irrigation system similar to that of northern India. Pakistani farmers typically respond to increased canal-water supplies by extending their cropped acreage rather than increasing the water-application rate. West Pakistan has a problem of increasing soil salinity due to underirrigation, and increasing the rate of water application would be beneficial (see Garth N. Jones and Raymond L. Anderson, *The Problem of Under-Irrigation in West Pakistan: Research and Study Needs* [Fort Collins: Colorado State University, 1971], p. 3).

10. *Cicer arietinum* or chick pea, a legume.

11. Gram is generally considered a dry-land crop, and its continued importance under Bhakra irrigation apparently came as a surprise to some evaluators of the project. They suggested gram's popularity might be due to tradition and the fact that it requires light, sandy soil and little labor and has had fairly good market prices (see Punjab Board of Economic Inquiry, *Report on Cropping Patterns of Hissar District* [Chandigarh: Economic and Statistical Organization, Punjab, 1965], p. 8). This report unfortunately completely neglected the problems of the canal-water supply itself. Tube-well–irrigated farms in the area with a completely controllable and certain water supply grow relatively little gram, as shown by local village record books. Many farms with separate tube-well–and canal-irrigated sections clearly show the contrast.

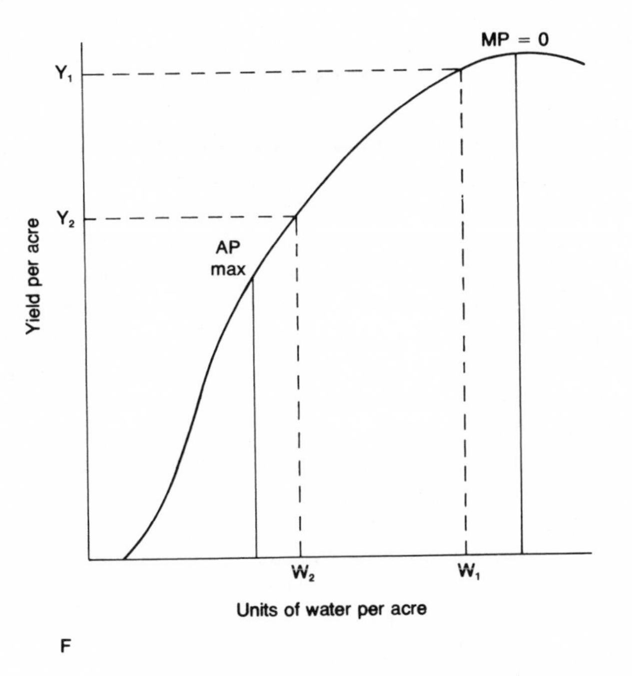

Figure 12.1. Production function for water.

what wheat can profitably use, the surplus can be applied to gram, which responds well to one or two irrigations. But if water is in short supply, the gram can generally produce a small profit without irrigation.[12]

General Characteristics of the System of Canal Irrigation

The system of canal irrigation—including operation, design, and administration—developed largely during British times, when objectives and conditions differed greatly from those of the present.[13] Both new Bhakra and old Western Jumna canals serve the study area, and the close similarity between the two systems indicates the extent to which the traditional system of canal irrigation has been institutionalized, or carried over into the modern system. The principal difference between them is the water source: Bhakra canals receive water from the 8-million-acre-foot reservoir of the modern, multipurpose Bhakra

12. Richard B. Reidinger (*Canal Irrigation and Institutions in North India: Microstudy and Evaluation* [Ph.D. diss., Duke University, 1971], esp. pp. 235–61) compares crop patterns and economics under canal and tube-well irrigation.

13. W. Eric Gustafson and Richard B. Reidinger ("Delivery of Canal Water in North India and West Pakistan," *Economic and Political Weekly*, December 25, 1971, pp. A151–62) examine the historical basis of modern irrigation in north India and West Pakistan.

Dam, whereas Western Jumna canals depend on the unregulated flow of the Jumna River. The same traditional system of irrigation very likely exists in much of northern India (and Western Pakistan), where modern canal irrigation developed largely under the Northern India Canal and Drainage Act of 1873, where the engineering traditions of the Thomason Civil Engineering College at Roorkee remain strong, and where canal irrigation facilities are generally owned and operated by the government.

A prominent characteristic of both systems is the lack of water-supply flexibility and control by both the farmers and the operating engineers. Engineers operating the system have very little descretionary control over the irrigation-water supply. Physical control facilities, such as gates and checks built into the system, are minimal. The engineers are further bound by the operating rules, and they must generally follow a predetermined schedule approved by their superiors. The system is apparently intended to operate more or less automatically, just as during British times it was to provide "equitable distribution [of water] without any interference by the canal establishment—which is an important advantage to the irrigating community."[14] Each individual farmer receives a share of the available water supply according to the number of acres he owns. But he has no control over the timing and quantity of this water—neither when he will receive his share nor how much of his total share he will receive in any given turn. The system thus provides no formal facility through which a farmer can match available water supply to crop water needs, and no way for him to compensate for the largely unpredictable changes in water use by his crop due to climatic variations. Crop damage from soil moisture deficits is generally irreversible and can substantially reduce yields below potential levels.

The rationing system itself is strictly administrative. A fixed water levy is charged per acre irrigated, regardless of the quantity of water applied; the amount paid for water does not reflect its marginal value. A set of institutionalized decision rules governs the sharing of the available water supply at all levels within the canal system, and no farmer can legally obtain more than his share. The quantity of each share varies directly with landholding size. In contrast, electricity generated at the dam is sold on a per unit basis, and more power sold brings in additional revenue for the project. Consideration of the power market will likely affect management decisions to release water according to electricity demand versus irrigation needs.

14. J. Clibborne, *Irrigation Work in India,* rev. G. T. Anthony (Roorkee: Thomason College, 1924), p. 146. This work formed a part of the well-known "Roorkee Treatise on Civil Engineering" of the Thomason College. Lack of discretionary control implied by this quote may have had a benefit, for instance, in reducing corruption, when a major purpose of irrigation was to achieve for a colonial government a maximum amount of famine protection at minimum cost in supervisory manpower and capital. But such a system may not match the needs of modern India and the new agricultural technologies.

The total amount of water available to each farmer is severely limited; each season most farmers can irrigate only about one-third of their land included in the canal service area.[15] In fact, the planned irrigation intensity for the Bhakra system is 62 percent total for winter and summer seasons combined.[16] The engineers interviewed indicated that a factor of 2.4 cusecs[17] per thousand acres served or commanded was used in designing the system.[18] This figure is empirical, and is roughly equivalent to 400 acres per cusec. In contrast, a similar factor used in the western United States averaged 75–100 acres per cusec.[19] Augmenting the shortage of available water is the high seepage from the generally unlined canals.[20] One common estimate of seepage losses is that on an average 47 percent of the water entering at the head of the system is lost before reaching the farmers' fields, with losses of 20 percent in main canals and branches, 6 percent in distributaries, and 21 percent in the watercourses leading to the fields.[21]

15. Reflecting a common justification for the low water availability, John Mellor states that "water will generally produce a higher return if it is spread relatively thinly. As more and more water is given per acre, the yield per acre rises, but returns per acre-inch of water decline. Thus the protective policy of making relatively small quantities of water available per acre ensures the highest returns" (John W. Mellor et al., *Developing Rural India: Plan and Practice* [Ithaca, N.Y.: Cornell University Press, 1968], p. 43). This apparently neglects net returns. And, in any event, if the farmer is a profit maximizer and can choose the amount of land to irrigate with his fixed quantity of water available, he will choose an amount to irrigate such that each acre is at the profit-maximizing level. A smaller quantity of available water simply results in fewer irrigated acres. Furthermore, in any given soil and climatic region, the technical range of water application possible with a particular crop should be relatively limited, if indeed "water requirements for all crops must be about the same, if they are grown on the same soil and for the same growing season" (Colin Clark, *The Economics of Irrigation* [London: Pergamon Press, 1970], p. 4). As indicated above (n. 9), however, lower application rates and thus higher production per unit of water could easily result from water-supply uncertainty, though at a sacrifice of yields and net returns from the irrigation. Similarly, increasing the marginal cost of irrigation by, for instance, volumetric water charges should also reduce the application rate. Volumetric charges without certainty would not, however, aid the farmer in increasing yields or maximizing net returns from irrigation.

16. Ansari, p. 135. Intensity is percentage irrigated out of total irrigable service area, combining both winter and summer seasons.

17. A rate of flow, cubic feet per second. Roughly, 1 cusec flowing for 24 hours provides 2 acre-feet of water; similarly, 1 cusec for 1 hour provides 1 acre-inch.

18. Oral communication, S. P. Malhatra, superintending engineer, Hissar Bhakra Circle, Hissar.

19. Duty is the commonly used term for this design factor, though its use is now out of fashion in the United States. In the western United States, with plentiful water, the duty ranged from 60 to 120 acres per cusec; with moderate water scarcity it ranged from 100 to 200 acres per cusec; and with extreme scarcity, such as in southern California, it sometimes reached 300 acres per cusec (see John A. Widtsoe, *The Principles of Irrigation Practice* [London: Macmillan, 1914], pp. 337–46).

20. Unlined canals do not necessarily result in high seepage losses if they are not in porous soil and if they are kept well puddled and wet. The method of operation for the canals under discussion, however, indicates that the canals will be empty repeatedly during the season. The repeated drying and wetting should produce substantial cracking in the soil lining of the channels.

21. This is the estimate of R. G. Kennedy in the early 1900s, cited in D. V. Jogleker, *Irrigation Research in India* (New Delhi: Central Board of Irrigation and Power, 1965), p. 56.

Rationing at Higher Administrative Levels within the System

At the level of releases into the canal system, the principal elements of the rationing system are the reservoir factor, the capacity factors, and the competition between electricity and irrigation for use of water stored in the reservoir.[22] The reservoir factor governs the general magnitude of releases from the reservoir during a given season. It remains constant for the season and represents an estimate of how much water will be available for irrigation in a given season relative to the maximum possible amount available. If, for example, the water available for a given season is 6.1 million acre-feet in a reservoir with a maximum irrigation storage capacity of 6.8, the reservoir factor is about 0.9. For the cultivator, the reservoir factor can roughly indicate the amount of land to sow in irrigated crops in any given season.

In contrast, the capacity factor changes several times during the season. It indicates how near to full the canal system will operate at any given time relative to designed capacity. It thus determines the general time distribution of the available water during the year. Capacity factors generally follow the same pattern every year and are shown in Table 12.1 for the Bhakra and Western Jumna canal systems.

Especially in the winter, changes in the capacity factor do not match the water need of the crops. The highest winter capacity factors occur in the early part of the season, when crops in seedling stages need less water than later, when they are approaching maturity. In contrast to the capacity factor, water requirements of wheat, for example, increase substantially as the season progresses, as shown in Figure 12.2.

The results for the individual cultivator will likely be a substantial waste of water during the early part of the season, and possibly a shortage later in the season during the grand growth period, when water stress can be most damaging to yields. On the basis of his past experience, the cultivator knows he will receive too much water when he needs it least and too little when he needs it most. The extent of the resulting waste and damage depends largely on how conservative he is in his farm planning at the beginning of the season; or spe-

---

The water lost, and thus production sacrificed, from this seepage is undoubtedly substantial. If, for instance, 50 percent of the water is lost to seepage, then reducing seepage to zero (admittedly impossible) should double the area irrigated from the canal. But perhaps the largest long-run cost results from damage to the land due to salt accumulation from rising groundwater levels. Each year seepage and the resulting high water table renders about 2 percent of land in northern India unfit for cultivation. Damage often appears first as strips of salt-affected land alongside the canals. Width of the damaged strips varies from a few hundred yards to a mile or so (see J. F. Mistry, "Water Conveyance Efficiency, with Particular Reference to Lining Channels" [paper presented at Water Management Symposium, Punjab Agricultural University, Hissar, March 11-13, 1969], p. 2; and Evans, p. 39).

22. The reservoir factor and electricity generation apply only to the Bhakra system, since the Western Jumna system has no storage reservoir or hydroelectric facilities.

Table 12.1.

Capacity factors for Bhakra and Western Jumna canal systems

Month	Bhakra canals	Western Jumna canals
Summer		
April	0.50	0.28
May	0.90	0.51
June	0.90	0.51
July	0.80	0.52
August	0.80	0.52
September	0.90	0.57
Mean summer	0.80	0.51
Winter		
October	0.90	0.45
November	0.90	0.45
December		
1-15	0.90	0.45
15-31	0.50	0.29
January		
1-15	0.50	0.25
15-31	0.50	0.29
February		
1-10	0.50	0.29
11-28	0.75	0.39
March	0.75	0.39
Mean winter	0.72	0.38
Yearly average	0.76	0.45

Source: Office of the Chief Engineer, Irrigation Department, Government of Haryana, Chandigarh.

cifically, the number of irrigated acres he sows with his fixed and small (relative to his landholding size) irrigation supply.

The reservoir and capacity factors are forms of interseasonal and intraseasonal rationing which are scheduled beforehand, and are reasonably consistent during the season and from year to year. In contrast, water releases for generating electricity are determined on a day-to-day basis. The irrigation system can absorb some limited fluctuation in releases due to changes in demand for electricity. But if the magnitude of the fluctuations is greater than the system can absorb, unscheduled releases must either be made into the irrigation system or passed on downstream. In the latter case the water is lost for this irrigation system, although it may be used further down the river. In the former, the water is unexpected and thus has a low value to cultivators. Similarly, engineers should find it difficult to distribute equitably such unscheduled releases.[23]

The extent to which irrigation is treated as a by-product of reservoir

23. Achieving an equitable distribution (supplying each acre with its proportionate share of the available water) was the criterion consistently mentioned by the engineers interviewed and in the quote from Clibborne (see n. 14).

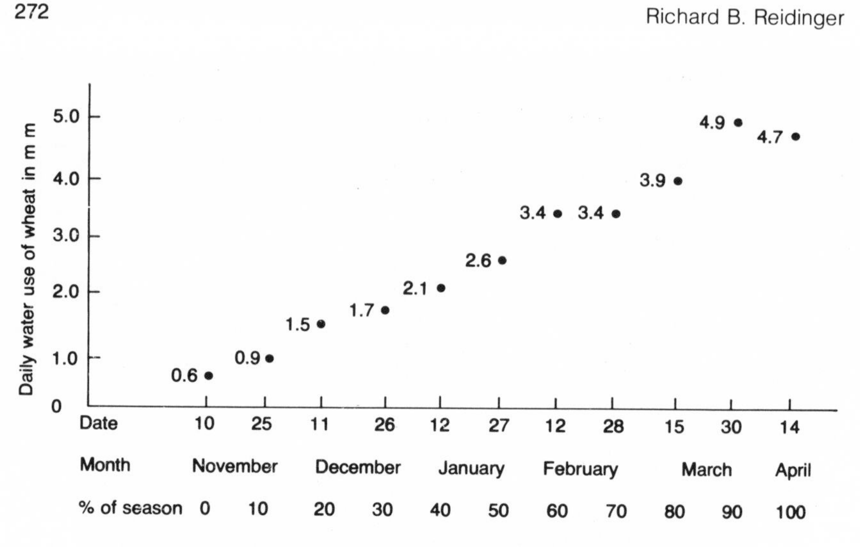

Figure 12.2. Estimated daily water requirements for wheat at various stages of growth in Hissar District. Estimates based on weekly pan evaporation, averaged for 1968-69 and 1969-70; data for previous years on a monthly basis. Period 1968-69 was roughly normal, while 1969-70 had substantial precipitation in late February and March; this accounts for the relatively low water use then. Data from Richard B. Reidinger, "Canal Irrigation and Institutions in North India: Microstudy and Evaluation," Ph.D. dissertation, Duke University, 1971, pp. 128-30.

storage cannot be quantitatively estimated due to lack of data, but it should be substantial for several reasons. Demand for electricity by industrialists for their factories, and by wealthier farmers for their tube wells and machinery, is strong and growing. These groups have much influence and political power. In addition, a fertilizer factory established in the area when the Bhakra Dam electricity was surplus uses much power in electrolysis, although the opportunity cost of that power has increased substantially.[24]

Perhaps most important in deciding the scheduling of releases are the financial returns from power relative to those from irrigation. Every unit of power sold brings in additional revenues for the system. In contrast, additional water supplied for irrigation brings in no additional revenue. Whether the irrigation supply matches the needs of agriculture makes little or no difference in revenues. Revenue from irrigation would be maximized by encouraging farmers to plant as many irrigated acres as possible (through a high initial-capacity factor, for instance) and avoiding crop failure from lack of irrigation.[25] The engineers operating the system are under great pressure to make the system appear as financially successful as possible. In this situation, irrigation understandably may often receive secondary consideration.

24. When operating at capacity, the fertilizer plant at Nangal utilizes about one-third of the firm generating capacity of Bhakra Dam, which is a major source of power for North India.

25. The Revenue Department remits the irrigation levy for farmers whose crops fail.

The water available for irrigation is divided among the user states on a day-to-day basis according to a politically fixed ratio. Within each state, the water is shared among irrigation circles[26] according to acreage served or commanded. Every circle includes several divisions, each of which receives a share of the available water according to the size of its command area. Fatehabad Division, which includes the case-study area, is the largest of four divisions in the Hissar Bhakra Circle, one of two Bhakra circles which serve Haryana state. Below the division level, water is no longer divided according to area served. Rather, the water rationing results from the interaction of two separate but simultaneous rotations which most directly determine the patterns of each farmer's water supply. It is this interaction which virtually guarantees a highly predictable water supply from the individual cultivator's viewpoint.

The Rotation among Channels.

Within Fatehabad Division, the channels are divided into three groups. During the season these groups change in preference status (rotate) to receive the available water, generally after every eight days. Each group down to the smallest distributaries and minors[27] must therefore be operable independently of the other two. Table 12.2 shows the rotation schedule designating the changing preference status for the 1969–70 *rabi* season. Groups A, B, and C have capacities of 596, 582, and 539 cusecs, respectively.[28] Each season the schedule and channel groupings are revised by the operating engineers and approved by the chief engineer.

Starting at the beginning of the schedule in Table 12.2, group B had first, group A second, and group C third preference for receiving the available water from October 14, 1969, to October 21, 1969. During that period, if the available supply were 900 cusecs, for example, B would be full, A with second preference would receive the remaining 318 cusecs, and C would receive nothing. The channels to be opened (completely) in group A are chosen such that available capacity in A equals the 318-cusec flow available to it, and thus the open channels tend to operate at capacity. Only with the available flows at capacity levels (1,717 cusecs) would all three groups operate at capacity. On October 22 the channels rotate in priority, with C having first, B second, and A third preference, and so on through the season.

The schedule also divides the winter season (*rabi*) into three stages: *rabi* sowing, *rabi* growing, and *rabi* maturing. Each stage is several weeks long

26. The circles are the highest administrative units below the office of the chief engineer, irrigation, for the state. They may include service areas of, for instance, 2 million acres.

27. The minors are the smallest controllable channels of the system. A single minor may serve a single village, though not all irrigated villages are served by a single minor. A minor may serve an area of several thousand acres.

28. An equitable distribution of water requires the capacities to be as equal as possible.

Table 12.2.

Hissar Bhakra canal circle, Hissar-Fatehabad Division (Sirsa branch system): grouping of channels under rotation program from 14.10.1969 to 8.4.1970

	Preference turns		
Dates	I	II	III
Rabi sowing			
14.10.69–21.10.69	B	A	C
22.10.69–29.10.69	C	B	A
30.10.69–6.11.69	A	C	B
7.11.69–14.11.69	B	A	C
15.11.69–22.11.69	C	B	A
23.11.69–30.11.69	A	C	B
1.12.69–3.12.69	Balancing period		
Rabi growing			
4.12.69–11.12.69	B	A	C
12.12.69–19.12.69	C	B	A
20.12.69–27.12.69	A	C	B
28.12.69–4.1.70	B	A	C
5.1.70–12.1.70	C	B	A
13.1.70–20.1.70	A	C	B
21.1.70–28.1.70	B	A	C
29.1.70–5.2.70	C	B	A
6.2.70–13.2.70	A	C	B
14.2.70–16.2.70	Balancing period		
Rabi maturing			
17.2.70–24.2.70	B	A	C
25.2.70–4.3.70	C	B	A
5.3.70–12.3.70	A	C	B
13.3.70–20.3.70	B	A	C
21.3.70–28.3.70	C	B	A
29.3.70–5.4.70	A	C	B
6.4.70–8.4.70	Balancing period		

Source: Canal Headquarters, Hissar Bhakra Circle, Hissar.

Note: During balancing period, channels will run at the discretion of the regulation in charge according to demand and share. Annual closure will be held in January or April 1970 if considered necessary for emergent repairs, and other closure works and the duration of the closure period will depend upon the extent of the nature of repairs considered necessary. Approval dates shown on this schedule were October 10 and November 11, 1969.

and is followed by a short balancing period. The operating engineers maintain complete water accounts for all channels. Channels receiving relatively less water during a stage have first preference for water available in the balancing period the follows.

The large designed capacity of the canal system relative to the flows normally available during the year necessitates the rotation. The large capacity is needed to transport the large quantities of water available in the monsoon season.[29] The rotation enables the canal system to provide each farmer with a flow of useful size at intervals during the season, even with the severely limited supplies available at the head during the winter dry season. At the same time, the rotation lets a large-capacity canal system operate at much less than optimal state, namely, full capacity, with minimum seepage losses in the largely unlined channels. The policy of adjusting capacity to match flows available should minimize the wetted perimeter per unit of discharge, and thus seepage for all channels in total.[30] Obviously, matching capacity and flows becomes more difficult with increasing variability and unpredictability of releases into the system.

The actual effect of the rotation on the cultivator's water supply depends on the level of releases into the system; as the level of releases relative to capacity decreases, the irrigation interval and timing unpredictability increase. For each farmer, the preference status of the group containing the channel that serves his watercourse provides a rough indication of his water supply at any date during the season. When that preference status is first, he is virtually certain of receiving a full supply; with second preference, he may receive a supply ranging from nothing to full; and with third preference, he will likely receive no water.[31] The preference status changes every eight days. Each group should therefore receive a full supply at least once every twenty-four days, and possibly some supply after sixteen days, were it not for the *warabandi* rotation.

The Warabandi Rotation among Farmers

Occurring simultaneously with the eight-day channel rotation is the seven-day *warabandi* rotation among farms on the same watercourse. The *warabandi* rations the water among farms served by a particular watercourse through a roster of fixed turns to receive water. A watercourse may extend one or two miles and may serve an area of over 1,000 acres containing 100 or more

29. Most of the large-scale canal-irrigation projects built during British times depended on river flows, which varied greatly between wet and dry seasons. The canal system in its operation had to transport to relatively dry areas large amounts of water available during the monsoon season, and at the same time provide flows of useful size during the dry season. Whether this same type of system makes most effective use of a modern dam with 8.0 million acre-feet of storage capacity is questionable. But its continued use probably indicates a high variability in releases from the dam.

30. A large channel running partially empty wets a larger surface relative to its rate of flow than when it is running full. Seepage losses are generally proportional to wetted surface area. Similarly, a partially full canal exposes more water surface for evaporation, relative to the rate of flow. Seepage and evaporation losses per unit of flow therefore increase as flow relative to capacity decreases.

31. For this situation to hold, releases into the system need only vary between about 30 percent and 70 percent of capacity because each group includes roughly one-third of the total capacity.

separate farms.[32] Each farm has a turn for receiving water once a week, and the length of that turn is directly proportional to the farm size. The day of the week on which the turn occurs depends primarily on the location of the farm on the watercourse. The quantity of water received by a farmer during his turn depends on the level of supply in the main channel serving his watercourse.[33]

Suppose a farmer owns ten acres in a 1,000-acre watercourse command or service area. Every week has 168 hours during which each acre in the command area, from the watercourse head to its tail, must receive its allotment of time. Each acre therefore has a time allotment of 0.168 hours, and the ten-acre farm thus has a turn 1.68 hours long.[34] With the command area of 1,000 acres and the design factor of 2.4 cusecs per 1,000 acres, a correctly designed outlet for this watercourse will provide a flow of 2.4 cusecs when the main channel runs at full supply. During a time allotment of 0.168 hours, 2.4-cusec flow would provide approximately 0.4 acre-inches for each acre. The farmer with ten acres would thus receive a total of about four acre-inches of water each week if the main canal is at full supply during his turn.[35] He can apply the water anywhere on his farm; he might concentrate it on a couple of acres or spread it thinly over several acres.

The permanent order of turns in the *warabandi* rotation depends somewhat on agreement and consultation among the cultivators. But the primary responsiblity for determining a just and equitable rotation lies with the local canal officers. Once fixed, it remains relatively immutable. A farmer can have his turn changed only if he can prove to the relevant canal officer that the *warabandi* is unjust, inequitable, or improper, or that it caused legal injury. The canal official may alter it if a change is necessitated by engineering changes in the watercourse or its alignment.[36]

The *warabandi* rotation of turns among farmers rations time, or the opportunity to receive water, rather than water itself. It is essentially permanent and

32. Fragmented holdings for separate plots belonging to the same farmer are treated as separate farms in the *warabandi*.

33. This obviously neglects seepage from the watercourse, condition and slope of the watercourse, height of the farmer's land, distance of the land from the watercourse head, and distance from the watercourse itself, all of which greatly affect the quantity of water received by the farmer.

34. These calculations neglect the adjustment made for filling the watercourse, which reduces the total week's time by one minute per acre of watercourse length (203 feet). Since watercourse length is generally limited to about two miles, the total adjustment would rarely exceed one hour for the whole week.

35. For any particular cultivator, the supply he actually receives during his turn depends primarily on his farm's location in the watercourse command area, his distance from the watercourse head, its condition between his land and the outlet, the height of his land relative to the outlet and the watercourse, and the cooperation of those farmers ahead of him on the watercourse.

36. J. L. Jain, *Northern India Canal and Drainage Act, Act No. VIII of 1873, Punjab State Wells Act, 1954; with Rules, Notifications, and Latest Case Laws,* 2d ed. (Chandigarh: Jain's Law Agency, 1968), sec. 68, pp. 45–48.

continues automatically regardless of the water flow in the channel. If the channel and thus the watercourse were constantly full, the farmer during his weekly turn would receive 0.4 acre-inch of water per acre owned, neglecting seepage from the unlined watercourse. The preceding discussion, however, clearly indicates a highly variable flow in the channels. In fact, the rotation among channels guarantees that flows will range from full to empty repeatedly during the season. The rotation periods for the *warabandi* and the channel rotations differ, and the interaction due to this difference in periods causes a major element of uncertainty in the farmer's water supply.

Interaction of Canal and *Warabandi* Rotations

The canals when empty require some time to fill. During that time only a partial and uncertain water supply is available to the farmers. The channel-rotation schedule therefore allows one extra day for filling, in addition to the seven-day *warabandi* rotation. This results in a total of eight days for each channel rotation. This eight-day period, however, causes an overlap of the two rotations during each following period. Tables 12.3 and 12.4 show the effect of this overlap or interaction. To indicate the effect of the channel rotation on the farmer's water supply, the patterns previously mentioned are assumed in the tables. The farmer receives a full supply when his group has first preference, an unpredictable supply during second preference, and no supply during third preference. The eight-day canal rotation and the seven-day *warabandi* rotation are compared in Table 12.3. Each farmer's turn always occurs on the same day of the week. To indicate the intraseasonal pattern of water supply for individual farmers as a result of the interaction of the two rotations, the data in Table 12.3 are transposed to a day-of-the-week basis in Table 12.4.

Table 12.4 indicates a common, though not dependable, pattern: a turn of full supply followed by turns of unpredictable supply and no supply. The pattern of water supply for a particular day of the week, however, may violate this pattern several times during the season. Because of the interaction of the two rotations, two consecutive turns of the same preference status commonly occur. For example, Sunday, Tuesday, Thursday, Friday, and Saturday all have two consecutive empty turns; for Sunday and Thursday this occurs twice in the season. Two such turns guarantee the farmer an interval of at least three weeks without water, even if by chance he gets a full supply when his group has second preference. If he receives no effective supply during that second preference turn, the interval will extend to four weeks. Further, if the farmer's turn occurs on the first day of first-preference status for his group, his supply is unpredictable because the canals are filling. Such a turn occurs at least once in the season for all farmers except those on Monday and Saturday. Similarly, the occurrence of any balancing-period turns introduces additional uncertainty

Table 12.3.

Interaction of *warabandi* and channel rotation in group B: hypothetical water supplies under channel rotation according to day of month and day of week, *rabi* 1969

	October		November		December		January		February		March		April	
Day of Month	Channel flow level	Day of week	Channel flow level	Day of week	Channel flow level	Day of week	Channel flow level	Day of week	Channel flow level	Day of week	Channel flow level	Day of week	Channel flow level	Day of week
1	...	...	0	Sa	B	M	+	Th	?	Su	?	Su	0	W
2	...	...	0	Su	B	Tu	+	Fr	?	M	?	M	0	Th
3	...	...	0	M	B	W	+	Sa	?	Tu	?	Tu	0	Fr
4	...	...	0	Tu	±	Th	+	Su	?	W	?	W	0	Sa
5	...	...	0	W	+	Fr	?	M	?	Th	0	Th	0	Su
6	...	...	0	Th	+	Sa	?	Tu	0	Fr	0	Fr	B	M
7	...	...	±	Fr	+	Su	?	W	0	Sa	0	Sa	B	Tu
8	...	...	+	Sa	+	M	?	Th	0	Su	0	Su	B	W
9	...	...	+	Su	+	Tu	?	Fr	0	M	0	M	...	...
10	...	...	+	M	+	W	?	Sa	0	Tu	0	Tu	...	...
11	...	...	+	Tu	+	Th	?	Su	0	W	0	W	...	...
12	...	...	+	W	?	Fr	?	M	0	Th	0	Th	...	...
13	...	...	+	Th	?	Sa	0	Tu	0	Fr	±	Fr	...	...
14	±	Tu	+	Fr	?	Su	0	W	B	Sa	+	Sa	...	...
15	+	W	?	Sa	?	M	0	Th	B	Su	+	Su	...	...
16	+	Th	?	Su	?	Tu	0	Fr	B	M	+	M	...	...
17	+	Fr	?	M	?	W	0	Sa	±	Tu	+	Tu	...	...
18	+	Sa	?	Tu	?	Th	0	Su	+	W	+	W	...	...
19	+	Su	?	W	?	Fr	0	M	+	Th	+	Th	...	...
20	+	M	?	Th	0	Sa	0	Tu	+	Fr	+	Fr	...	...
21	+	Tu	?	Fr	0	Si	±	W	+	Sa	?	Sa	...	...
22	?	W	?	Sa	0	M	+	Th	+	Su	?	Su	...	...
23	?	Th	0	Su	0	Tu	+	Fr	+	M	?	M	...	...
24	?	Fr	0	M	0	W	+	Sa	+	Tu	?	Tu	...	...
25	?	Sa	0	Tu	0	Th	+	Su	?	W	?	W	...	...
26	?	Su	0	W	0	Fr	+	M	?	Th	?	Th	...	...
27	?	M	0	Th	0	Sa	+	Tu	?	Fr	?	Fr	...	...
28	?	Tu	0	Fr	±	Su	+	W	?	Sa	?	Sa	...	...
29	?	W	0	Sa	+	M	?	Th	...	...	0	Su	...	...
30	0	Th	0	Su	+	Tu	?	Fr	...	...	0	M	...	...
31	0	Fr	...	...	+	W	?	Sa	...	...	0	Tu	...	...

Note: Symbols and assumptions: +—first preference for group B, assume full flow; ?—second preference for group B, assume unpredictable flow; 0—third preference for group B, assume no flow; ±—canal filling on first day of first preference, unpredictable flow; and B—balancing period turn, unpredictable flow.

Table 12.4.

Pattern of water supplied to farmers under group B command during the season, according to interaction of *warabandi* and channel rotation

Week of the season	Day of the week on which farmer's turn occurs						
	Sunday	Monday	Tuesday	Wednesday	Thursday	Friday	Saturday
1	. . .	. . .	±	+	+	+	+
2	+	+	+	?	?	?	?
3	?	?	?	?	0	0	0
4	0	0	0	0	0	±	+
5	+	+	+	+	+	+	?
6	?	?	?	?	?	?	?
7	0	0	0	0	0	0	0
8	0	B	B	B	±	+	+
9	+	+	+	+	+	?	?
10	?	?	?	?	?	?	0
11	0	0	0	0	0	0	0
12	±	+	+	+	+	+	+
13	+	?	?	?	?	?	?
14	?	?	0	0	0	0	0
15	0	0	0	±	+	+	1
16	+	+	+	+	?	?	?
17	?	?	?	?	?	0	0
18	0	0	0	0	0	0	B
19	B	B	±	+	+	+	+
20	+	+	+	?	?	?	?
21	?	?	?	?	0	0	0
22	0	0	0	0	0	±	+
23	+	+	+	+	+	+	?
24	?	?	?	?	?	?	?
25	0	0	0	0	0	0	0
26	0	B	B	B	. . .	. . .	. . .

Note: Symbols and assumptions same as in Table 12.3.

in future water supplies for all except Thursday and Friday. Even the engineers cannot know the allocation of water in those turns until very late in the particular stage of the season. Finally, the closure of the canals for annual repairs provides another element of uncertainty. As indicated in Table 12.2, the closure dates are only generally specified.

Thus, although the farmer's water supply often repeats the pattern of full followed by unpredictable and no supply, he cannot depend on this pattern. The interaction of the two rotation systems, along with turns when the canals are filling and when balancing periods occur, constantly alters this pattern. The farmer can, however, be fairly certain of receiving his water at least after 4 weeks. Most farmers interviewed verified this. They felt reasonably certain of receiving water twice a month, but they never knew when during the month they would receive it. While permitting the survival of irrigated agriculture in an arid region, such a water supply makes precise crop and water management difficult, if not impossible.

Table 12.5.

Discharge as percentage of capacity by days of the week, *Rabi* 1969–70

Week of season	Sunday			Monday			Tuesday		
	BSB	BD	DD	BSB	BD	DD	BSB	BD	DD
1. Oct. 12–18	...	...	...	...	...	...	93	100	43
2. Oct. 19–25	100	100	94	100	100	94	100	100	89
3. Oct. 26–Nov. 1	77	57	96	78	59	100	78	59	100
4. Nov. 2–8	0	0	0	0	0	0	14	0	0
5. Nov. 9–15	78	81	94	100	74	98	100	100	100
6. Nov. 16–22	78	100	60	76	100	60	76	100	52
7. Nov. 23–29	0	29	0	5	0	0	5	0	18
8. Nov. 30–Dec. 6	4	0	11	100	100	84	54	14	100
9. Dec. 7–13	100	100	100	100	100	100	100	99	100
10. Dec. 14–20	27	44	0	49	81	0	53	78	0
11. Dec. 21–27	0	0	0	0	0	0	0	0	0
12. Dec. 28–Jan. 3	100	90	79	100	87	97	100	88	92
13. Jan. 4–10	100	88	97	59	33	100	34	0	60
14. Jan. 11–17	79	75	67	68	66	60	10	29	0
15. Jan. 18–24	0	0	0	0	0	0	0	0	0
16. Jan. 25–31	39	8	100	49	20	100	49	20	100
17. Feb. 1–7	79	78	70	79	84	65	79	85	60
18. Feb. 8–14	0	0	0	0	0	0	0	0	0
19. Feb. 15–21	100	88	100	100	90	95	100	88	94
20. Feb. 22–28	100	88	94	100	89	100	100	89	97
21. March 1–7	83	91	100	87	85	81	87	71	95
22. March 8–14	12	0	21	9	0	21	34	16	39
23. March 15–21	85	91	74	100	90	70	100	90	100
24. March 22–28	77	78	74	79	85	56	77	78	56
25. March 29–April 4	0	0	56	0	0	0	0	0	0
26. April 5–11	0	0	0	46	78	0	84	88	76

Source: Original data from Canal Daily Discharge Records, Fatehabad Division, Canal Headquarters, Hissar Bhakra Circle, Hissar.

Note: BSB = Balsamand Sub Branch, capacity 512 cusecs; BD = Balsamand Distributary, capacity 292 cusecs; DD = Dewan Distributary, capacity 168 cusecs. Flow rates below 80 percent are in brackets.

Resultant Canal Flows under the Institutional Rationing System

Flows in the canals serving the farmer's watercourses integrate all the aspects of canal-water rationing discussed above. Table 12.5 shows flow levels in three primary canals serving the case-study area. Canal flows as shown are based on canal record books, but modified to reflect the farmers' viewpoint. Whereas canal records show flows in cusecs for every day of the month, the table shows flows relative to capacity arranged by day of the week, with flows less than 80 percent in brackets.[37]

Balsamand Sub Branch (BSB) includes all of group B. Of the three chan-

37. When canal flows are less than 80 percent, farmers cannot effectively use their water supply (oral communication, B. S. Bansal, deputy commissioner for the Indus Water Treaty, Ministry of Irrigation and Power, Government of India; formerly superintending engineer, Hissar Bhakra Circle and later chief engineer for Haryana on the Bhakra Management Board). Flows of less than 80 percent are largely ineffectual for irrigation use by the farmers because of the watercourse outlet design. As mentioned previously, the watercourse outlets magnify the effect of

Table 12.5.—*continued*

Wednesday			Thursday			Friday			Saturday		
BSB	BD	DD	BSB	BD	DD	BSB	BD	DD	BSB	BD	DD
91	100	70	83	100	60	98	100	74	100	100	94
72	103	100	78	49	100	79	57	100	79	54	100
72	49	100	5	51	0	0	0	0	0	0	0
12	19	0	14	15	0	78	78	0	88	91	94
100	100	100	39	65	100	49	5	94	62	24	100
76	100	47	76	100	47	76	100	47	76	100	47
1	1	0	1	0	0	4	0	0	1	0	0
78	45	94	100	94	100	100	103	104	100	94	100
100	100	100	100	100	100	62	100	60	27	84	0
59	84	10	24	49	0	27	27	0	0	29	0
0	0	0	0	0	0	0	0	0	0	0	0
100	88	92	100	90	89	100	92	85	100	92	87
37	28	65	40	36	58	40	33	56	37	26	56
0	0	0	14	29	0	0	0	0	0	0	0
93	64	100	100	85	100	59	36	100	39	12	100
69	41	100	59	52	74	65	68	60	78	71	74
79	85	60	81	85	70	34	49	0	0	0	0
0	0	0	0	0	0	0	0	0	95	85	89
98	89	86	100	89	97	100	89	94	100	88	100
83	89	89	91	81	100	87	74	100	87	74	100
74	49	95	0	0	60	10	0	49	14	0	21
27	39	56	19	31	0	100	88	21	100	88	97
100	90	100	100	92	100	100	92	95	81	81	89
72	71	56	65	71	56	73	78	56	73	69	56
0	0	0	0	0	0	0	0	0	17	0	0
85	81	72									

nel groups (A, B, and C), only group B has a single contact point where it receives all its water. Balsamind Sub Branch later divides into Balsamand Distributary (BD) and Dewan Distributary (DD). All are relatively large and serve many watercourses and several minors.

The table clearly indicates a high degree of irregularity and uncertainty for the farmers' water supply during the season with respect to both timing and quantity. Several weeks of no effective supply often follow two consecutive weeks of essentially full supply. Furthermore, water-flow patterns in the three channels have no easily recognizable or consistent relationship, although the channels are physically connected and the two distributaries share all the available water from the subbranch.

Thus, canal-irrigation supplies have at least three types of uncertainty or unreliability with which the farmer must contend: the timing of water supplies during the season, the quantity of water to be received at various times during the season, and in total, and the timing and quantity of water received at various locations. This uncertainty is in direct contrast to the farmer's allo-

canal-level fluctuations on the water supply received by the farmer. The magnification becomes especially severe below an 80 percent level of canal supply. In balancing the water supplied to their various canals, engineers do not count flows below 80 percent.

Table 12.6.

Effect of varying schedule and frequency of irrigations on yield of dwarf wheat (Sonora 64)

Treatments	Schedule of irrigation (days after sowing)						Irrigation frequency	Grain yield (quintal per hectare)	Relative yield (percent)
	25 (crown root)	45 (tillering)	65 (jointing)	85 (flowering)	105 (milk)	120 (dough)			
A	–	–	–	–	–	–	0	9.3	100
B	+	–	–	–	–	–	1	30.4	327
C	–	–	+	–	–	–	1	20.6	222
D	–	–	–	–	+	–	1	10.5	113
E	+	–	+	–	–	–	2	34.2	368
F	–	–	+	–	+	–	2	26.1	280
G	+	–	–	–	+	–	2	31.6	340
H	+	–	+	–	+	–	3	35.4	381
I	+	+	+	–	+	–	4	41.8	450
J	+	–	+	+	+	–	4	42.8	460
K	+	–	+	–	+	+	4	37.6	404
L	+	+	+	+	+	–	5	47.8	514
M	+	+	+	–	+	+	5	43.3	466
N	+	–	+	+	+	+	5	43.5	468
P	+	+	+	+	+	+	6	51.1	550
Rainfall (mm)	–	–	–	–	0.8	12.0	. . .	. . .	. . .

Source: N.G. Dastane et al., *Review of Work Done on Water Requirements of Crops in India* (Poona: Navabharat Prakashan, 1970), p. 43.

Note: + and – indicate irrigation and no irrigation. C.D. or significant difference in yield at the .05 level is 6.0 quintals per hectare. Many similar experiments have been done, and while their results do not all agree quantitatively, the evidence is strong that the dwarf's respond well to irrigation at particular times, especially at crown-root initiation.

cated turn to receive water, which occurs weekly for a fixed period of time, and the water needs of crops, which generally increase as the season progresses.

Precise and accurate estimation of the effects of an uncertain water supply on production and yields requires the kind of detailed field and experimental data that are at present largely unavailable.[38] The recent research on critical irrigation periods for dwarf wheat, however, can provide a quantitative indication of the effect from varying timing and quantity (as indicated by the number or frequency of irrigations) of water applied. Such research is now fairly widespread in India, although results are sometimes contradictory. Table 12.6 shows the results of a recent critical-period experiment. Yields generally increase with increasing amounts (frequency in the table) of irrigation, but substantial yield differences also occur with different timings of the same number of irrigations. Clearly, both timing and quantity of water applied affect yields substantially.

Most farmers interviewed felt severely constrained by their canal-water supply, although they generally seemed anxious to take advantage of new crop varieties and to improve their economic situation when possible. Most wanted to install a tube well to supplement their canal water and improve the timeliness of their irrigations, but few could because of the saline groundwater common in the area. The substantial advantage in per acre profitability of tube-well- over canal-irrigated farms indicates the value of timeliness and precise water management, as well as the opportunity cost of not improving the canal system to provide a more controllable, predictable, and certain water supply.[39]

Summary and Recommendations

The rationing system tends to assure a high degree of uncertainty in the timing and quantity of water supply received by farmers. And for the individual farmer it virtually guarantees an interval between irrigations varying unpredictably from one to several weeks in length during the season. The competition between electricity and irrigation, the capacity factors, and the interaction between the canal and *warabandi* rotations appear as primary causes of this situation. At the same time, the nonmarket nature of the rationing system and its lack of flexibility and controllability largely preclude adjustments to compensate for the uncertainty and variability. A primary virtue of the rationing system is that all recipients apparently share both water and insecurity equally.

38. Available field data, especially for canal irrigation, generally do not include the measured amount of water used or the exact timing of the application. And experimental data unfortunately usually reflect the attempt to maximize yields with optimum conditions, rather than exploring the production surface.

39. Reidinger (pp. 208–311) provides a detailed comparison and analysis of performance and economics for private tube-well- and canal-irrigated (and dry-land) farms in the case-study area.

Through the uncertainty it produces, as well as the mismatch of water supply to crop needs, the rationing system can greatly limit the effectiveness of water use by farmers, and thus the overall returns to large-scale canal irrigation. Clearly, the new high-yielding varieties—the basis of the green revolution—cannot approach their potential without precise water management. This requires flexibility and control, or at least predictability of the water supply. The government recognizes the need for improved water management with canal irrigation.[40] As indicated above, however, the major difficulty lies not so much with the operating engineers or with the farmers, but rather with the traditional system of canal irrigation itself.

The canal system as an institution which rations water thus exemplifies the impact of economic or production institutions on agricultural development. Institutional changes to provide greater certainty or control of water supply would allow farmers to reduce risk adjustments, improve water and crop management, and thus increase yields and profits. The preceding analysis of the rationing system suggests several changes which should enable it more closely to meet the needs of modern agriculture and to increase the predictability, certainty, and control of the farmer's canal-water supply.

At the level of releases into the system, matching the capacity factors more closely to crop water needs could improve the farmers' efficiency of water use for both irrigation and stored soil moisture. For example, in October just after the monsoon, the soil-moisture reservoir should be near capacity. Nevertheless, the initial winter-season capacity factors appear relatively high compared to levels later in the season, especially during the grand growth period. To say that "cultivators will have to be taught—if indeed they do not know this already—that lesser quantities of water are needed for plants in seedling stages than when they are growing rapidly toward maturity"[41] neglects the root of the problem. Cultivators cannot change the capacity factors that guarantee that the largest water supplies of the season will occur during the seedling stages when the quantity of water needed is relatively low. Even

---

The study indicates a substantial yield and profit advantage with controllable tube-well irrigation compared to uncontrollable canal irrigation. Even though the water application (frequency) was higher under tube-well irrigation, the regression coefficient for water application (especially on dwarf wheat) was also larger compared to canal irrigation, indicating a higher marginal product for water use with tube-well irrigation. T. V. Moorti (*A Comparative Study of Well Irrigation in Aligargh District, India* [Ithaca, N.Y.: Cornell University, Department of Agricultural Economics, Occasional Paper no. 29, 1970], p. 53) reports similar results in comparing private tube wells and state-owned tube wells which are not controllable by the farmer. He attributed this to a combination of good water timing and control, use of high-yielding varieties, and high use of other inputs under private tube-well irrigation.

40. For example, at a meeting of the Central Board of Irrigation and Power in December 1969, G. A. Narshimha Rao, then chairman of the Central Water and Power Commission, stated: "I consider it the duty of the Irrigation Engineer to supply the required quantity of water at the right time on the farm."

41. Evans, p. 21.

during the stages of maximum consumptive use of water by the crop, the capacity factor never reaches its early-season level.

The competition between electricity and irrigation introduces a much more complicated problem. To explore this would require a detailed analysis of both private and social returns from improving the supply of water for irrigation, as compared to treating irrigation as a by-product and letting power production take precedence as may be happening at present. Putting the irrigation levy on a volumetric basis would provide decision makers with a rough indication of the marginal value of irrigation water, and with an incentive to give more emphasis to agriculture. A volumetric water levy should also encourage more judicious water use by farmers.[42] It would not, however, help them equate marginal revenues and marginal costs and attain economic efficiency in water use.

At lower levels of the rationing system, the policy of sharing the available water among farmers on the basis of their landholding size certainly violates the spirit and the letter of a major stated national objective: equalization of income distribution.[43] In a country like India, with its concentration of landholdings,[44] such a policy favors the relatively few wealthy landlords. And giving more water to the larger farmers should provide little or no economic

42. Much literature on Indian irrigation has emphasized the need for volumetric water charges. The *Report of the Indian Irrigation Commission of 1901–03* provides a cogent statement of the case. The situation with respect to volumetric water levies has changed little since then (see *Report of the Indian Irrigation Commission, 1901–03,* pt. 1, *General* [Calcutta: Superintendent of Government Printing, 1903], pp. 90–98); Ansari (pp. 109–13) presents a brief discussion of the report's conclusions in this regard. Inefficient use of water—overirrigation—is often attributed to the per acre levy as an argument for volumetric water charges. The marginal cost of applying additional water to the same acre is very low, and indeed the farmer who receives much more water than he planned for could easily overirrigate. A volumetric water levy for a highly uncertain water supply may possibly reduce the tendency to irrigate beyond the point of zero marginal product. It would not, however, help the farmer to reach the economic optimum and thus increase net returns per acre from irrigation; that is, while the levy would increase technical water-use efficiency (average product), it would not help in reaching economic efficiency without a more certain or controllable water supply.

43. "The basic goal is a rapid increase in the standard of living of the people, through measures which also promote equality and justice. . . . planning would result in greater equality of income and welath, and there should be progressive reduction of concentration of incomes, wealth, and economic power" (*Fourth Five-Year Plan,* p.4).

44. Of the agricultural families, 60.8 percent hold less than 2.5 acres and own 6.3 percent of the agricultural land, while 3.4 percent of the families hold more than 25 acres and own 34.5 percent of the land (see *National Sample Survey,* report no. 10 [Delhi: Government of India, 1958], p. 47, cited in P. V. John, *Some Aspects of the Structure of Indian Agricultural Economy, 1947–48 to 1961–62* [Bombay: Asia Publishing House, 1968], p. 91). Indeed, allocation of canal water in direct proportion to the number of acres owned should encourage concentration of holdings, contrary to national policy. By retaining their land, the large landowners receive a larger share of the available water. Canal irrigation may also affect the other major land-tenure policy—consolidation of fragmented holdings—which is necessary for efficient agricultural management. As indicated in the analysis, the canal-water rationing system can result in a water supply that is good at one location and bad at another nearby. A farmer's fragmented holdings can thus reduce the risk that all his crops will receive a poor canal-water supply.

advantage in terms of production and, in fact, may have a negative effect. Substantial evidence exists indicating that small farms have a per acre productivity at least as high as that of large farms under roughly comparable conditions.[45] Analysis of field data collected in the case study area verifies this. With the present low intensity of irrigation, allocating relatively more water to small farmers should therefore not sacrifice total production, and could result in increases. Finally, a policy that places the small farmers at a competitive disadvantage in resource availability could drive them back toward subsistence agriculture, and ultimately off the farm. In view of the large numbers involved and the limited opportunity in industry in the foreseeable future, a policy retaining labor in agriculture appears worthwhile. It seems much more beneficial to provide a more equal sharing of the scarce water, which provides the basis for raising income in arid agriculture.

Increasing the predictability of water supply would enable more effective water use by the farmer. More knowledge or forewarning of future flows should enable him to plan and manage his farm more precisely. News media, and especially radio broadcasting, could quickly spread news of expected water supplies for the season, for the system as a whole, and for particular times and areas. For example, water released from the Bhakra Dam requires about three days to reach Hissar; broadcasting information on releases would give at least a few days' forewarning. Broader dissemination of information supposedly already available—such as the channel-rotation schedule—should also receive more emphasis. No farmers questioned about it had seen a channel-rotation schedule in several years. Knowledge of the rotation schedule well before the season begins would also benefit the farmers; for the 1969 *rabi* schedule, the approval dates, October 10 and November 11, are too late to enable effective diffusion for the *rabi* sowing period. An irrigation extension agent could also play a vital role in dissemination and interpretation of information such as the rotation schedule and releases into the system.

The interaction of the rotation among channels and the farmers' *warabandi* rotation produces much of the uncertainty. As indicated above, much of the problem results from the unequal periods of the two rotation. Making the two periods equal in length would eliminate the interaction problems. A seven-day canal-rotation period, for instance, would match the seven-day *warabandi* rotation, thus eliminating the overlap.

45. A. M. Khusro ("Returns to Scale in Indian Agriculture," reprinted in A. M. Khusro, *Readings in Agricultural Development* [New Delhi: Allied Publishers Private, 1968], pp. 123–59), presents results indicating that, in fact, output per acre decreases as farm size increases, as well as references to research showing similar results. This also occurs in the Intensive Agricultural Development Districts. See Expert Committee on Assessment and Evaluation, Ministry of Food, Agriculture, Community Development and Cooperation, *Modernizing Indian Agriculture: Report on the Intensive Agricultural District Programme (1960–68)* (New Delhi: Government of India, 1969), 1:30, commonly known as the *Sen Report*.

The problem of the day required for filling, however, would still exist. Farmers with turns on the first day of the week would always face a very uncertain water supply and possibly always receive relatively less water. Compressing the turns of all the farmers on the watercourse into six days and allowing the water on the filling day to go to waste[46] (which should not amount to much until the canals reach full supply levels) would reduce each farmer's time allotment by one-seventh but might still result in production gains, depending on the quantity of water actually lost on the filling day and the value of the resulting increased certainty of water supply.

Alternatively, an association of farmers on the same watercourse could oversee the distribution of whatever water comes during the filling day, rather than wasting it. Such a water users' association would absorb the uncertainty of the water supply on the filling day because no farmer would depend on it. At the same time, it could provide more water control for the farmers and additional water to those for whom the water was most valuable, perhaps by auctioning the water available on the filling day to the highest bidders. A mutual company–type organization could provide for a just sharing or rebating of net income from water sales according to some specified formula.[47]

In the long run, devising an organization to replace entirely the *warabandi* system appears highly beneficial. Designing a workable system for Indian conditions presents many problems, but the likely returns from greater flexibility and control by the farmers certainly justifies further research and investigation. Field interviews indicated that some informal water trading—both interseasonal and intraseasonal—among neighboring farmers already occurs, even though the canal authorities apparently consider it illegal. Such trading exists in highly diverse situations in the world,[48] and has much to recommend it as an approach to economic efficiency. A water users' association could provide the institutional framework to encourage and expedite the transfer of water to those farmers for whom it has the highest marginal value, while protecting the rights of the economically and politically weak.

Certainly the development of some type of water users' association among the farmers represents only one of many possible alternatives. The method of providing more flexibility and control in farm-water supply remains an open question, but an undeniable need for both does exist. With further advancement of agricultural technology, the need for more precise management of

46. The law does not actually require use of the water (see N. D. Gulhati, *Administration and Financing of Irrigation Works in India,* publication no. 77 [New Delhi: Central Board of Irrigation and Power, 1965], p. 79).

47. Various types of effective local water users' associations operate, for example, in Japan, Taiwan, and Indonesia, especially Bali.

48. Substantial water trading apparently occurs in the western United States, though it has rarely received much emphasis. Wells S. Hutchins (*Mutual Irrigation Companies in California*

water, and the requisite flexibility and controllability of its supply, will increase. Lack of these characteristics in the canal system could greatly impede future agricultural development and leave untapped much of the potential of the new agricultural technologies.

---

and Utah, bulletin no. 8 [Washington, D.C.: Farm Credit Administration, Cooperative Division, 1936], pp. 123–24), briefly discusses it. For more recent analysis of such transfers, see Raymond L. Anderson, "The Irrigation Water Rental Market: A Case Study," *Agricultural Economics Research* 12, no. 2 (April 1961): 54–58, and B. Delworth Gardner and Herbert H. Fullerton, "Transfer Restrictions and Misallocations of Irrigation Water," *American Journal of Agricultural Economics* 50, no. 3 (August 1968): 55–71. Similarly, organized water trading as an integral part of a distribution system occurs in the traditional *huerta* (irrigation association among farmers) of Alicante, Spain, and dates back several hundred years; see the section on Alicante in the forthcoming book by Arthur Maass and Raymond Anderson dealing with irrigation water distribution in three areas of the western United States and three in Spain. Raymond L. Anderson and Arthur Maass (*A Simulation of Irrigation Systems: The Effect of Water Supply and Operating Rules on Production and Income on Irrigated Farms,* technical bulletin no. 1431 [Washington, D.C.: Department of Agriculture, Economic Research Service, 1971], p. 6) briefly mention the study.

13

Local Water Management in the People's Republic of China

James E. Nickum
SAN JOSE STATE UNIVERSITY

While there are several excellent studies on local irrigation management in pre-1949 China and post-1949 Taiwan Province, little attention has been given by outside analysts to the nature of the organization of water resource management in the People's Republic of China. This paper provides a brief introduction to the latter, as described in Chinese source materials and supplemented by a limited amount of personal observation as a member of the U.S. Water Resources Delegation in late summer 1974.

The management of irrigation, drainage, and flood-control facilities affecting more than one production team, once such facilities are constructed, is commonly placed in the hands of a separate administrative entity with a full-time specialist staff. This entity can be simply a handful of persons at the production brigade, or village level, whose primary responsibility is to oversee the management of pumps or checkdams; or it can cover the area governed by a major state reservoir, pumping station, or dike. Some of the most famous of these include the irrigation districts of the Red Flag Canal (control area: 40,000 ha.), the Kiangtu (Jiangdu) Pumping Stations (600,000+ ha.), the People's Victory Canal (in Honan) (40,000 ha.), and the Shaoshan Irrigation District at Mao's birthplace in Hunan (50,000+ ha.). A reservoir used primarily for irrigation, with a storage capacity of over 10 million cubic meters (ca. 80,000 acre-feet), may have separate management organs for the reservoir and its irrigation district (Chou 1965:7–8). Here I will focus on the medium-sized irrigation district, such as that for the Lin Xian portion of the Red Flag

Previously published as "Local Irrigation Management Organization in the People's Republic of China," *China Geographer,* no. 5 (Fall 1976), pp. 1–12. Reprinted by permission.

Canal, which we visited, with an irrigated area of 40,000 hectares; and the Meich'uan (Meiquan) Reservoir Irrigation District in Hupeh Province, a national model, which has an irrigated area of 8,000 hectares.

An irrigation district is commonly established directly by the lowest administrative level whose boundaries encompass the benefited area. For example, the *xian* (county) founded the irrigation district for the Red Flag Canal, which affects several communes (townships) within Lin Xian. A small portion of land outside the *xian* is benefited by the Red Flag Canal, notably in Anyang and P'ingshun Xian. The portion of the canal system in Anyang is managed by a local body there. This encompasses only a small portion of the entire system, however, and was a later accretion. The Red Flag Canal originates outside the Lin Xian boundaries, but yields the bulk of its benefits after it crosses into the *xian*.

While most irrigation districts appear to be formed around a single key installation when it is built, they may also be consolidations of preexisting structures and districts (*Meiquan* 1974:15–16). A district may also be organized by several of the next lowest administrative units, such as communes, as a joint enterprise under the leadership of the next highest level. This type of genesis requires a higher degree of horizontal integration, however, and does not appear to be very frequent.

Once established for a year, a state-run district (i.e., above the commune level) is operated as an autonomous financial entity with its own profit-and-loss accounts, much like other state run economic enterprises. It is supposed to become self-sufficient in funds for current operation and extension of basic projects. It may do so by one of three means: the levying of water fees, the development of economically productive sidelines, or a reduction in staff and partial decentralization of certain management functions to the collective sector. These will be taken up in turn.

Water fees are set once a year and charged to the user production teams. For the Red Flag Canal, the amount is determined by the canal irrigation administration office before the May wheat harvest, and approved by the *xian* government. The fees are essentially to cover the cost of maintenance and repair, a large proportion of which is labor cost. Original capital costs are absorbed at the time of construction by the developing agency and the beneficiary teams. In addition, the aggregate ability to pay appears to play a role, in that the amount charged is commonly reduced when the harvest is bad.

According to State Council regulations, funds collected for irrigation are to be used exclusively for water conservation, and may not be levied for local general revenue purposes. While irrigation districts are supposed to make ends meet, and if possible accumulate some surplus, the basic policy is to keep the charges as low as possible.

Both the amount charged and the basis on which it is levied vary from district to district. Charges may be based on irrigated acreage, which may or may not be zoned according to productivity; or on the amount of water used, measured in cubic meters; or on some combination of fixed and variable price bases (Nickum 1974:106–8).

Each of these methods has its advantages and its drawbacks. An acreage charge, such as used at the Red Flag Canal, is quite common. It is the simplest to administer, inasmuch as it does not require monitoring of actual water use. By the same token, it does not encourage frugality by water users, and may lead to a neglect of smaller, team-owned projects which could provide alternative sources of water. Disputes over priority of use are thereby exacerbated. The problem of overutilization of canal water was acknowledged at the Red Flag Canal project, where they are considering a dual price basis incorporating actual water usage as well as acreage.

A pure water-usage charge, while less likely to promote waste than an acreage charge, also requires more complex administration, particularly in monitoring and calculating the amount used in each watering. It encourages production teams to make greater use of their own internal water sources, and in at least one instance in the early 1960s, teams were encouraged to market their own internal water surpluses among themselves, either directly or through the reservoir administration (*Renmin Ribao,* 10 March 1963:2).

On the other hand, variations in natural conditions from year to year lead to wide fluctuations in the amount of supplemental irrigation water used. This could lead to major instability in the water rates or, more likely, in the district's revenue. In addition, since prices are set on a cost basis and not in the free and easy markets of our textbooks, certain dysfunctional allocational and behavioral consequences are likely to occur. The Meich'uan Reservoir, which initially levied water charges according to the amount used, abandoned that system within two years in favor of an acreage charge, since the "simplistically economic" approach led to the following consequences:

> A small number of communes and teams asserted that no matter what, if I pay you the money, you release the water. They wanted to use too much water. . . . A small number of communes and teams upstream along the canal would not request water at the appropriate time, but waited to use the water which seeped through as it passed by. Others did not request reservoir water at all and only used pond water, causing their fish to die and affecting agricultural and fishery production. [*Meiquan* 1974:47]

The third scheme of calculation, which was in use in the majority of irrigation districts in Kiangsu Province in 1965 and which appears to be a future possiblity for the Red Flag Canal, is a combination of the above two methods. Charges are divided into two components: a fixed charge based on acreage, used to cover fixed, noncapital expenditures, and a "production charge" based on water use, which goes to cover variable costs. Little information appears to be available on how this scheme has worked in practice, however.

Map 13.1. Sketch map of water conservancy projects of Lin Xian.

The actual amount charged varies widely, depending upon the efficiency of management, the amount of fuel oil or electricity used and, presumably, the local wage rate. The Red Flag Canal, which relies on gravity irrigation, charges a rather low 5.70 yuan per hectare ($2.85 at the current exchange rate of $0.50 = 1 Yuan). In Kiangsu Province in 1964, electrical irrigation averaged 27 yuan per hectare; mechanical (diesel) irrigation cost somewhat more, 46.5 yuan per hectare (Yang 1965). The latter figures are comparable to those obtaining else-where in Asia where irrigation is highly developed (Asian Agricultural Survey 1969:539). Meich'uan Reservoir originally charged 0.008 yuan per cubic meter with an average application of 150 cubic meters, costing 18 yuan per hectare (*Meiquan* 1974:47). In some cases the bulk of the fees are paid in labor services and materials rather than in cash (*Renmin Ribao,* 13 December 1963:2).

One means of holding down costs to water users is summarized in the slogans "Use the canal to sustain the canal" and "Use the reservoir to sustain the reservoir." In addition to their water-management duties, the employees of an irrigation district and their dependents engage in other economically productive activities, both for their own consumption and for market. For instance, the thirty-three staff members of the Meich'uan Reservoir and their dependents reclaimed a small portion of land and by 1972 had become more than self-sufficient in grain, cotton, oil-bearing crops, pigs, and fish, the latter being their principal subsidiary occupation. In their spare time, they planted over 20,000 trees in one year (1972). In addition to providing themselves with the basic means of subsistence, they turned a profit of 40,000 yuan from their economic activities (*Meiquan* 1974:54–55).

While the Meich'uan Reservoir Irrigation District may be particularly successful in its supplemental economic endeavors, a certain amount of such activity appears to be nearly universal. Most common for reservoirs is the establishment of a fishery, and afforestation is universal. Another economic activity in recent years is the establishment of small-scale hydropower stations, which may provide grain processing for a fee for nearby production teams.

Another method of reducing the cost of running the irrigation district management office is to reduce the size of the full-time specialist staff, or at least maintain it at certain prescribed levels. In 1965, suggested staffing levels (not including project maintenance crews or staff of the experimental stations) ranged from one to three per 10,000 *mu* effectively irrigated (three to nine per 2,000 hectares) for large districts to about triple that for smaller districts (Chou 1965:8). The Red Flag Canal—irrigating 40,000 hectares—has a maintenance crew of 300, but we did not get figures for the size of the management office staff.

Staff size is kept down by extensive reliance on the decentralization of certain functions to user groups, called "democratic management organizations," or "mass organizations." These consist of full or part-time personnel whose wages are paid within the collective system.

The actual form of user organizations and the mix of full- and part-time workers varies considerably from district to district. The main canal and three principal branches of the Red Flag Canal are managed by the central irrigation office, but they are subdivided into nine administrative divisions and, consequently, into forty-one administrative subsections. In some other districts such subsections appear to be determined by political boundaries, and their maintenance and repair are placed under separate management committees related to the communes. At the Red Flag Canal, however, they are not partitioned along commune boundaries, and are staffed directly by the 300 full-time workers belonging to the irrigation district management office. These workers, each assigned to a specific subsection, are responsible for overseeing the smooth running of their segment of the irrigation system and the delivery of water. Once the water leaves one of the branch canals, however, its management is the responsibility of the user unit.

In Meich'uan Reservoir's irrigation district, the main reservoir and five smaller reservoirs are managed by specialist organs; sixteen very small reservoirs (under 1 million cubic meters, or 8,000 acre-feet) are managed separately by specialists sent from beneficiary communes or brigades; and ponds are managed by the production teams. Labor for reservoir and canal maintenance and repair is provided mainly by user units, but allocated by the irrigation district (*Meiquan* 1974:48–49).

As would be expected, there are some problems in China in balancing "unified planning" with member unit autonomy, especially in the collection of water fees and in water allocation. These problems may be compounded by the position of the irrigation district administration outside the main direct channels of political authority.

The problem of insufficient collection of revenue by a significant proportion of irrigation districts was alluded to in a *Renmin Ribao* editorial on 13 December 1963, which criticized the "supply of water without collecting charges while extending one's hand to the State for money year after year," and insisted that irrigation districts "make efforts to strive for self-sufficiency in management expenses in the next one or two years."

Clearly other factors are involved here than the dichotomy between irrigation district management and local administration, such as the availability of state subsidy and the inexperience or incompetence of management, due partly to the newness of many districts. On the other hand, allusions are made in Chinese materials to a rift between supplier and user, characterized by a lack of ability on the part of the district to enforce the collection of water fees and by an absence of active involvement in each other's affairs.

One of the primary mechanisms to overcome the involvement gap is the "irrigation district representatives' conference," held once or twice a year, at which are discussed the repair and construction plans, water utilization plans, and water-fee collection. That the communication is sometimes one-way is indicated by a report in 1965 which claimed: "In the past, when a meeting of representatives of the irrigated areas was called, either a demand was made for repairs to engineering projects, or a collection of water charges was asked for. The problem of production was basically ignored" (Yuan 1965).

The most effective irrigation districts appear to be those which have direct ties to the regular administrative apparatus, and where there is open exchange in terms of communication and involvement with beneficiary teams, although water allocation plans and fee schedules are still set from above. The Meich'uan Reservoir Irrigation District has a close symbiosis with the local political apparatus. The (political) district (*qu*) party committee, which provides overall leadership, appoints a secretary to serve as head of the management committee of the irrigation district, and assigns another of its committee members to oversee the work of Meich'uan District. A reservoir management office party branch secretary participates on the *qu* party committee. The irrigation district's work is a regular part of the agenda of the *qu* committee.

On the other side, commune party secretaries at Meich'uan also participate on the irrigation district management committee. The management sections which are the administrative subdivisions of the irrigation district make regular reports to the commune, and assiduously heed the latter's leadership. At a more basic level, ties are established between section personnel and production brigade party branch secretaries. Clearly, the channels of authority are open to the Meich'uan Irrigation District, and as a consequence it is able to be quite active in matters such as the collection of water fees and the integration of all local water sources into a single system.

The more successful irrigation districts tend to become involved in local agricultural production problems beyond simply providing water. They provide some planning and leadership for ancillary farmland capital construction, and expertise, including surveying for leveling land and training courses for local water technicians. They operate scientific experiment stations which may assist and direct local units in developing their own scientific experimentation activities, including the testing of new crop varieties. In one reported instance, an irrigation district not only supplied technology to collectives to develop well irrigation, but contributed funds as well. In another, a commune-level irrigation station operated a free machine repair service (*Yikao* 1973:24–27, 84–88).

One mark of a successfully integrated irrigation district is that it may establish a so-called watermelon water system, wherein the small projects owned by the production teams, such as ponds, are linked up with each other and with the main canal system. They are then included in the overall water utilization plans, providing supplemental irrigation during busy times and receiving additional supplies from the main reservoirs during slack periods.

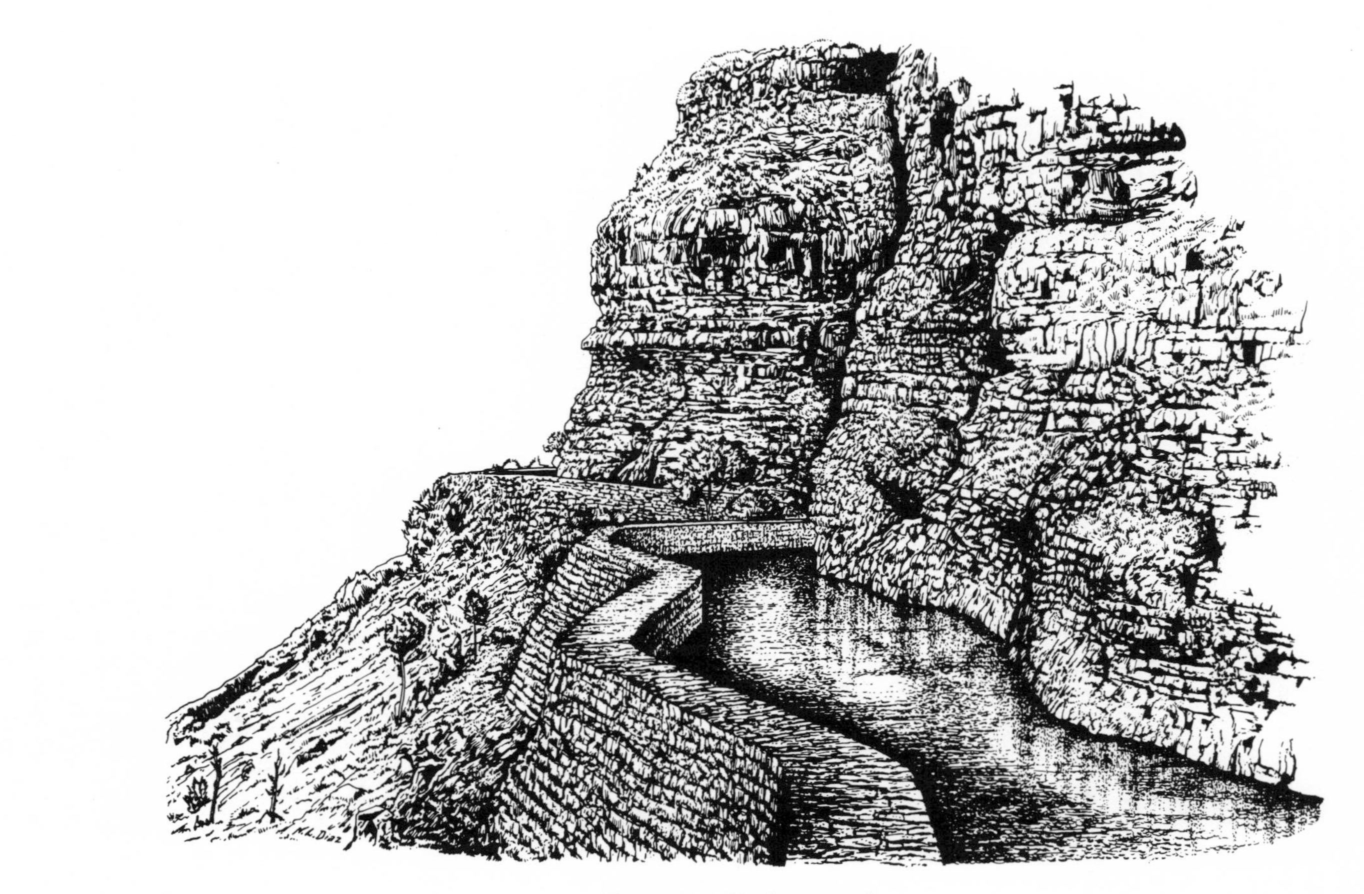

Figure 13.1. Distributory canal.

The operation of a watermelon water system is rather sophisticated, and requires a high degree of cooperation between the irrigation district management and the user teams.

Finally, a brief note on water allocation. Conflicts over water use between upstream and downstream, between major state-owned and minor collective-owned projects, and between old and new districts are frequently alluded to in materials published in China. In fact, cooperation between units in water development and use is the theme of a major didactic Peking opera, *Ode to Dragon River,* indicating that more of such cooperation would be desirable.

Disputes such as those between upstream and downstream are commonly mediated by the next higher administrative level; for example, a quarrel between the brigades would be handled by the commune party committee or revolutionary committee. The irrigation district appears to defer to the regular administrative apparatus in such affairs.

It is unclear whether or not there is a universal water code to provide specific guidance to dispute resolution. It is clear, however, that the first developer and upstream units are normally given priority in obtaining sufficient water to maintain their yields. For instance, in Hsuchiang (Xujiang) Xian, Kiangsi, before a new canal was built which would tap the same water as that of two other canals, the *xian* called a "new and old irrigation canal representatives conference," where an accord was signed in which representatives of the old canals agreed that a new canal was necessary, and representatives from the new canal area "resolutely guaranteed that in a year of great drought, the old irrigation canals would certainly have priority in using water" (*Renmin Ribao,* 8 November 1974:4).

At Meich'uan, the following approaches are taken by the irrigation district management and the *qu* political administration to provide downstream units with sufficient water during a drought:

1. "Ideological work" is carried out among upstream and midstream units to encourage them to be as frugal as possible and supply "Dragon River water" downstream.

2. Downstream units are encouraged to use water rationally and to open up alternative water sources such as ponds.

3. Water allocation procedures are altered, such as releasing water upstream only during the day and releasing it during the night for downstream use.

4. Upstream and midstream cadres are organized by the *qu* party committee and the reservoir management to go to the lower reaches to witness conditions there (*Meiquan* 1974:24–25).

To sum up, medium-scale irrigation facilities in China, once built, are placed in the hands of irrigation districts which are separate accounting entities. These districts administer the major works within their domain and, when successful, establish a comprehensive irrigation and drainage network incorporating the smaller projects owned by user units. The success of a district appears to be determined by its ability to tie into the local political apparatus and by the effectiveness of its outreach programs to beneficiary units. In terms of specific forms of organization and procedures, China exhibits its customary flexibility in producing a variety of ways to adapt to local conditions.

References

Asian Agricultural Survey. 1969. University of Washington Press.

CHOU CHIN-YUAN and LIANG YUNG-SHUN, eds. 1965. *Guan'gai Guanli* [Irrigation Management]. Peking: Zhongguo Gongye Chubanshe.

MEIQUAN SHUIKU GUANQU DI GUANLI [Management of the Meich'uan Reservoir Irrigation District]. 1974. Peking: Shuili Dianli Chubanshe.

NICKUM, JAMES E. 1974. "A Collective Approach to Water Resource Development: The Chinese Commune System, 1962–1972." Ph.D. dissertation, University of California, Berkeley.

YANG LIU-WU. 1965. "Reduction of Irrigation and Drainage Costs." *Nongtian Shuili yu Shuitu Baochi* [Farmland water conservancy and erosion control], August (JPRS 32,970).

YIKAO QUNZHONG, GUANHAO SHUILI [Rely on the Masses to Manage Water Conservancy]. 1973. Peking: Shuili Dianli Chubanshe.

YUAN LUNG. 1965. "Agricultural High Yield in Irrigated Areas." *Nongtian Shuili yu Shuitu Baochi,* July (JPRS 32,970).

14

Local Consequences of a Large-Scale Irrigation System in India

Edward J. Vander Velde
NORTH-EASTERN HILL UNIVERSITY (INDIA)

Irrigation projects have often formed a part of development programs aimed at economic and social change in rural India. A major irrigation development project in southern Fatehabad tehsil, Hissar District, Haryana, initiated important changes in the spatial pattern of agricultural land use in the area. Certain design and operational features of the irrigation system tended to concentrate the advantages of irrigation at specific locations within the area it serves, and the resulting patterns of agriculture substantially reflect the differential access of farmers to the advantages and benefits of irrigation. Moreover, differential access to irrigation water has had important implications for the various social groups resident in villages in the area, particularly when examined within the context of governmental action to implement a land-reform program intended to have major social and economic benefits.

The temporal coincidence of perennial irrigation from the Bhakra-Nangal project and the application of the Punjab Occupancy Tenants and Security of Land Tenures Act, 1953, and Rules, 1956, may have been accidental. The general success of the Bhakra-Nangal project in bringing water to agricultural lands in Punjab and Haryana is widely acknowledged. Yet the general failure of land-reform legislation to alter significantly the concentration of land-

Field research on which this paper is based was carried out in 1964–67, supported by a junior fellowship from the American Institute of Indian Studies and a research grant from the Center for South and Southeast Asian Studies, University of Michigan. The advice and assistance of Jhujar Singh throughout my field research is gratefully acknowledged. The comments and criticisms of Gerard Rushton, Canute VanderMeer, and Nancy Hammond are deeply appreciated. An earlier version of this paper was presented at the 22nd annual meeting of the Association for Asian Studies, April 1970, San Francisco.

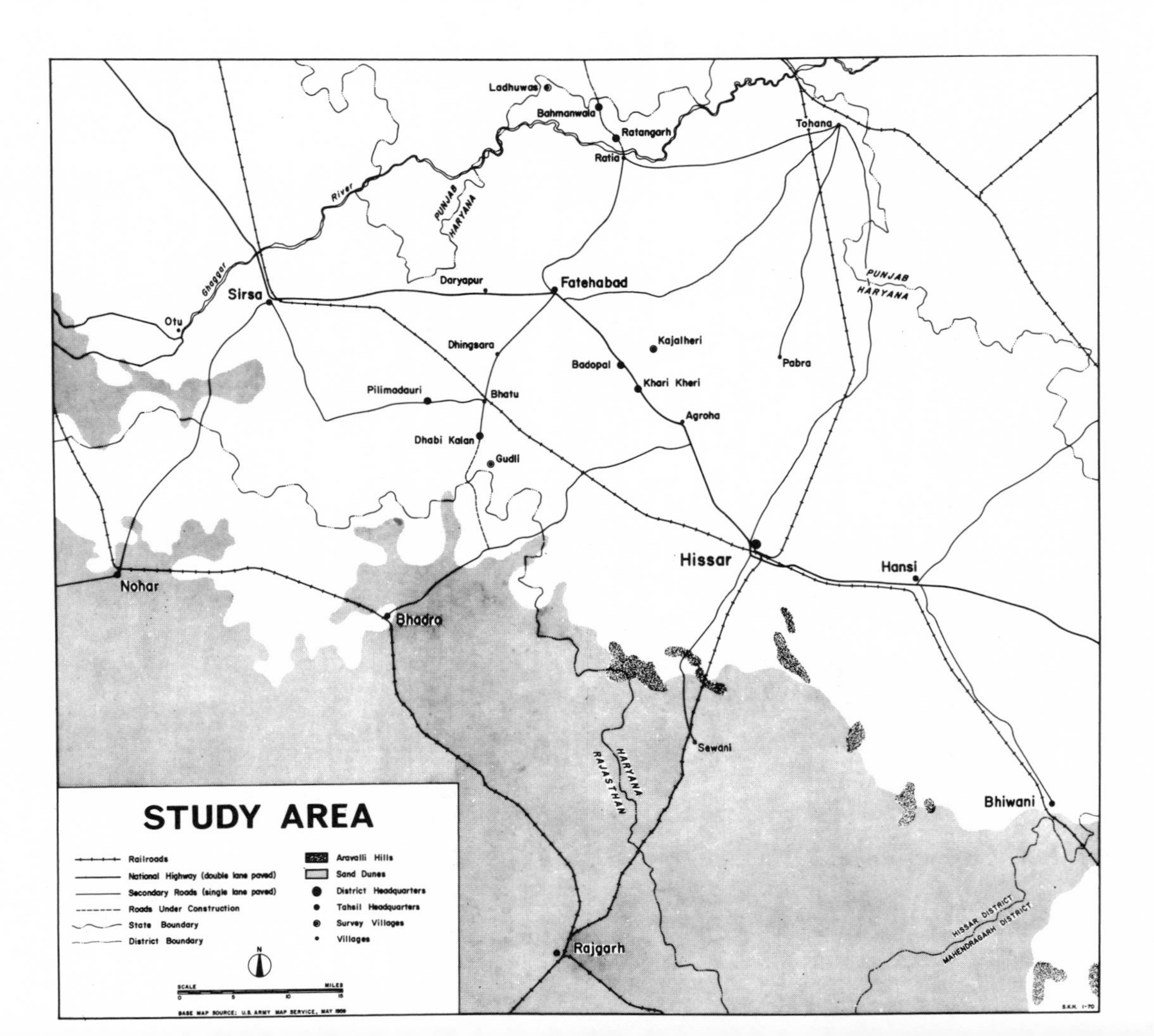

STUDY AREA
Railroads
National Highway (double lane paved)
Secondary Roads (single lane paved)
Roads Under Construction
State Boundary
District Boundary
Aravalli Hills
Sand Dunes
District Headquarters
Tahsil Headquarters
Survey Villages
Villages
N
SCALE
MILES
0
5
10
15
BASE MAP SOURCE: U.S. ARMY MAP SERVICE, MAY
S.K.H. I-70
Ladhuwas
Bahmanwala
Ratangarh
Ratia
Tohana
PUNJAB
HARYANA
River
Ghaggar
Otu
Sirsa
Daryapur
Fatehabad
Dhingsara
Kajalheri
Badopal
Khari Kheri
Pabra
Pilimadauri
Bhatu
Agroha
Dhabi Kalan
Gudli
Hissar
Hansi
Nohar
Bhadra
Sewani
RAJASTHAN
Bhiwani
Rajgarh
HISSAR DISTRICT
MAHENDRAGARH DISTRICT

ownership in the hands of rural elites within the environment of rapid technological change in agriculture facilitated by the Bhakra-Nangal project appears to have substantially frustrated efforts to redress the severe social and economic inequalities of rural life in this region.[1]

The Research Area

The Bagar region of Haryana, in which the survey village for this study, Dhabi Kalan, is located, was long recognized as one of the least productive agricultural areas in India (Fig. 14.1.).[2] Recurring famine conditions dramatically emphasized the environmental severity and agricultural marginality of Hissar District, particularly its southern and western parts. Occasionally entire village populations would temporarily emigrate for the wetter and agriculturally more prosperous districts of Punjab and Uttar Pradesh, to the north and east, where they searched for work as daily or seasonal agricultural labor and for fodder supplies for their animals. Many also sought and found work as casual laborers in the Delhi urban area or in the other towns and cities north and east of Hissar District.[3]

This region and the bordering area of Rajasthan have an average rainfall of only sixteen inches, and it varies substantially annually, seasonally, and monthly. Most of the annual rainfall comes during the July-August monsoon season, when evapotranspiration rates are close to maximum. Highest temperatures occur in May and June, but especially in June, when mean daily maximums often exceed 105° F. The coldest month is January, when mean daily minima in the low forties are common; frost occurs occasionally in January and February. The rates of evapotranspiration in the winter months partially offset the low probability of significant rainfall during that period.[4]

1. The Punjab Occupancy Tenants and Security of Land Tenures Acts, 1953, and Rules, 1956, were applicable throughout the former Punjab. See Government of Punjab, Legislative Department, *Land Code* (Chandigarh: Controller of Printing and Stationery, Punjab, 1963), vol. 2.

2. "Bagar" is the indigenous term for the sandy portion of western and southern Haryana.

3. As recently as 1965–66, such conditions prevailed in parts of southern Haryana, including Hissar District. See "Famished Villages Cry for Water," *The Tribune* (Ambala), November 16, 1965, p. 3. See also R. Humphreys, "Report on the Famine in the Hissar District," in Punjab Government, *The Punjab Famine of 1899–1900* (Lahore: The "Civil and Military Gazette" Press, 1901), vol. 4; Punjab Government, *Final Report on the Operations for the Relief of Distress in the Hissar District (Ambala Division), during the Year 1932–33* (Lahore: Superintendent of Government Printing, Punjab, 1934), pp. 2–15; Punjab Government, *Reports on the Famine Relief Operations in the Southeast Punjab (Hissar, Rhotak, and Gurgaon), 1938–1940* (Lahore: Superintendent of Government Printing, Punjab, 1946); Sir Malcolm Darling, *The Punjab Peasant in Prosperity and Debt* (London: Oxford University Press, 1947), pp. 82–88.

4. Annual rainfall at Hissar for the past seventy years has ranged from 34.3 inches in 1917–18 to 5.4 inches in 1951–52. The erratic seasonal distribution of rainfall is well illustrated by the fact that in 1965 the monsoon "average" for the study area came in two forty-eight-hour periods separated by nearly a month of rainless weather. L. Labroue, P. Legris, and M. Viart ("Bioclimats du Sous-Continent Indien," *Travaux de la Section Scientifique et Technique* (In-

The soils in the Bagar region are alluvial in origin, potentially very fertile, but sandy in many places and highly permeable.[5] Shifting sand dunes are scattered throughout the area, and it is plagued by severe dust storms from April until the onset of the monsoon. Groundwater is found at depths commonly exceeding 100 feet, but often it is too brackish for humans, beasts, or crops. The occasional tank or pond has been and remains a major focal point for settlement in this area.

Before the completion of the Bhakra-Nangal project, farmers in the Bagar region depended on the scanty and variable rainfall, except in small areas in the northwest and in the east-northeast where limited irrigation was possible from the seasonal Ghaggar River and the Western Jumna and Sirhind Canal systems, respectively. Thus cereals and pulses with low moisture requirements and a high tolerance for drought were the predominant crops. Bajra (*Pennisetum typhoideum*) was the principal summer season (*kharif*) crop, and gram (*Cicer arietinum*) was the most widely grown crop in the winter season (*rabi*). If soil moisture conditions were favorable, a less drought-resistant but better yielding and more remunerative crop such as jowar (*Sorghum vulgare*) in *kharif* or wheat in *rabi* was sown intermixed with bajra or gram.

To counter the low yields usually produced under these conditions, large holdings under cultivation were necessary; 15 to 20 acres was a common holding for a family of the dominant, landowning Jat caste, and holdings that exceeded 50 or even 100 acres were not unusual.[6] In the one good year in five that might normally be expected, these holdings would produce large, marketable surpluses of gram, bajra, or wheat as well as abundant fodders to support the extensive livestock production of the region. The more persistent pattern of pre-irrigation agriculture, however, was bare subsistence, punctuated by a year or two of genuine famine.

Haryana first received irrigation though the Bhakra system during the summer of 1952; supplies were gradually increased as Bhakra Dam and the network of irrigation channels were constructed. In the agricultural year 1961–62, the Bhakra irrigation system was finally capable of supplying water on a year-round basis to a large portion of the *barani* or dry crop tract in

stitut Français de Pondichery) 3, no. 3 [1965], p. 25) classify Hissar as "tropical dry," calculating 237 physiologically dry days. See also S. L. Duggal, *Agricultural Atlas of Punjab* (Ludhiana: Punjab Agricultural University, 1966), pp. 30–31 (Plate 14).

5. The pre-irrigation soil survey of the area south of Fatehabad and Sirsa, which includes the research area, found that 29 percent of the soils surveyed were "excessively sandy," and as a consequence, recommended that canal irrigation be carefully controlled to avoid water losses through excessive seepage. This important conclusion has proved to be prophetic. See C. L. Dhawan and others, "Pre-irrigation Soil Survey of Some Districts of the Punjab," *Indian Journal of Agricultural Science* 27, no. 4 (1957): 375–94.

6. Large holdings still prevail; 32 percent of all landowning households in Hissar District own 15 to 30 acres, and 21 percent own in excess of 30 acres (Census of India, 1961, *Hissar District, Punjab,* District Census Handbook no. 1 [Chandigarh: Government of Punjab, 1966], p. 32).

Hissar, Fatehabad, and Sirsa tehsils in Hissar District. The arrival of fairly regular perennial irrigation fundamentally altered the resource base of agriculture for a considerable area of cropland in the study area. Wheat quickly became the most important winter crop for many farmers, though gram remained a significant crop in unirrigated fields. The cultivation of profitable cash crops such as cotton and sugarcane also spread rapidly throughout the area served by the Bhakra Canal system. Most significant, however, wherever it was available, year-round irrigation sharply reduced the dependence upon rainfall and brought to the agriculture of many farmers a degree of dependability and security previously unknown in this region.[7]

Western and central Hissar District is the very core of the region comprised of parts of Punjab, Haryana, and Rajasthan in which the Jat community is numerically dominant. In the 1931 census, the last census to enumerate caste populations, the Jats constituted over 24 percent of the population of Hissar District. In a more recent study, the geographer Joseph Schwartzberg found that the Jat population in villages in Hissar District typically exceeded that of the next most numerous caste in the village by over 150 percent.[8]

Dhabi Kalan, the research village, is fifteen miles southwest of Fatehabad, the headquarters of tehsil and subdivisional administration, and five miles south of the large village of Bhatu, a station on the narrow-gauge railway line between Hissar and Bhatinda. A single-lane tarmac road connects Bhatu and Dhabi Kalan, and continues on to the Haryana–Rajasthan border, six miles further south; regular bus service along this route between Fatehabad and Bhadra, Rajasthan, facilitates travel through the region.

Dhabi Kalan was colonized by Jats of the Beniwal clan (*got*) immigrating from neighboring Rajasthan about 150 years ago, and many cultural links persist between the village and that region. Numerous families in the Jat community still maintain close relations with the parent Rajasthan villages by exchanging daughters in marriage with families there. Although it is a multicaste village, Dhabi Kalan is numerically dominated by the Jats, who make up nearly 50 percent of the population, more than double the population of the next largest caste, the Kumhars (Table 14.1).

Dhabi Kalan has an area of 4,222 acres or about 6.6 square miles (Fig.

7. See Economic and Statistical Organisation, *Effects of Bhakra Dam Irrigation on the Economy of Barani Villages in the Hissar District, 1964–65* (Chandigarh: Government of Haryana, 1967), for a partial account of the cropping pattern changes associated with the supply of Bhakra Dam irrigation to Hissar District since 1952–53. The Board of Economic Inquiry, Punjab, *Report on Cropping Pattern of Hissar District* (Chandigarh: Economic and Statistical Organisation, Government of Punjab, 1965), is another useful though occasionally inaccurate source of data concerning the shifts in the cropping pattern of Hissar District. Duggal, *Agricultural Atlas of Punjab*, is a convenient reference for the spatial distribution and relative significance of various crops grown in this region.

8. Joseph E. Schwartzberg, "The Distribution of Selected Castes in the North Indian Plain," *Geographical Review* 55, no. 4 (1965): 477–95 and map; Census of India, 1931, *Punjab*, vol. 17, pt. 1 (Lahore: The "Civil and Military Gazette" Press, 1933), pp. 339–42.

Table 14.1.

Dhabi Kalan: Caste population and households, 1966[a]

Caste	Population	Households	Percent of Population
Jats			
Beniwal	479	53	28.4%
Dhindwal	40	6	2.4
Jakhar	149	18	8.8
Lamba	32	3	1.9
Netar	10	3	0.6
Puniya	23	3	1.4
Other *gots*	82	12	4.9
All Jats	8.15	98	48.4
Caste Hindus			
Brahmin	21	4	1.2
Aggarwal	19	4	1.1
Sunar	9	1	0.5
Kumhar	369	48	21.8
Khati	26	5	1.5
Nai	44	6	2.6
Gujar	3	1	0.2
"Refugee"	2	2	0.1
All caste Hindus	493	71	29.0
Other castes			
"Muslim"	10	1	0.6
Lohar	28	5	1.7
All other castes	38	6	2.3
Harijans			
Chamar	141	23	8.3
Dhanak	171	25	10.1
Other	31	5	1.9
All Harijans	343	53	20.3
Total population	1689	228	100.0%

[a]Excluding Patwaris and their families.

14.2), of which only 6 percent is uncultivable. Of the 3,958 acres of cultivable land, 2,065 acres are officially classified as irrigable and the remainder as nonirrigable. Irrigation is supplied from the Dhabi Minor Canal, which traverses the village area from northeast to southwest for a distance of about three miles. The headworks of Dhabi Minor are on Kheri Distributary, and the last outlets of the Minor irrigate fields of the neighboring village of Dhabi Khurd.

Before the completion of the Bhakra irrigation system, agriculture in Dhabi Kalan was totally dependent upon the capricious rainfall. Consequently, drought-resistant gram was by far the most widely grown winter crop and bajra dominated agriculture in the summer season. Sowing mixed crops was the usual method of diversifying agriculture to meet environmental uncertainty; bajra-jowar-pulse, bajra-pulse, and bajra-guar (*Cyamopsis psoraloides*) were frequent summer crops mixes, and gram-wheat, gram-

Table 14.2.

Dhabi Kalan: agricultural land ownership by caste (acres)

Caste	All land	Classed Nehri	Classed Barani	Acres of cultivated area per household
Jats				
Beniwal	2152.0	1282.5	869.5	40.6
Dhindwal	84.6	37.5	47.1	14.1
Jakhar	403.0	201.0	202.0	22.4
Lamba	207.1	134.1	73.0	69.0
Netar	4.7	1.9	2.8	1.6
Puniya	55.1	9.9	45.2	18.4
Other *gots*	159.4	52.0	107.4	13.3
All Jats	3065.9	1718.9	1347.0	31.2
Caste Hindus				7.6
Brahmin	27.0	8.4	18.6	6.8
Aggarwal	85.4	56.2	29.2	21.4
Kumhar	319.8	133.7	186.1	6.7
Khati	20.8	20.8	0.0	4.2
Nai	69.7	16.9	52.8	11.6
All caste Hindus	522.7	236.0	286.7	7.4
Other castes				
Lohar	21.1	16.8	4.3	4.2
Harijans				
Chamar	75.0	41.4	33.6	3.3
Dhanak	51.5	16.9	34.6	2.1
All Harijans	126.5	58.3	68.2	2.4
Other owners				
Panchayat	87.6	6.8	80.8	
Temple	20.7	0.5	20.2	
Reside outside	113.0	27.3	85.7	
All other owners	221.3	34.6	186.7	
All owners	3957.5	2064.6	1892.9	16.7

mustard, and barley-wheat were often planted in winter. Nevertheless, partial crop failures were common, and occasionally crops failed totally, irrespective of the location of the field within the village area, the caste of the farmer, or the tenure of the cultivation unit.[9] Since the perennial operation of Dhabi Minor in 1960–61, agriculture in Dhabi Kalan has ceased to depend exclusively on rainfall, and crop failures have been reduced—perhaps more accurately, confined—to those areas of the village lands that remain outside the area served by the irrigation system.

Jat households own over three-quarters of the agricultural land in Dhabi

9. In two consecutive winter seasons, 1938–39 and 1939–40, all sown acreage failed to mature, and in the winter season 1965–66 the rate of failure of rainfall-dependent crops exceeded 50 percent; rain-fed agriculture remains risky (*Lal Kitab, Mauza Dhabi Kalan*, Tehsil Records Office, Fatehabad, Hissar District, n.d., and personal communication, District Commissioner, Hissar, August 1966).

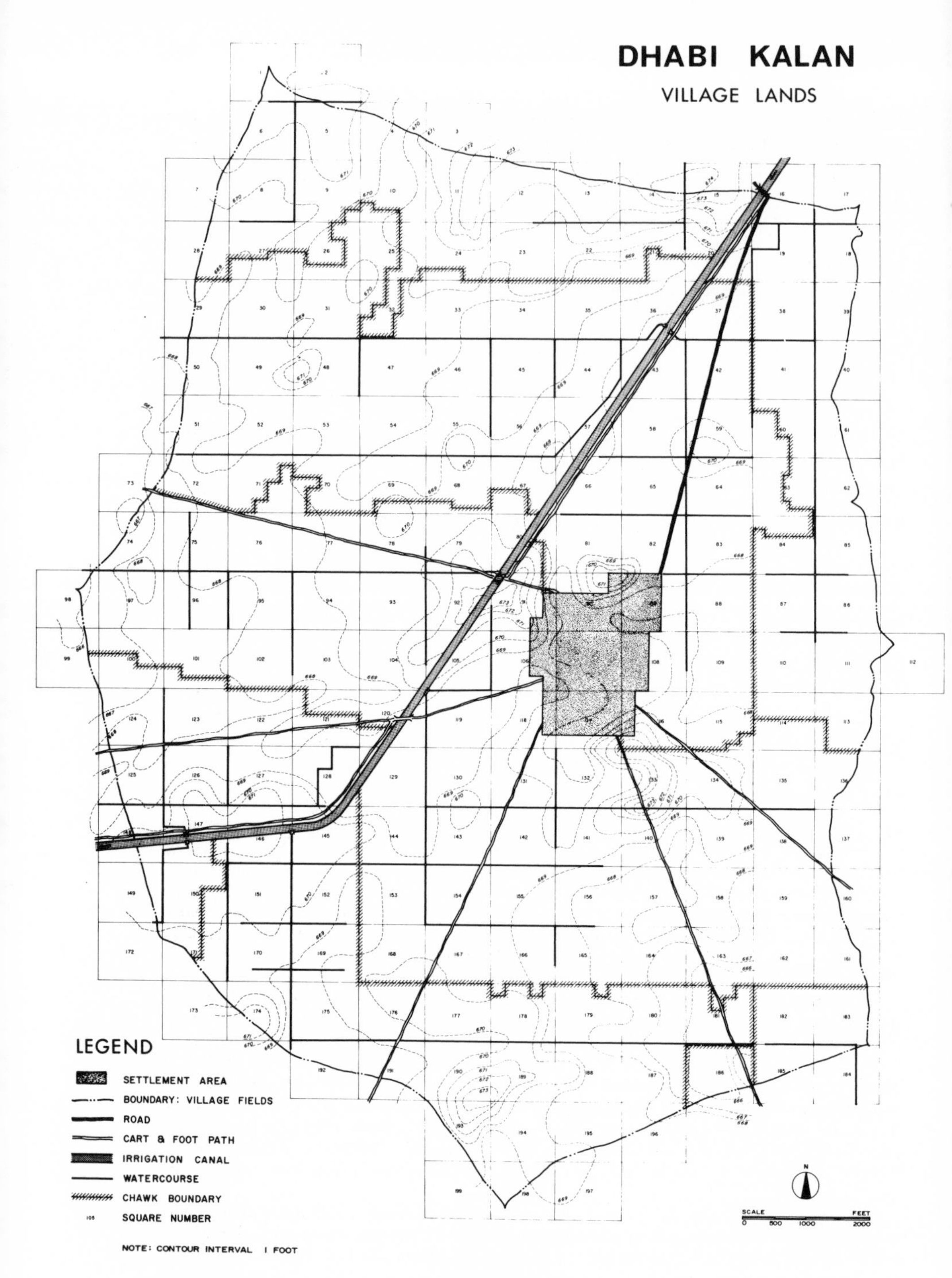
DHABI KALAN
VILLAGE LANDS
LEGEND
SETTLEMENT AREA
BOUNDARY: VILLAGE FIELDS
ROAD
CART & FOOT PATH
IRRIGATION CANAL
WATERCOURSE
CHAWK BOUNDARY
SQUARE NUMBER
NOTE: CONTOUR INTERVAL 1 FOOT
SCALE
FEET
0 500 1000 2000
N

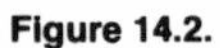

Figure 14.2.

Kalan and, significantly, 83 percent of all irrigable land (Table 14.2).[10] The amount of land owned by most Jat households exceeds 30 acres and holdings larger than 50 acres are common. All of these landholdings contain variable amounts of irrigable and nonirrigable land, and they are usually located at two or three sites within the village area. Most Jat landowners have always tried to maximize their ownership or control over agricultural land, and now their competition for control over more irrigable land or for additional water rights is the principal cause of the feuds and faction fights that erupt from time to time within the Jat community of Dhabi Kalan.

Although it is true that most households in Dhabi Kalan own at least a bit of land, only the Jats, and a few other families ranking immediately below them in the social hierarchy, own enough land to support the individual households. Thus, the basis of the Jat dominance of all important levels of society in Dhabi Kalan is not simply their numerical superiority but their extensive control of the agricultural base of the village economy.[11]

The Irrigation System

Dhabi Minor and the network of field watercourses it sustains are but a small part of the Second Bhakra Main Line Circle, a major component of the Bhakra canal system built as a part of the multipurpose Bhakra-Nangal project (Fig. 14.3).[12] The planned intensity of irrigation for Dhabi Minor, as for all perennial Bhakra canals, was 62 percent of the cultivable commanded area (CCA),

10. In all, fourteen Jat *gots* are represented in Dhabi Kalan. Following the Beniwal *got* in both numbers and importance is the Jakhar *got*. The population and caste data for Dhabi Kalan used here were compiled from a census conducted by the author in 1965 and a resurvey in 1966. Two sources are of particular value in any analysis of the tribes and castes of this region of Haryana: J. Wilson, *General Code of Tribal Costume in the Sirsa District of the Punjab* (Calcutta: Superintendent of Government Printing, India, 1883), especially pp. 1–66; and Sir Denzil Ibbetson, *Panjab Castes* (Lahore: Superintendent of Government Printing, Punjab, 1916), a reprint of the chapter, "The Races, Castes, and Tribes of the People" in his *Report* on the Census of the Punjab, 1881.

11. Although caste hierarchies and caste dominance have been studied extensively by anthropologists, frequently in villages in the North Indian plain, the spatial context of caste hierarchy and dominance has yet to attract much systematic attention by geographers. Kashi N. Singh, "The Territorial Basis of Medieval Town and Village Settlement in Eastern Uttar Pradesh," *Annals of the Association of American Geographers* 58, no. 2 (1968): 203–20, and Joseph E. Schwartzberg, "Caste Regions of the North Indian Plain," in *Structure and Change in Indian Society*, ed. Milton Singer and Bernard S. Cohn (Chicago: Aldine Press, 1968), pp. 81–113, are important exceptions to this criticism. See Ralph W. Nicholas, "Structures of Politics in the Villages of Southern Asia," in ibid., pp. 243–84, for definitions of caste dominance by an anthropologist.

12. The Second Bhakra Main Line Circle is comprised of nearly 2,000 miles of large and small irrigation channels providing perennial irrigation to a cultivable commanded area of 1.9 million acres in Hissar District, Haryana, and Bhatinda and Sangrur districts, Punjab. The system was declared perennially operational beginning in the winter season of 1961–62 (Superintending Engineer, Second Bhakra Main Line Circle, Hissar, "Note on the Development of Irrigation in Second Bhakra Main Line Circle, Hissar" [Hissar, 1964], typescript).

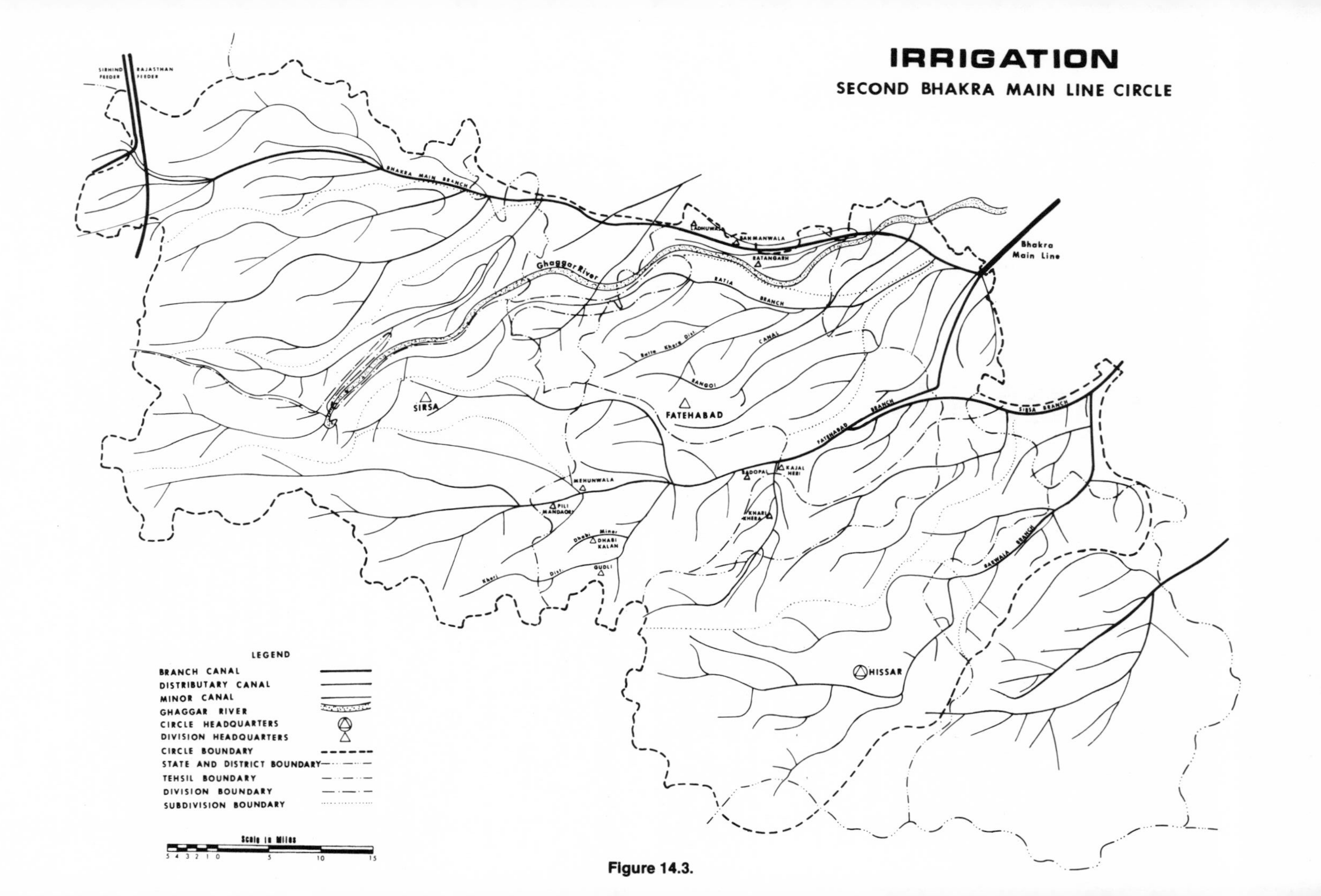
IRRIGATION
SECOND BHAKRA MAIN LINE CIRCLE
SIRHIND FEEDER
RAJASTHAN FEEDER
BHAKRA MAIN BRANCH
Bhakra Main Line
Ghaggar River
RATIA BRANCH
Ratia Khera Disty.
CANAL
RANGOI
SIRSA
FATEHABAD
FATEHABAD BRANCH
SIRSA BRANCH
BARWALA BRANCH
HISSAR
LADHUWAL
BAHMANWALA
RATANGARH
MEHUNWALA
PILI MANDAORI
Dhabi Minor
DHABI KALAN
GUDLI
Kheri Disty.
BADOPAL
KAJAL HERI
KHARI KHERA
LEGEND
BRANCH CANAL
DISTRIBUTARY CANAL
MINOR CANAL
GHAGGAR RIVER
CIRCLE HEADQUARTERS
DIVISION HEADQUARTERS
CIRCLE BOUNDARY
STATE AND DISTRICT BOUNDARY
TEHSIL BOUNDARY
DIVISION BOUNDARY
SUBDIVISION BOUNDARY
Scale in Miles
5 4 3 2 1 0 5 10 15

Figure 14.3.

divided between 28 percent in the summer season and 34 percent in the winter.[13] The acreage actually irrigated in the command of Dhabi Minor, however, has consistently fallen well short of this target, as it has for the entire area supplied by Kheri Distributary in the Fatehabad irrigation subdivision and in many other areas irrigated by the channels in the Second Bhakra Main Line Circle.[14]

Part of this failure is doubtless due to large water losses through seepage in the unlined channels traversing the sandy and highly permeable soils of the Haryana Bagar region and the high rates of evaporation that are common in the region. Part of the failure, however, also rests with certain weaknesses in the design of the project itself. When the irrigated and highly productive areas of western Punjab were lost to Pakistan in 1947, the Bhakra irrigation system was conceived as a way to offset this loss by supplying water to the semiarid areas of the southeastern Punjab plains. Simply put, the project was planned to cover too large an area given the supplies of water that would be available. Apparently the desire to expand agricultural production and at the same time to reduce the risk faced by the farmer in the sparsely populated and potentially productive Bagar region led to an overly optimistic assessment of project capability.[15] In the operations of Dhabi Minor during the first six years of perennial water supplies, this design weakness led to annual deliveries as low as 56 percent of the promised amount (Fig. 14.4). The shortages of water supplies have been particularly acute in the winter crop season.

A second weakness of the Bhakra project was that the water allowance was predetermined and rigidly fixed, and the new crop patterns then had to be worked out within those limits.[16] This arrangement had major implications for

13. Intensity of irrigation is the ratio of the area actually irrigated by the canal to the total cultivable commanded area of the canal.

14. The mean annual intensity of irrigation for the period 1961–62 to 1966–67 in the area served by Dhabi Minor was 35.5 percent, 15 percent in the summer season and 20.5 percent in the winter season (Edward J. Vander Velde, Jr., "The Distribution of Irrigation Benefits: A Study in Haryana, India," unpublished Ph. D. dissertation, University of Michigan, 1971, pp. 102–12). Robert C. Repetto (*Time in India's Development Programmes* [Cambridge: Harvard University Press, 1971], pp. 73–80) reports a similar pattern in other areas of Hissar District served by Bhakra canals.

15. K. N. Raj, *Some Economic Aspects of the Bhakra Nangal Project* (New York: Asia Publishing House, 1960), p. 53, notes that in 1953 it was recognized that "there will be a shortage of 9% even in years when the reservoir is full, and about 25% on an average over a longer period," in relation to the requirements of the Bhakra irrigation system. See his discussion, pp. 36–37, for a fuller exposition of the technical implications of the Bhakra project design.

16. Ibid., p. 92. The water allowance or duty of water is the relation between the area to be irrigated and the quantity of water available or required to irrigate it, often expressed in cusec per unit area. The Bhakra-Nangal project design set a water allowance of 2.75 cusecs of capacity at the head per 1,000 acres of CCA in the unrestricted perennial irrigation area of Zone III, in which the study area is located. Dhabi Minor's full supply discharge at the head per 1,000 acres CCA is 0.35 cusecs less than the duty of water established for this area. Thus, at full supply, this canal carries 3.65 cusecs *less* than it should for its assigned CCA; functionally, this has meant that the duty of water in the CCA of Dhabi Minor is even greater than planned for channels providing perennial irrigation (Sub-Divisional Officer [Irrigation], Fatehabad, personal communication, August 1966).

DHABI MINOR: IRRIGATION DELIVERIES

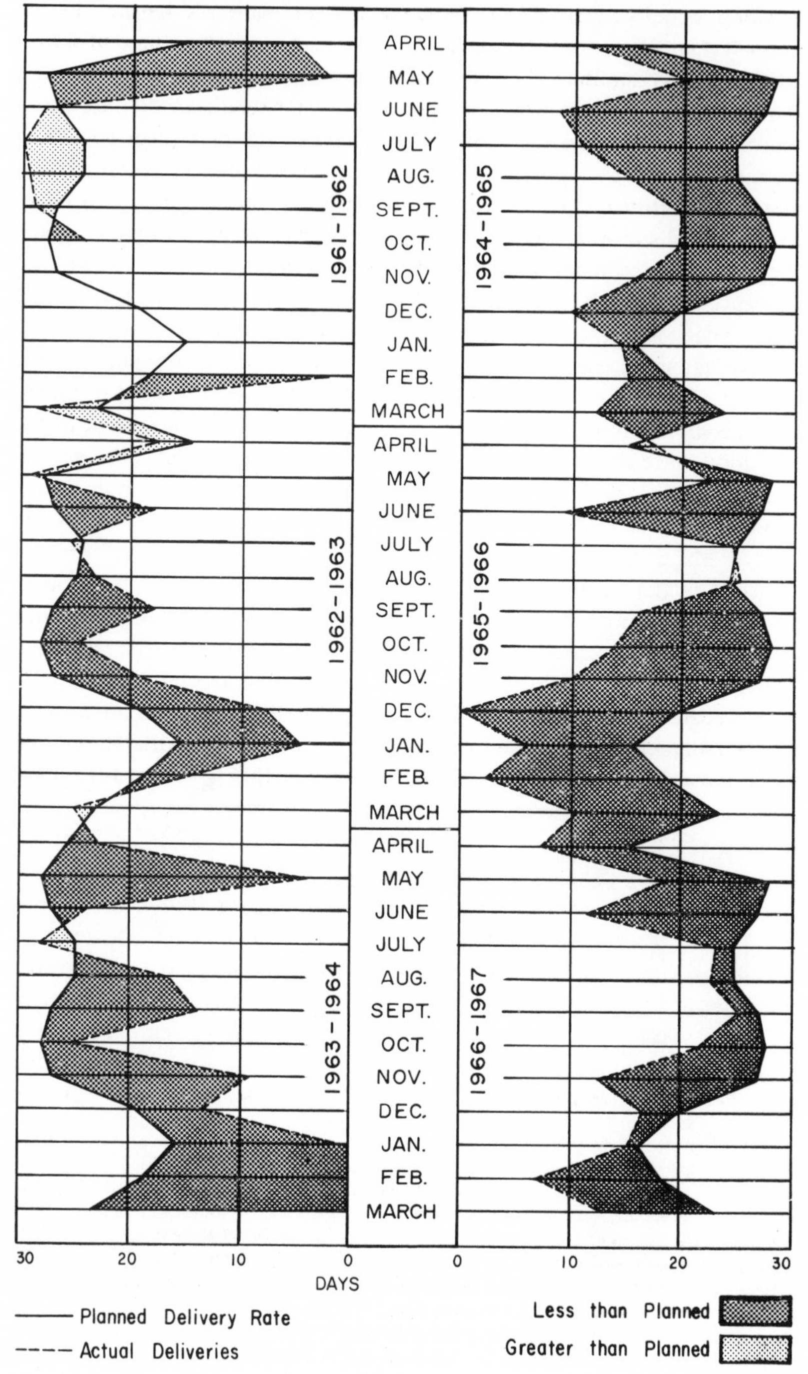
Actual vs. Planned Full Supply Days, Perennial Irrigation 1961-62 To 1966-67
APRIL
MAY
JUNE
JULY
AUG.
SEPT.
OCT.
NOV.
DEC.
JAN.
FEB.
MARCH
1961-1962
1962-1963
1963-1964
1964-1965
1965-1966
1966-1967
30
20
10
0
0
10
20
30
DAYS
Planned Delivery Rate
Actual Deliveries
Less than Planned
Greater than Planned

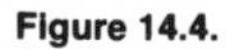

Figure 14.4.

the resulting patterns of agricultural land use in the area served by the Bhakra irrigation system. If maximizing agricultural productivity per volume of water provided by the system had been the goal, the water allowance for particular areas would have been adopted on the basis of water requirements per crop, the seasonal ratio of irrigation, and the area planted to each crop per 1000 acres CCA. Under such a plan, the effects of particular environmental factors in an area, primarily relief and soils, on a planned crop pattern would have been minimized. Adoption of a fixed water allowance *regardless* of areal differences, however, magnified the significance of relief and soils, as well as field location with respect to the water distribution system, as factors in the resulting patterns of land use in the command of the Bhakra project.

The irrigation allowance for each cultivable acre in the command or *chak* of an outlet on the canal is determined by dividing the total amount of water that an outlet can deliver at full supply for a week by the total cultivable acres in the *chak*. The resulting irrigation allowance of so many minutes per acre is then multiplied by the number of acres in a farmer's cultivation unit in the *chak* to determine the total length of his irrigation turn during the rotation of turns in an irrigation week. Again, a fixed water allowance has been established irrespective of environmental variation and field location in the *chak*.

Seen in this context, the implicit goal in the perennial irrigation area of the Bhakra project surely was to *protect* agriculture from the effects of a failure of the monsoon rains and subsequent drought, a common occurrence in the Haryana Bagar. The duty of water was set to ensure that a maximum amount of CCA would be established and supplied with water, but the productivity of irrigated lands would not be maximized in these areas.[17] Presumably, most farmers would perceive the irrigation system as providing a "floor" below which their productivity would not fall and they would adjust their cropping patterns accordingly. That a few farmers would shift to growing higher yielding, more profitable, *and* more water-consumptive crops must have been anticipated. In view of the weaknesses in the final project design, however—known at the time of its adoption—and the methods adopted to determine water allowances and farmers' irrigation turns, the Bhakra system obviously could supply water in sufficient amounts and at suitable times over only a small portion of the total CCA to sustain these highly water-consumptive crops.[18] The subsequent pattern of concentration of such crops at specific

17. Others have concluded similarly. For example, Hari Lal Sally has noted that the objective in distributing irrigation supplies on the Bhakra canals was to provide irrigation to the maximum area *regardless of the adequacy of supplies for essential crops*. . . . The available supplies have been utilized to the full and on the maximum area by reducing the water allowances to the barest minimum and without adjusting them on the basis of rainfall, soil and other characteristics of the tracts concerned" (emphasis mine) (Hari Lal Sally, *Irrigation Planning for Intensive Cultivation* [New York: Asia Publishing House, 1968], p. 85).

18. Using officially supplied data and other published studies, Raj (*Some Economic Aspects*, pp. 94–102) attempted to assess the likely impact of Bhakra irrigation on agricultural production

locations in the irrigation system less obviously conflicted with developmental policies and goals.

Irrigation and Agriculture in Dhabi Kalan

As previously noted, irrigation opened up new agricultural options for farmers in Dhabi Kalan. In so doing, however, it also produced an unintended concentration of benefits by creating a pattern of differential access to the water essential for the cultivation of the new crops. The spatial pattern of irrigated and dry crops as well as fallows and crop failures in Dhabi Kalan suggests that local relief is a significant land-use variable. The 670-foot contour is a frequent barrier to irrigation, occurring throughout the lands of the village. Seldom are fields at this height or higher actually irrigated, even though they may adjoin frequently irrigated fields, and then only with difficulty in this gravity flow system. Increasing distance from the head of irrigation and correspondingly lower volumes of flow in the watercourse network also means that the maximum elevation of fields sown to irrigated crops diminishes.[19]

Often minor fluctuations in field and watercourse elevations pose substantial problems for the irrigation farmer in a gravity-flow irrigation system. Leveling fields, improving the layout of field channels, and maintaining these improvements are not readily and easily accomplished tasks, and many farmers in Dhabi Kalan had not eliminated such obstacles to more efficient and effective delivery of water to their crops.

Much more significant in determining the spatial pattern of irrigation agriculture in the various *chaks* of Dhabi Minor are the rates of water loss through seepage, rates that increase with increasing distance from the watercourse outlets. Throughout Dhabi Kalan, soils are sandy and permeability is high. Progressive losses through seepage in the unlined watercourses considerably reduce the velocity of water flow in the more distant reaches; generally poor levels of watercourse maintenance, siltation of the watercourses, and the varying dimensions of the channels from place to place combine to reduce further the velocity of flow in the watercourses, while raising seepage losses.

In *chak* 16000 Left, Dhabi Minor (Fig. 14.5), the average rate of water loss in the first 2000 feet of watercourse was 20 percent; it increased to 40

---

throughout the commanded area; at the time, no specific field studies had been conducted for such a purpose.

19. This problem was anticipated by the irrigation engineers. The formula used to determine the area designated in the CCA of an outlet reduces the maximum elevation at which a field is classified as irrigable by 0.2 feet per 1,000 feet from the head of irrigation.

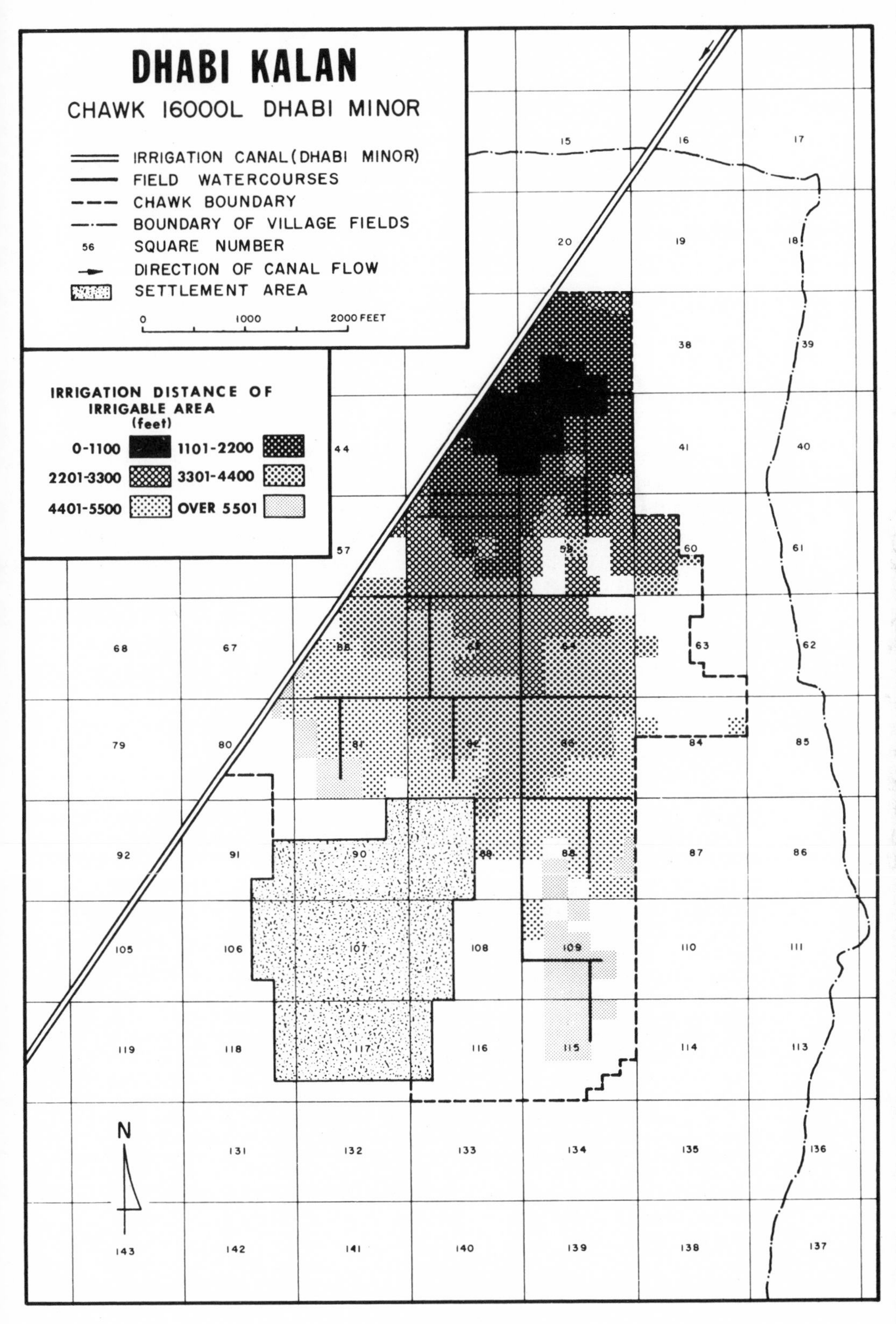
DHABI KALAN
CHAWK 16000L DHABI MINOR
IRRIGATION CANAL (DHABI MINOR)
FIELD WATERCOURSES
CHAWK BOUNDARY
BOUNDARY OF VILLAGE FIELDS
56 SQUARE NUMBER
DIRECTION OF CANAL FLOW
SETTLEMENT AREA
0
1000
2000 FEET
IRRIGATION DISTANCE OF IRRIGABLE AREA (feet)
0-1100
1101-2200
2201-3300
3301-4400
4401-5500
OVER 5501
N

Figure 14.5.

Table 14.3.

Land use on irrigable area, *chaks* 16000 left and 21000 right, Dhabi Minor, 1964-66 (percent)

Irrigation distance of fields (feet from outlet)	Irrigated		Dry Cropped		Fallow	
	16000 L	21000 R	16000 L	21000 R	16000 L	21000 R
Less than 1101	64%	41%	3%	13%	33%	46%
1101–2200	59	32	16	25	25	43
2201–3300	43	26	28	33	29	41
3301–4400	34	24	29	33	38	43
4401–5500	29	16	30	38	41	47
More than 5500	24	19	38	30	38	51

percent over the next 2000 feet of watercourse.[20] Farmers in this and all other *chaks* in Dhabi Kalan were keenly aware of the water-loss/distance relationship, for they altered the amount of time used to irrigate an acre. For example, a farmer whose field was located one mile from the outlet for *chak* 16000 Left typically took twice the time to irrigate an acre of cotton or wheat that a farmer did for a field located within 1100 feet of the outlet.[21]

The general pattern of land use in *chaks* 16000 Left and 21000 Right, Dhabi Minor, confirms that the frequency and intensity of irrigated agriculture in Dhabi Kalan is directly related to the distance of fields from the source of

20. Parshall flumes were used to measure field watercourse discharges at the head of irrigation, at the midpoint of the watercourse, and at three tail locations during a complete irrigation week. Measured values were averaged for the week to reduce the effect of slight alterations in the discharge of the Minor. The resulting rates of water loss were somewhat higher than Bhakra project estimates (Raj [*Some Economic Aspects*, p. 92] cites a 20 percent loss rate for watercourses), but within the range of estimates made by knowledgeable local Irrigation Department personnel. Certainly the higher rates of loss are closely related to the more permeable sandy soils of this region.

21. The time needed to irrigate an acre increased rapidly with distance from the outlet. Farmers with fields one-half mile from the outlook took as much as 70 percent more time to irrigate an acre than those close to the outlet. Of course, these measurements assume that farmers were attempting to deliver similar irrigation purposes: this assumption was generally validated by direct field observations, though a few minor differences were noted (Vander Velde, "Distribution of Irrigation Benefits," pp. 144–53). Similar adjustments to the distance/water loss relationship were made by farmers in other villages in Fatehabad tehsil under the command of the Bhakra irrigation system. The difference in time taken by farmers to irrigate one acre in two other villages served by Kheri Distributary, which also supplies Dhabi Minor, was as great as four hours per acre in a distance interval of 1,500 feet (Repetto, *Time in India's Development Programmes,* p. 77).

Without doubt, the difficulties experienced by farmers in making an adequate adjustment to water losses in the watercourse network were compounded by the interpenetration of canal rotational operations and the *warabandi*, the roster of irrigation turns that governs the distribution of water among co-shares in each *chak*. This problem is cogently explained in Richard B. Reidinger, "Institutional Rationing of Canal Water in Northern India: Conflict between Traditional Patterns and Modern Needs," *Economic Development and Cultural Change* 23, no. 1 (1974): 79–104.

Table 14.4.

Frequency of irrigated agriculture, *chaks* 16000 left and 21000 right, Dhabi Minor, 1964-66

		Percentage of irrigable area				
		Irrigated crops				
Irrigation distance (feet from outlet)	Total irrigable area (acres)	4 seasons	3 seasons	2 seasons	1 season	Dry crops
0–2200	160	8%	29%	32%	16%	14%
2201–4400	255	4	16	28	20	22
More than 4401	251	4	10	15	14	36

irrigation. As distance from the watercourse outlet increases, the percentage of irrigable area on which irrigated crops are grown decreases while that devoted to dry crops increases substantially, and the area in fallows increases slightly (Table 14.3).

During the four seasons of 1964–66, 37 percent of the irrigable area located within 2200 feet—that is, within ten acres—of the watercourse outlet was planted to an irrigated crop three or four times; more than two-thirds of this area had an irrigated crop at least twice. By contrast, 57 percent of the irrigable area at distances greater than 4400 feet or more than twenty acres from the outlet had no irrigated crop in any of those four seasons (Table 14.4). Indeed, with few exceptions, fields at those distances that carried an irrigated crop three or all four seasons were located adjacent or very near to the watercourse; more distant fields served by laterals were rarely if ever irrigated (Fig. 14.6).

Three broad patterns of land use can readily be identified. Within an irrigation distance of about one-half mile, an average of 50 percent of the irrigable area will be planted to irrigated crops in any season. Fallowing is frequent here, too; about one-third of the area in any season is in fallow. This pattern reflects dependence upon fallowing as a technique to reduce the rate of nutrient losses in irrigation agriculture and partially to restore soil fertility, still necessary given the low levels of commercial fertilizer use in agriculture in Dhabi Kalan, while adjusting seasonal and yearly agricultural labor requirements to available labor supplies. In the middle-distance range, a well-mixed pattern of irrigated fields, dry crop farming, and fallow land prevails, though fallow land now exceeds 40 percent of the irrigable area in any season. In the lower reaches of the watercourse network, particularly at irrigation distances greater than one mile from the outlet, less than one-quarter of the irrigable area typically is planted to irrigated crops in any season, whereas more than one-third of this area is dry farmed, virtually a reversal of the pattern for these two land-use catagories in the middle-irrigation-distance locations. Here marginal agricultural conditions prevail for most farmers, and

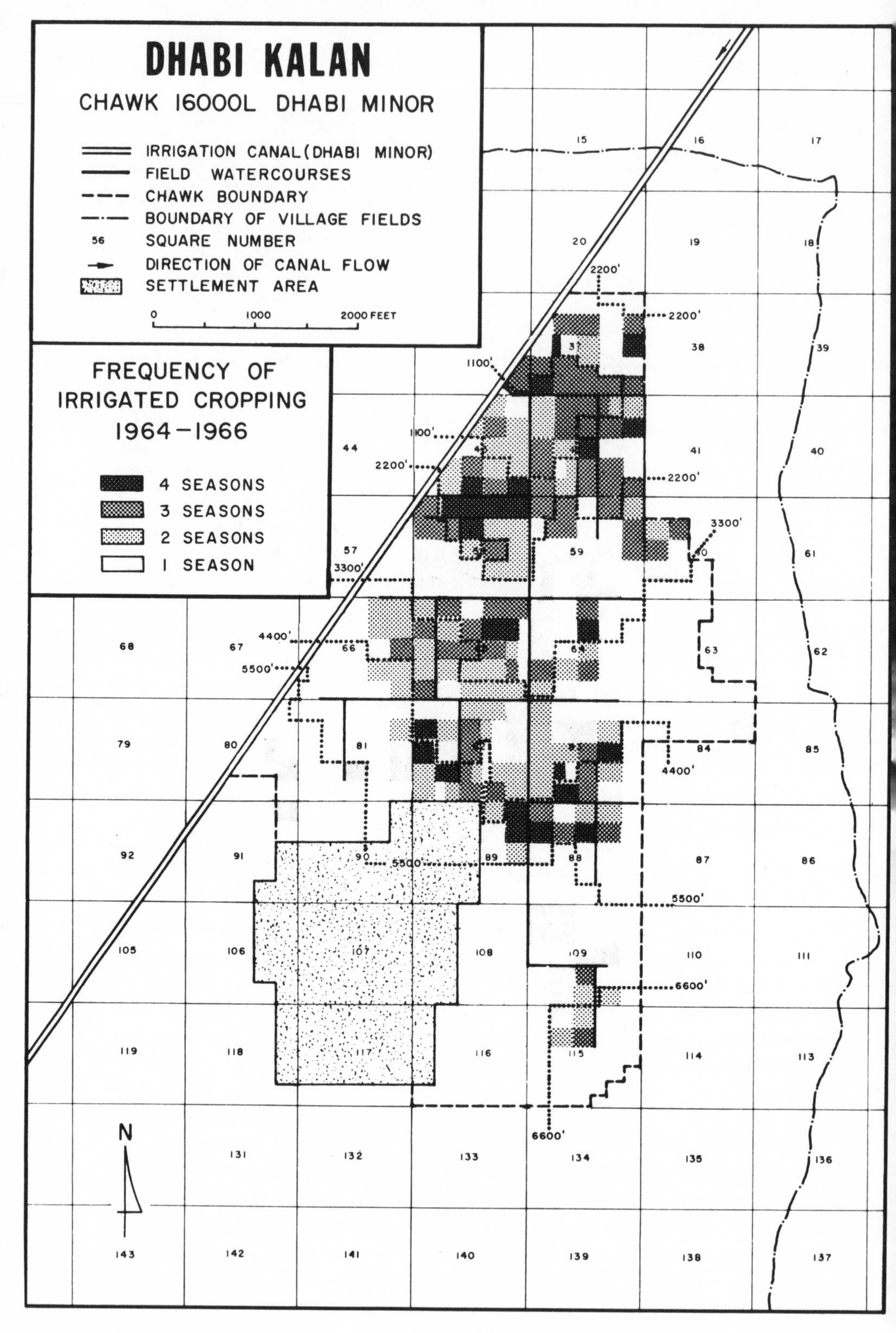
DHABI KALAN
CHAWK 16000L DHABI MINOR
IRRIGATION CANAL (DHABI MINOR)
FIELD WATERCOURSES
CHAWK BOUNDARY
BOUNDARY OF VILLAGE FIELDS
56 SQUARE NUMBER
DIRECTION OF CANAL FLOW
SETTLEMENT AREA
0 1000 2000 FEET
FREQUENCY OF IRRIGATED CROPPING 1964–1966
4 SEASONS
3 SEASONS
2 SEASONS
1 SEASON
N

Figure 14.6.

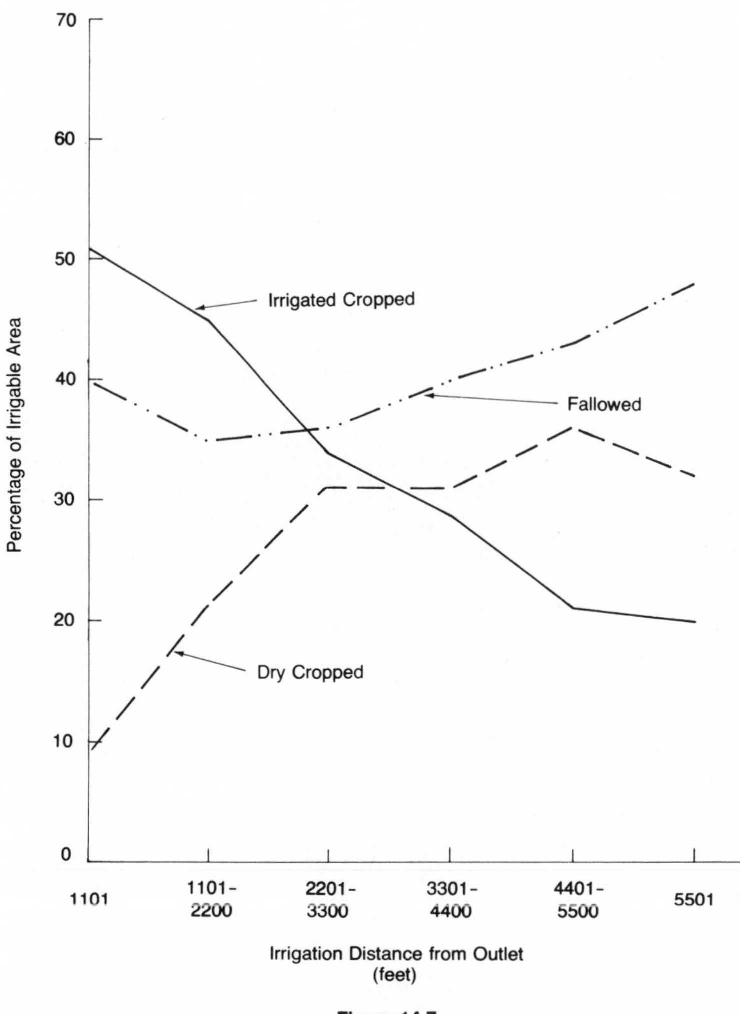

Figure 14.7.

many are nearly as dependent upon rainfall to supplement the marginal amounts of water delivered by the irrigation system or for dry crop farming as they were in the pre-Bhakra period (Fig. 14.7).

Land Tenure and Land Reform in Dhabi Kalan

Only the Jats and a few other families high in the social hierarchy of Dhabi Kalan own sufficient land to support the individual household (Fig. 14.8). Most Kumhars and virtually all Harijan families whose livelihood is agriculture must either lease land on shares or for cash rents, usually from their Jat

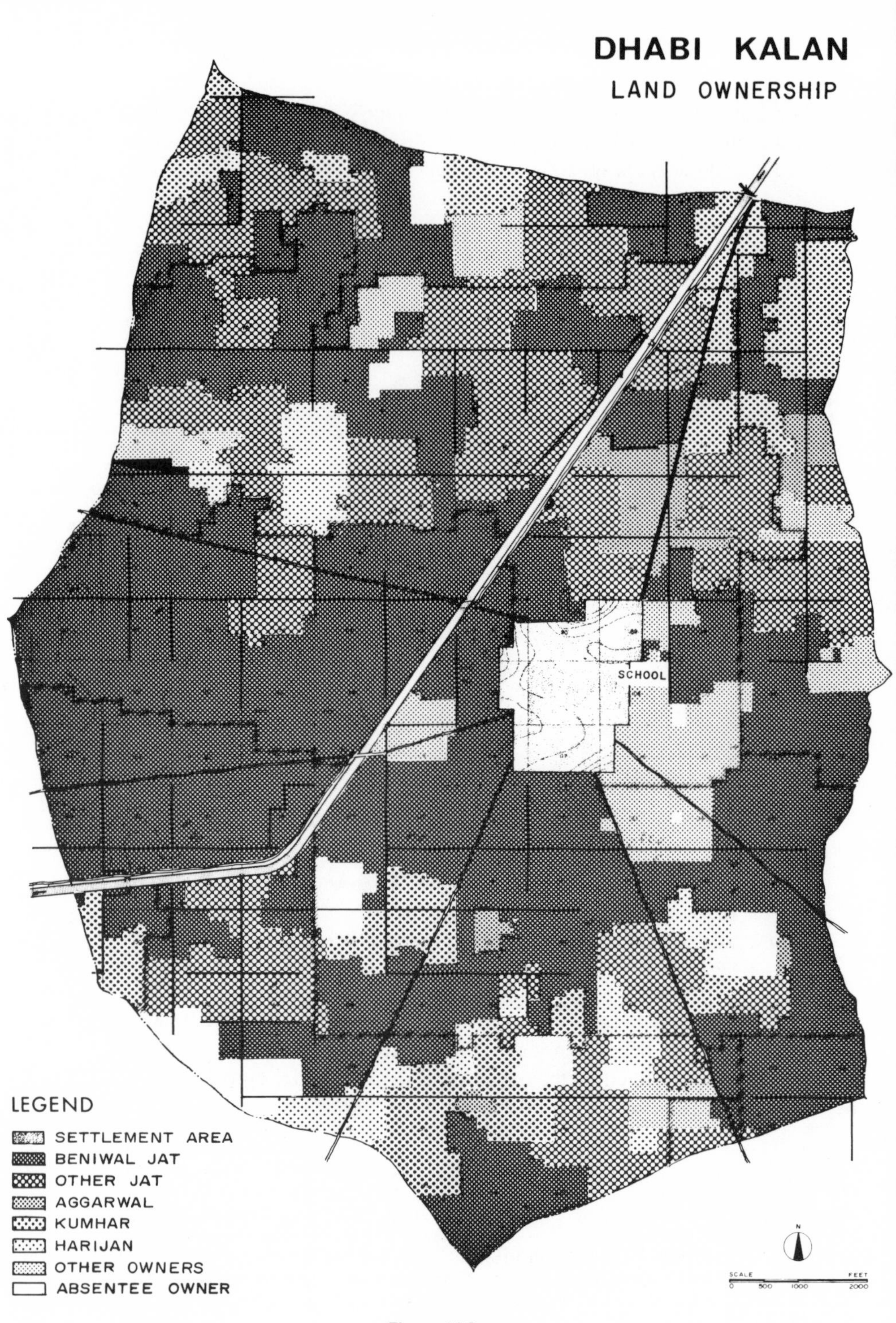
DHABI KALAN
LAND OWNERSHIP
SCHOOL
LEGEND
SETTLEMENT AREA
BENIWAL JAT
OTHER JAT
AGGARWAL
KUMHAR
HARIJAN
OTHER OWNERS
ABSENTEE OWNER
N
SCALE
FEET
0
500
1000
2000

Figure 14.8.

neighbors, or find work as seasonal or permanent agricultural laborers. Most of these families successfully piece together bits of agricultural land from two or three Jat landowners or obtain an occasional large holding from a single Jat family; in virtually all such cases, sharecropping or *batai* is the leasing arrangement. The Punjab Security of Land Tenures Act (1953) limited the maximum amount of produce to which a landowner was entitled as rent under any *batai* arrangement to a one-third share.[22]

Four distinct kinds of sharecropping leases are used in Dhabi Kalan:

- *Tisra batai*: Two-thirds of the produce is kept by the cultivator and one-third goes to the owner of the land. Commonly, all inputs—seed, fertilizers if used, animal power and labor—are supplied by the cultivator, who also typically decides on the crop to be cultivated.
- *Paanjaa batai*: Three-fifths of the produce goes to the cultivator and two-fifths is given to the landowner. Again, all inputs usually are provided by the cultivator, and he also decides on the crop to be cultivated. Of the four sharecropping arrangements, *paanjaa* leases are the least common.
- *Aadh batai*: The production is divided evenly between cultivator and landowner. Generally inputs are also shared on a fifty-fifty basis, although exceptions do occur; typically the crop to be cultivated is jointly agreed. This is the single most common tenure arrangement in Dhabi Kalan.
- *Tijiya hissa*: One-third of the production is kept by the cultivator and two-thirds is turned over to the landowner, who also supplies seed, fertilizers if used, and animal power. The cultivator supplies only his labor. The owner of the land selects the crop to be grown.

Tenure agreements under any of these four types of *batai* are made on a year-to-year basis, and while they are often renewed for a succession of years between the same landowner and tenant, more often than not the fields cultivated by the tenant in one year are not those he cultivated the previous year or

22. The land-reform legislation referred to here and earlier established a ceiling on the ownership of agricultural land by a single owner at thirty standard (that is, irrigated) acres or sixty ordinary acres, provided security of tenure under certain conditions for occupancy tenants, set the maximum limit of produce to which a landowner is entitled as rent under any sharecropping arrangement at one-third share, and provided for the redistribution of land held in excess of the ceiling limitation among the landless or small landowning peasants (Government of Punjab, Legislative Department, *Land Code*, vol. 2, pp. 275–89, 305–96). That shares or rents exceeding one-third of the crop remain common in Punjab and Haryana has been noted elsewhere, though their caste distribution has not been discussed. See W. Ladejinsky, *A Study on Tenurial Conditions in Package Districts* (New Delhi: Government of India, Planning Commission, 1965), pp. 33–37; Board of Economic Inquiry, Punjab, *Resurvey of Village Birampur, District Hoshiarpore*, no. 88 (Chandigarh: Punjab Government, Economic and Statistical Organisation, 1963), pp. 16–17; Board of Economic Inquiry, Punjab, *Resurvey of Village Suner, District Ferozepore*, no. 89 (Chandigarh: Punjab Government, Economic and Statistical Organisation, 1963), p. 45; J. S. Uppal, "Implementation of Land Reform Legislation in India: A Study of Two Villages in Punjab," *Asian Survey* 9, no. 5 (1969): 359–72.

the ones he will cultivate the following year. In this way, the landowner ensures that the tenant cannot legally claim permanent occupancy of and ultimately proprietary rights over a portion of the owner's land. If a tenant can prove continuous cultivation of the same plot or plots for six years or more, under the provisions of the Security of Land Tenures Act (1953) he completes the first step necessary to substantiate a claim to occupancy of that land on a permanent basis. For similar reasons, landowners also insist upon oral leases; only one written lease existed among all the tenancies studied in Dhabi Kalan.

For the lessee, the favorable-to-unfavorable hierarchy of the four types of sharecropping tenures is immediately apparent. Certainly a one-third/two-thirds (*tijiya hissa*) lease extracts from the tenant a very high (and illegal) rent that makes him little more than an agricultural laborer. In fact, many landowners leasing land on such an arrangement do not verbally distinguish between their tenant and yearly contract agricultural labor. The word used to refer to both, *siṛi*, literally means "servant" or "helper."

Among the nonrandom sample of tenancies studied, 76 belonged to one of the four classes of sharecropping leases (Tables 14.5 and 14.6).[23] Thirty-six Jat households, who owned more than 55 percent of the cultivable land (2191 acres) in Dhabi Kalan, leased out 36 percent of their agricultural land through 71 sharecropping leases, primarily to members of the Harijan castes (29 tenancies) or to Kumhar families (21 tenancies). The preeminent position of the Beniwal clan as landowners among the Jats in Dhabi Kalan was reflected in the 62 tenancies held by 28 Beniwal households; they accounted for 86 percent of all agricultural land leased out by Jat landowners in the survey.

Not a single Jat or caste Hindu household, other than Kumhar households, in the survey cultivated land under a *tijiya hissa* lease, but 72 percent of all land leased to Harijans and 47 percent of that leased to Kumhars were leased through *tijiya hissa* tenure. By contrast, 80 percent of the land leased by Beniwal Jats to other Jat farmers and 53 percent of that leased to other caste Hindus were let on the much more favorable *paanjaa* and *tisra* leases, suggesting a close correlation between position in the social hierarchy and an ability to acquire a favorable sharecropping arrangement. Only three Harijan tenancies and eight Kumhar leases with Jat landowners, totaling a very modest 4 percent and 23 percent, respectively, of the land leased to members of these communities were under such favorable terms of tenure.

Clearly, the implementation of land reform in Haryana simply has not

23. This discussion is based on a detailed examination of 109 separate tenancies in Dhabi Kalan for which it was possible to collect data. Not surprisingly, securing information about land-tenure arrangements proved to be extremely difficult, so much so that data collection had to be halted before the survey was completed. Most of the tenancies studied cover both seasons of the 1965–66 agricultural year, though they are not a complete enumeration of all such arrangements operative throughout that period. There is reason to believe, however, that about two-thirds of the existing tenure agreements were identified, and the data collected do include every type of tenancy said to exist in Dhabi Kalan. Therefore, they are reasonably representative of the tenancy situation in the village.

Table 14.5.

Sharecrop tenancies in Dhabi Kalan: Jat lessors

Lessees	Total	*Tisra*	*Paanjaa*	*Aadh*	*Tijiya*	Season[a]
	Number of leases by tenure					
Jats	9	3	1	5	0	
Kumhars	21	5	3	7	6	
Other caste Hindus	13	3	1	9	0	
Harijans	28	2	1	13	12	
All lessees	71	13	6	34	18	
	Total acreage					
Jats	64	6	30	28	0	K
	76	8	30	38	0	R
Kumhars	218	29	22	69	98	K
	222	29	22	71	100	R
Other caste Hindus	61	17	13	31	0	K
	61	17	13	31	0	R
Harijans	437	7	12	88	330	K
	441	7	12	92	330	R
All lessees	780	59	77	216	428	K
	800	61	77	232	430	R
	Irrigated acreage					
Jats	25	1	15	9	0	K
	14	3	5	6	0	R
Kumhars	43	1	0	17	25	K
	31	3	2	10	16	R
Other caste Hindus	15	2	0	13	0	K
	15	2	0	13	0	R
Harijans	92	3	0	24	65	K
	55	3	0	15	37	R
All lessees	175	7	15	63	90	K
	115	11	7	44	53	R

Note: All acreage values to nearest full acre.
[a] K = *kharif*; R = *rabi*.

been successful in delivering security of tenure to tenants or, insofar as it represented an effort to diminish the concentration of wealth among the caste-based peasant elite through extractive land tenures, in limiting the rent they could be charged for the use of the land. Indeed, given that forty Jat households have holdings of thirty acres or more, and that twenty-one of these holdings contain more than thirty irrigable acres each, its failure in limiting the amount of agricultural land that individuals could own also must be acknowledged. The failure of public policy to reform either concentrated ownership in rural Haryana or to redress the unequal balance of social power and economic wealth in village society, however, was not solely the result of poorly designed legislation or subsequent legal and illegal activities to circumvent the ceilings on landholdings and rents. Nor did failure stem merely from poor enforcement of the new laws. It also failed, at least in Haryana,

Table 14.6.

Sharecrop tenancies in Dhabi Kalan: Jat Beniwal lessors

Lessees	Number of leases by tenure					
	Total	*Tisra*	*Paanjaa*	*Aadh*	*Tijiya*	Season[a]
Jats	7	3	1	3	0	
Kumhars	19	5	3	5	6	
Other Hindu castes	11	3	1	7	0	
Harijans	25	1	1	12	11	
All lessees	62	12	6	27	17	
	Total acreage					
Jats	40	6	30	4	0	K
	52	8	30	14	0	R
Kumhars	211	29	22	62	98	K
	213	29	22	62	100	R
Other Hindu castes	56	17	13	26	0	K
	56	17	13	26	0	R
Harijans	366	2	12	87	265	K
	371	2	12	92	265	R
All lessees	673	54	77	179	363	K
	692	56	77	194	365	R
	Irrigated acreage					
Jats	19	1	15	3	0	K
	8	3	5	0	0	R
Kumhars	37	1	0	11	25	K
	28	3	2	7	16	R
Other Hindu castes	13	2	0	11	0	K
	13	2	0	11	0	R
Harijans	85	1	0	23	61	K
	51	1	0	15	35	R
All lessees	154	5	15	48	86	K
	100	9	7	33	51	R

Note: All acreage values to nearest full acre.
[a]K = *kharif*; R = *rabi*.

because the resource environment was, albeit coincidentally, fundamentally altered by the provision of irrigation to a large area of the state.

Even the government fully recognized the superior quality of irrigable acreage over its rainfall-dependent dry crop counterpart when it set the land-holding ceiling at thirty standard acres, or sixty ordinary (i.e., dry crop) acres, thereby implicitly establishing the value of irrigable land as twice that of dry crop land. Cultivators in this semiarid region quickly came to hold a similar regard for the worth of irrigable land in comparison with dry crop land.[24] In the erratic rainfall environment of the Haryana Bagar, acquiring and maintain-

24. To obtain a rough estimate of the relative values of irrigable and nonirrigable land as they were perceived by farmers in Dhabi Kalan, 40 Jats, 32 Kumhars, and 40 Harijans were asked the maximum price they would pay, if they could pay any price, for "good" irrigable and "good" nonirrigable land. The responses of these farmers suggest that nonirrigable land is "worth" about 60 percent of the value of irrigable land. Jat farmers tended to ascribe a slightly higher price to both irrigable and nonirrigable land than did Kumhar and Harijan farmers.

ing a large, extensively cultivated holding was now no longer essential "survival insurance." Instead, the possession of irrigable acreage became the key to agricultural success and prosperity, so much so that nonirrigable land might be considered scarcely worth the effort to farm, particularly if the holding was very large and distant from the canal or poorly served by the watercourse network. A more satisfactory alternative, however, was readily available.

Land Reform and Irrigation: The Coalesence of Time and Location

To understand the meaning and future implications of present land-reform legislation and irrigation development in Haryana, we must shift our focus from the failure of land reform to accomplish its stated goals to the consequences of its structural weaknesses and its implementation coincident with the development of perennial irrigation. In the tenancies surveyed, in Dhabi Kalan, about 36 percent of the total acreage leased out by Jats was actually planted to an irrigated crop during the agricultural year, irrespective of season. Nearly 50 percent of this irrigated acreage was leased under the most exploitative terms of tenure by Harijan and Kumhar cultivators; no such lease was negotiated between a landowner and any Jat or other caste Hindu lessor. This is the price that most Harijan farmers and many lower-caste farmers must pay to participate in the changing and more remunerative cropping patterns made possible in Dhabi Kalan by the Bhakra irrigation system, given that 80 percent of Harijan households and 35 percent of lower-caste houesholds own no irrigable agricultural land whatsoever. Equally significant, about 60 percent of the dry crop area leased out by Jat landowners was also under the most exploitative terms of tenure to Harijan and Kumhar cultivators, accounting for very nearly 25 percent of all nonirrigated cultivated area in Dhabi Kalan.

A more detailed examination of the pattern of landownership by caste and of cultivation units with tenants in *chak* 16000 Left, Dhabi Minor, makes the emerging pattern more explicit. In this *chak* Jat farmers own 61 percent of the CCA, typically in fairly large—ten acres or more—and compact blocks, distributed throughout the *chak*; a large and compact unit is also owned by an Aggarwal household (Fig. 14.9). Altogether, these farmers control nearly 75 percent of the CCA in the *chak*. Kumhar farmers control a small, compact block of land immediately east of the outlet, virtually all the fields of which are located within 3000 feet of the watercourse outlet. This advantageous location with respect to irrigation is somewhat negated, however, by the fact that this block of land is divided into ten separate cultivation units, the largest of which is about six acres.[25]

25. A study of irrigation agriculture in three other villages in Hissar District revealed some similar patterns. Cultivators of small operational holdings (fewer than fifteen acres) were located at a maximum of 2.5 miles from the canal outlet; the maximum distance at which farmers of large holdings (over twenty-five acres), however, were found was only 1.5 miles from the outlet. Jats were most often farmers of large holdings in these villages, too (C. R. Kaushik, "Farm Adjust-

The cultivation units in which tenants sharecrop irrigated land and the locations of these units in the *chak* are shown in Figure 14.10; details of these tenancies are given in Table 14.7. Harijan farmers (Chamar and Dhanak) were able to lease land in the nearer irrigation distance zone (cf. Fig. 14.5) only on the usurious *tijiya hissa* terms. A Kumhar farmer, however, obtained irrigable land at a similar location on a fifty-fifty share basis. At the more extreme irrigation distances, irrigable fields were farmed by both Harijan and Kumhar farmers on *aadh* leases, and another Harijan farmer had a tenancy on a *paanjaa* agreement. Land is available on the legal *tisra* terms or on cash rents averaging 50 rupees per acre per season only at maximum irrigation distances in the *chak* and provided the farmer has bid successfully for its use in the seasonal "auction" of cultivation rights to the village panchayat land. In fact, not a single field in this area was actually irrigated in any season from 1964 to 1966. Although this sample of tenancies was very small, it is notable that the area of the tenure actually irrigated also declined sharply as the irrigation distance of the cultivation units from the outlet for the *chak* increased, a function both of the difficulty in securing an adequate irrigation supply at greater distances from the outlet and of the tendency of the Jat landowners to retain for themselves part or all of the irrigation turn the tenanted unit would otherwise obtain.[26] Indeed, dry crop farming and large seasonal fallows characterized agriculture in the two most distant tenancies.

Despite its weaknesses and its failure to accomplish its stated goals, land-reform legislation in Haryana did in fact spur the redistribution and reconsolidation of agricultural land among those who already controlled it.[27] The structural weakness of the legislation facilitated the retention of large land-holdings within the family. Coincidentally, irrigation development and the methods of operation of the irrigation system transformed these large holdings, now comprising mixed amounts of highly valuable irrigable land and much less desirable dry crop land, into an even greater asset than they had

ment on the Introduction of New Irrigation Facilities in Canal-Irrigated Area of Hissar District," unpublished M.Sc. thesis, Punjab Agricultural University, Ludhiana, 1966, p. 23.

26. Estimated land values in *chak* 16000 Left follow a similar hierarchial pattern. A number of Jat farmers, all recognized as excellent farmers, were asked the maximum price they would "bid" for specific irrigable fields at various locations in the *chak*. Fields rather close to the watercourse outlet were "worth" about 2,500 rupees per acre, those in middle irrigation distance locations were "worth" 500–700 rupees per acre less, and theoretically irrigable fields at maximum irrigation distance locations were thought to be "worth" only 1,500 rupees per acre. This procedure had to be adopted to determine land values in Dhabi Kalan because very little land had been bought or sold since the beginning of perennial irrigation. Official records of land sales proved to be useless for this purpose since, for a variety of reasons, land sale prices are either inflated or deflated.

27. In fact, Jat households secured 54 percent of the total of 272 acres to which tenants obtained permanent occupancy rights of tenancy in Dhabi Kalan under the terms of the land-reform legislation, including slightly more than two-thirds of all irrigable area within this acreage (Vander Velde, "Distribution of Irrigation Benefits," pp. 52, 249).

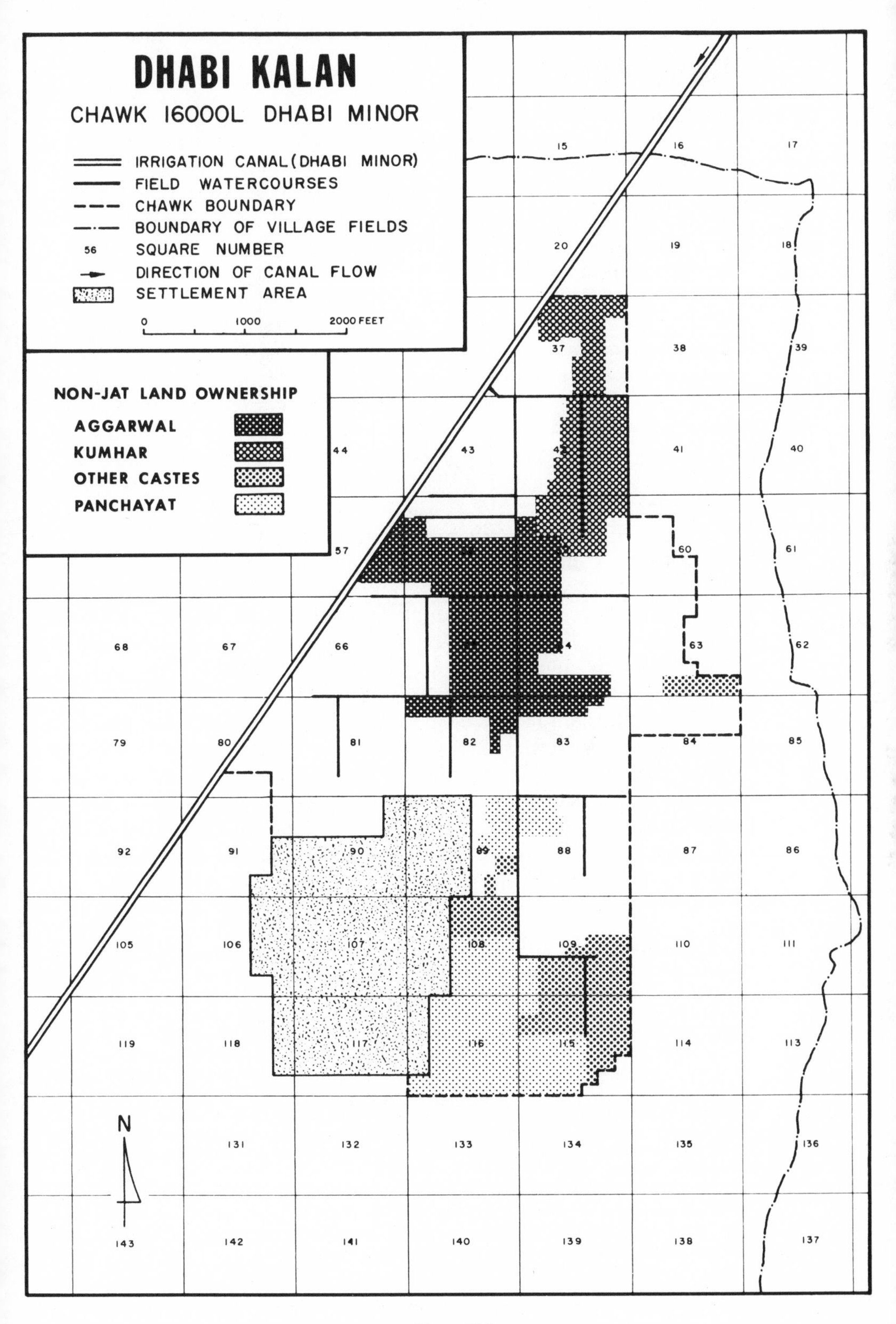
DHABI KALAN
CHAWK 16000L DHABI MINOR
IRRIGATION CANAL (DHABI MINOR)
FIELD WATERCOURSES
CHAWK BOUNDARY
BOUNDARY OF VILLAGE FIELDS
56 SQUARE NUMBER
DIRECTION OF CANAL FLOW
SETTLEMENT AREA
0 1000 2000 FEET
NON-JAT LAND OWNERSHIP
AGGARWAL
KUMHAR
OTHER CASTES
PANCHAYAT
N

Figure 14.9.

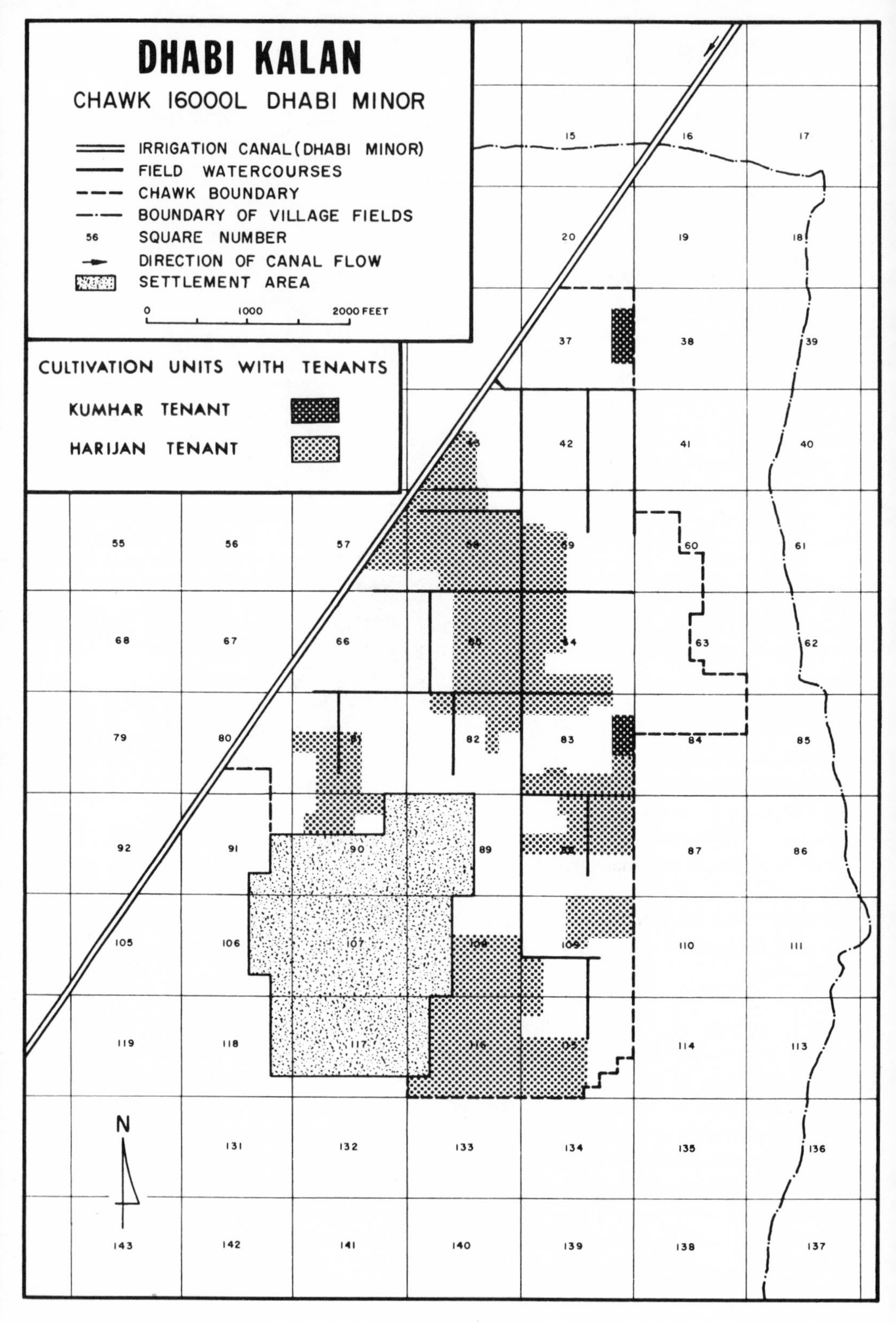
DHABI KALAN
CHAWK 16000L DHABI MINOR
IRRIGATION CANAL (DHABI MINOR)
FIELD WATERCOURSES
CHAWK BOUNDARY
BOUNDARY OF VILLAGE FIELDS
56 SQUARE NUMBER
DIRECTION OF CANAL FLOW
SETTLEMENT AREA
0 1000 2000 FEET
CULTIVATION UNITS WITH TENANTS
KUMHAR TENANT
HARIJAN TENANT
N

Figure 14.10.

Table 14.7.

Tenancy, caste, and location in *chak* 16000 Left, Dhabi Minor

Area of tenure (acres)	Irrigated area (acres)		Irrigation distance (feet)	Tenure type	Tenant's caste
	Summer	Winter			
10[a]	9	2½	1101-2200	*aadh*	Kumhar
30[a]	6	5½	1101-2200	*tijiya*	Chamar
26	14	9½	2201-4400	*tijiya*	Dhanak
2	0	1	4401-5500	*aadh*	Kumhar
22	3	1½	4401-6600	*aadh*	Dhanak
20½	1	0	5501-6600	*aadh*	Chamar

[a]This tenure was not wholly in chak 16000 Left.

been. Because the length of farmers' irrigation turns and thus the amount of water to which they are entitled are determined by the size of the cultivation unit in the command of the system, there is even greater reason to retain title to the largest area possible because by doing so one maximizes access to the most scarce resource in this environment.

By shifting tenants and the locations of their tenancies, registering multiple owners in the family, and using oral leases, Jat landowners secured the bulk of their large holdings from the occupancy of present and former tenants and precluded the confiscation of any significant amount of land for redistribution among the landless and low-caste peasantry. The capability and operation of the irrigation system generally restricts irrigation agriculture to a segment of the cultivation unit in any season. Thus, maintaining a large holding is necessary to assure irrigation supplies for that segment in sufficient amounts to offset the water losses experienced in the distribution system as field distance from the outlet increases. The typical Jat cultivation unit, however, still contains a considerable area that can be planted only to nonirrigated crops or fallowed. The maintenance of the usual unirrigated cropping pattern—one dry crop per year—is relatively unprofitable for the large landowner when compared with the potential of irrigated land to yield two crops a year or three crops in four seasons; moreoever, the risk of failure remains high for dry crop agriculture.

The persistence of numerous landless farmers and small landowners as well as the traditional system of sharecropping tenures in the post-land-reform period have made it possible for large landowners to keep substantial amounts of dry crop land in production at a profit even though it is too expensive for them to cultivate it. The frequent acceptance of disadvantageous and illegal tenures by farmers who rank low in the social and economic hierarchy of rural society reveals that those who most needed security of tenure and security from crop failure—the objectives of land-reform legislation and irrigation development—still have it least. By contrast, for the Jat landowners, the returns from such sharecrop tenancies constitute a kind of guaranteed bonus, since their extensive holdings have been largely protected from crop failure through the irrigation system.

In the pre-irrigation Bagar region of Haryana, landowners, owner-cum-tenants, and the landless, despite the obvious maldistribution of the land resource, faced essentially the same level of environmental risk. The *location* of one's cultivation unit counted for relatively little, although those whose fields were at the periphery of the village lands did face a greater investment of labor time in agriculture. This "production cost," however, was relatively insignificant when compared with the need to survive in a harsh environment with few economic alternatives. Rather, large holdings were important in at least partially offsetting the high risk of crop failure in the region, and the landholdings of the Jat clans were by far the largest.

Following the completion of the Bhakra irrigation project, the *location* of one's landholding counts for a great deal, for in large measure it determines who can grow what, how much, and when. Differential access to water in the irrigation system, a result of the interaction of size of holding and location within the *chak*, has tended to concentrate its benefits—greater security from crop failure, the opportunity to grow more valuable crops, and enhanced land values—among a relatively small number of farmers. And the pre-land-reform, pre-irrigation pattern of economic and social dominance, which land reform directly and irrigation at least tangentially sought to break up, continues almost intact.

The recent development and rapid diffusion of higher yielding, hybrid crop varieties may facilitate the further concentration of benefits among these secure and increasingly wealthy farmers. In the Haryana Bagar, irrigation is a prerequisite to the cultivation of these crops, which are far more remunerative than traditional varieties when grown with appropriate amounts of fertilizer and proper pest-control measures. Thus, even more than before, water is the critical means of agricultural production, and when access to water is combined with the concentrated ownership of agricultural land, the functional dominance of local economy and society by the high-caste, landowning agricultural classes is reinforced. In Hissar District, and throughout much of Haryana, this is, as it has long been, the Jat community.

15

Local Organization and Bureaucracy in a Lao Irrigation Project

E. Walter Coward, Jr.
CORNELL UNIVERSITY

> The building of "middle rungs" between traditional society and more modern forms of social organization seems to be a characteristic activity of people caught up in the processes of social transformation. As a group, this family of institutions should be, consequently, of particular interest for students of social and cultural development, highlighting, as they do, some of the central tensions involved in such development and the sorts of mechanisms by means of which those tensions are resolved.
>
> CLIFFORD GEERTZ (1962: 263)

The Problem of Interest

Agricultural development is frequently implemented through irrigation development projects. These irrigation projects are usually based on complex

Previously published as "Indigenous Organisation, Bureaucracy and Development: The Case of Irrigation," *Journal of Development Studies* 13, no. 1 (October 1976): 92–105. Reprinted by permission.

While I assume the usual responsibility for various facts, ideas, and interpretations contained in this paper, there are several organizations and individuals whose assistance I wish to acknowledge. The Ford Foundation provided travel and field support for my visit to Laos during the summer of 1974. While in Laos, I received excellent cooperation from the Directorate of Agriculture and the Agency for Development of the Vientiane Plain, both agencies of the Lao government. Important assistance was also received from the offices of the Agency for International Development. In addition to the 1974 field work in Laos the author worked with International Voluntary Services in rural areas throughout Laos during the period 1958–1964. Previous field studies of irrigation systems in Laos were conducted during the summers of 1969 and 1971. The information reported in this paper is based on information collected during field work in 1969, 1971, and 1974. Three colleagues at Cornell—John Cohen, Jere Gilles, and Gilbert Levine—kindly read early drafts and provided helpful suggestions.

engineering works constructed and operated by agencies of the central government but intended to deliver water to local communities and individuals.

Appraising the successful delivery of irrigation water is a difficult matter because of the complexity of success criteria. A useful set of criteria for evaluating a water delivery system is suggested by Reidinger (1974): "... Its capability to supply water according to the needs of the individual cultivator, and in particular, the predictability, certainty, and controllability of the timing and quantity of his water supply."

One can add that while some systems do this for none of the water users, others do achieve these objectives for some; those close to the head of the canal, those with large landholdings, or those with political influence are often able to achieve the ideal that Reidinger suggests, at least for their own holdings.

A major point of this discussion is that the extent to which Reidinger's criteria are achieved system-wide is dependent, among other things, upon the mode of articulation between the administrative authorities and the local water users.

There have been numerous solutions to this articulation problem. To illustrate, consider the following: the Gezira solution (Gaitskell, 1959), in which the water authority is given strong control not only at the project level but also at the farm level, and individual water users operate within a highly centralized management scheme; the Taiwanese solution (VanderMeer, 1968), in which water users are represented at several levels of project operation through selected representatives; or the California solution (Bain et al., 1966), in which water users form an association to purchase water wholesale from the central authority and retail it to their individual members.

Here, I discuss an alternative solution, applicable where irrigation projects provide water to communities of irrigators already experienced in irrigated agriculture. This solution involves the adaptation and incorporation of indigenous irrigation leadership roles for the achievement of satisfactory system-user interaction.[1] To examine this solution, I first present a case example from western Laos in which a new irrigation system has been constructed by the central government. The project is administered by an agency of the central government but has incorporated an indigenous irrigation role as part of the administrative organization. From that case, several generalizations about the use of indigenous organizational patterns for the implementation of irrigation development will be discussed.

1. Any discussion of irrigation bureaucracy should acknowledge its relationship to the well-known propositions of "hydraulic society" elaborated by Wittfogel (1957). This discussion is not a further "test" of, or direct comment upon, the Wittfogelian theory that posits a causative relationship between large-scale irrigation and a highly centralized and despotic bureaucracy. For one thing, as detailed below, this is not a discussion of a large-scale irrigation system. On the other hand, the discussion is concerned with the interaction between water users and water authorities, and thus, in a most general way, deals with the bureaucracy–irrigation theme so central in Wittfogel's writing.

Description of the Physical Setting and System

The project (the Nam Tan irrigation system, located in the province of Sayaboury in western Laos)[2] is within the Muong Phieng Valley, which is about twenty-one kilometers long from north to south and about five kilometers from east to west at its widest point (Coward, 1972). The floor of the valley is generally flat and 350 meters above sea level. The mountains to the west rise first to about 750 meters and then rapidly to 2,000 meters. The Laos-Thailand border is located in these mountains. The eastern edge of the valley is formed by mountains which rise to over 1,500 meters and separate the valley from the Mekong River.

The main physical structure in the irrigation system is a concrete diversion dam on the Nam Tan River. From this diversion structure, two main canals convey water to a left- and right-bank command area. The system provides supplementary irrigation for wet-rice cultivation during the rainy season (approximately April–May through September–October), and water for a limited area of rice cultivation during the dry season. The irrigable area of the left bank is estimated to be 1,127 hectares; the right bank is estimated to be 919 hectares. Each is further divided into two subsections, or blocks. A varying number of watercourses, or laterals, is found in each of the four blocks of the system.[3]

The overall pattern of water distribution involves three different levels. First, water is continuously served to both the left- and right-bank main canals. Second, within the command area of each bank, water is rotated between blocks. On the left bank, water is delivered to a block for six days at a time. On the right bank, each block receives water for a three-day period. Third, within the block receiving water, water is distributed simultaneously to all laterals.

The approximately 2,000 hectares of the command area serve nearly 900 farmers who live in eleven different villages. Size of holdings is relatively equal in the command area, since at the time the irrigation project was initiated water users were allocated farm holdings of approximately three hectares each. New settlers were assigned plots of this size, and established residents with less than this size were assigned additional land to raise them to this minimum size. Resident farmers holding more than three hectares did not have their additional land confiscated. There are no very large landowners in

2. For selected photographs of the project side, see: *War on Hunger*, Washington: Agency for International Development, June 1971. The text that accompanies the photos is expectedly glowing. Political events in Laos during the period of this project's implementation were extremely volatile. Laos was very much a part of the Vietnamese war, because of the activities of various Lao factions in alignment with the North Vietnamese on the one hand, and the U.S. government on the other. Throughout this period, the western province of Sayaboury remained relatively free of the direct effects of air and ground combat.

3. For a presentation of the basic terms used to identify the engineering components of gravity irrigation systems, see: Asian Development Bank (1973).

the area, resident or absentee. Subsequent discussion will deal with the organization of this population of small rice cultivators in relation to the physical layout of the irrigation system.

Project Administration

The project was initiated in 1968 as a joint program of the Lao government and the Agency for International Development (AID). Since then, there has been more or less joint administration of the program, with heavy American control in the early years and an increasing control by Lao administrators in the last few years. American support was due to be phased out by the end of June 1975. The discussion here will concentrate on the Lao central administration of the project.

The Nam Tan project is the direct responsibility of the Directorate of Agriculture of the Lao government. This is a relatively modest agency of the Lao government, with a small total staff and not much specialization. Special divisions within the Directorate, such as the Irrigation Division, when they exist operate with a very small headquarters and field staff.

Staff from the Directorate of Agriculture are assigned to Nam Tan to serve as central administrative staff. In addition to the project director, there are staff members working in irrigation, extension, research, and grain processing. The project operates a rice mill, managed and staffed by the Directorate. Likewise, as a part of the total project, a small experimental farm has been established, and is managed by the Directorate. The project also has an extension staff, to work with farmers in the irrigation system on technology diffusion and farmer organization, and an irrigation staff of engineers and various skilled craftsmen for operation and maintenance.

In the summer of 1974, this project staff comprised twenty-eight persons. The extension and irrigation groups each had nine staff members: about one staff member each per 100 farmers; or about one each per 225 hectares of land.

There is another group whose responsibilities for implementation of the project have been increasing: the project-wide Farmer's Association.[4] The

4. The long-range role of the Farmer's Association in project management has been an issue for several years. In general, the American project directors have urged a strong role for the Association and have envisioned a point in time when the project could be completely turned over to the farmers themselves: the ultimate rice-roots project. On the other hand, the Lao project directors have viewed a more subsidiary and complementary role for the Association; one in which government would still perform selected overall management and administrative functions while the Association accepted more responsibility for various implementation activities.

The memorandum of agreement setting a date for the termination of direct AID support seems to leave this issue sufficiently unresolved to obtain the necessary signatures of both parties. The specified intent of the memorandum is to deal with the transfer of physical property associated with the project. The wording indicates that the property is to be given by AID to the Directorate internal organization of the Farmer's Association has been made congruent with the physical layout of the irrigation system in the following way: The Association is segmented into twenty-one small groups (twelve on the left bank and nine on the right bank). These small groups are composed of water users whose fields are contiguous and who (in most cases) receive their water from a common watercourse in the system.[5]

Each of these small groups has a rather elaborate leadership structure selected by the members of the small group. These officers include the usual president, vice-president, secretary, and treasurer, as well as other specialized officers, including a water headman. This latter role will be discussed more fully below.

Water-User Organization

The water headman is an adaptation of a leadership role found in this area of Laos prior to the present irrigation development activities. This section of the paper discusses that role in both its historical and present contexts.

Traditional Models of Irrigation Leadership

Irrigation roles analogous to the water headman are reported by several writers who have examined the traditional organization of irrigation management in Southeast Asia. Of particular interest are the materials from Burma (Nash, 1965), Indonesia (Geertz, 1967), Laos (Taillard, 1973), the Philippines (Lewis, 1971) and Thailand (Bruneau, 1968; Frutchey, 1969; Moerman, 1968; Wijeyewardene, 1965, 1973).

The traditional form of irrigation leadership can be characterized as an accountability model. Notwithstanding the novel and idiosyncratic characteristics of irrigation organization in various local contexts, several basic features, which act to make the leader responsible to his irrigation group, are common to this role in numerous settings.[6]

---

of Agriculture for specific use on the Nam Tan project and that the Directorate will, at a future date, consider transferring this material to the Farmer's Association. It contains a final reiteration of the American preference for the role of the Association.

As this was being written (May 1975), the issue of American aid in Laos was very much in the headlines. All American aid, at least in the form it then took, was terminated in Laos prior to the date of 30 June 1975 stated in the memorandum to which I refer.

5. Information gathered in the project area in the summer of 1974 indicates the following size characteristics of these small groups:

(a) They range in membership size from 30 to 60 persons; on the left bank the average size is 38.6 members; on the right bank, 46 members.

(b) The land area of the groups ranges in size from 70 to 138 hectares; on the left bank the average is 93.9 hectares, on the right bank 102.1 hectares.

6. This issue of accountability between water users and the water authorities is fundamental, and has impacts on irrigation organization other than the leadership roles emphasized in this

The actual responsibilities of traditional irrigation leaders vary in accordance with the size and physical aspects of the irrigation system and the location of ultimate control of the system. In the case described by Moerman (1968), the irrigation system is a community-constructed, -maintained, and -operated entity. Here we see that the major duty of the irrigation leader is the mobilization of labor and materials for the annual repair of the diversion structure. In contrast, Nash (1965) describes a local role which is organized around the function of canal maintenance. The village studied obtains its water from a large, government-built irrigation system and thus has no responsibility with regard to maintaining the diversion structure(s).

Maintenance and repair of either the diversion structure or the canals is, of course, a basic activity for which all groups must arrange. However, there are instances in which the water headman's responsibilities have included activities considerably more complex and controversial. One example is the *klian subak* described by Geertz (1967). In the *subak* organization, the irrigation headman and his assistants are also involved in the distribution of water through the assignment of planting times for the various subsections of the *subak*.[7] While it is not clear what determines this more complex leadership role, one might hypothesize that the higher population density in Bali requires greater coordination of water use among various groups obtaining water from the same stream.

While this range of activities exists, several common denominators persist, as reported in the literature. Three are of importance for this discussion. First, traditional irrigation leaders serve relatively small groups of water users, not necessarily from the same village, and small areas of irrigated land surface. Second, they are selected, in some manner, by the members of the local group which they serve. While leaders often serve for long periods of time, perhaps a lifetime, they nevertheless are subject to review and replacement, as is clearly illustrated in Wijeyewardene's report (1973). Third, they receive compensation directly from the group members whom they serve: sometimes in the form of exemption from labor (Moerman, 1968), sometimes in the form of land for cultivation (Jay, 1969), sometimes in the form of a proportion of the crop harvested (Harriss, 1974), or a fixed amount of the crop (Coward, 1971).

---

discussion. For example, both Birkelbach (1973), discussing the *subaks* in Bali, and Bain et al. (1966), discussing water districts in California, refer to the propensity of irrigation units to strive to maintain relatively small entities even though various economies of scale could be gained by consolidation of some sort.

7. The assignment of planting times historically has been fixed and did not require the repeated attention of the *subak* leaders. However, Birkelbach (1966: 155) notes that the introduction of new rice varieties with shorter growing periods does create the need for some adjustment of planting schedules.

These three elements—small scale, local selection, and direct compensation—are the basis of the accountability model of irrigation leadership. The accountability required of leaders through the operation of these organizational principles seems to shape their performance in directions acceptable to the irrigation group.

Irrigation roles similar to those described above were found in various villages of the command area which practiced irrigated agriculture prior to the construction of the Nam Tan project (Coward, 1971). This pre-project organization was of two major types. First, in villages that irrigated their fields by constructing very small diversion dams on shallow streams, such small dams requiring the labor of perhaps two or four water users, no village leadership role for irrigation existed. Leadership resided in each of the small work groups.[8]

However, in villages that constructed one large diversion dam to serve nearly all farmers in the village, specialized roles for administering irrigation activities were present. These villages selected an individual to act as water headman (*nai nam*). This role was separate from the role of village headman (*nai ban*).[9] In some instances, the water chief was assisted by a group who acted as a water committee or council.[10]

In the summer of 1969, villagers reported that previously the water headman provided the leadership for the annual reconstruction of the diversion dam and the canals and also served in settling disputes regarding water allocation: major functions in any operational system. It was also reported that the water headman was paid for his service by each water user, who annually contributed approximately ten kilos of unmilled rice.

In this pre-project organization, the water-headman role required little or no articulation with external authorities. The village irrigation systems which they administered were self-contained, local systems, not dependent upon or linked with entities outside the village. The traditional water-headman role primarily focused on the mobilization of intravillage action and typically did not involve the articulation of village interests with those of some external

8. This miniscale irrigation may be similar to that described by Kunstadter (1966) at another location in the general region.

9. Moerman (1968: 50) also notes this separation between the role of the irrigation leader and the role of the village leader. He indicates that in 1960 none of the dam chiefs nor the overall dam chief was an official or a village headman. For those with an interest in the sociolinguistic perspective, the difference in local terminology between dam chief and water chief may be suggestive.

10. The organizational pattern of leader-cum-council seems to occur frequently in the context of Thai-Lao villages. Moerman (1969: 542) in his analysis of the Thai village headman refers to a group of elders who "... advise the headman, especially on matters of village policy toward official demands, the mobilization of village-wide activities, and important trouble cases that threaten to involve the police."

organization. As we shall see below, this aspect of the water headman's responsibility has been fundamentally changed.

Contemporary Adaptation

As mentioned previously, each of the 21 small groups of water users that has been organized selects an individual to serve as water headman. His basic responsibility is to see that water is actually delivered to the members of the small group during their scheduled period.

Thus, the role expectations in the project setting are significantly different from those found in the indigenous systems. The responsibility has shifted from a concern with maintenance, primarily the rebuilding of the temporary diversion structure, to a concern with water distribution. Furthermore, to perform this new role satisfactorily, the water headman must act as a link between his water-user group and the external project administration. He also must act as a water guard, protecting the group's delivery from other water users who are near to the supply channel.

The new role finds support from both the water bureaucracy and the water users. The project staff prefers to communicate with the water headman because of the relative simplicity of dealing with a small number of individuals ($n = 21$). Likewise, the project staff tends to conceptualize the command area in terms of the small group areas and not in terms of individual farm units. They plan water deliveries to "Group A" and not to "Farmer A." The difficulties of distance to be traveled and time lost are important incentives for the individual water users to issue complaints to their local water headman rather than going directly to the central staff.

Nevertheless, there are occasions when the role of the water headman is not performed. When this occurs, it can be complicated by the relative inexperience of both Lao officials and Lao peasants with modern irrigations systems. These difficulties are well illustrated by the following incident, which occurred during the author's 1974 field visit:

On a particular day, water was to be rotated from one block to another in the left-bank portion of the command area. At the appropriate time, a staff member of the project administration went to a specific turnout, closed the gate delivering water to a lateral, and allowed the water to continue in the main canal in the direction of the lower block, whose turn had now arrived.

Some time after that, evidently soon after, an unknown person came to the same turnout and reopened the gate, thus delivering water to the original lateral and preventing water from going to the new block. In addition, this individual removed a small part from the gate, which prevented the easy correction of this illegal action.

When the water did not begin arriving at the lower block as expected, a number of the farmers from that block went to the project authorities and to the site of the water theft to complain. Some farmers arrived with guns; all arrived with insulting words for the water bureaucracy and its agents. The water authorities discovered the broken gate, and since the day was Sunday, indicated they would not be able to repair it until the next day. They also indicated to the farmers that because of the delay in delivering water, their block would be provided water for one additional day beyond their assigned time. Farmers were dissatisfied with the delay but reluctantly agreed to wait. Some wanted to try to repair the gates themselves, but this was not attempted.

Other details related to the incident are not necessary here. Let us pick up with a staff meeting held a few days later to discuss this incident. This meeting was attended by leaders of the project administration and several local officials, including the police. No farmers were present at the meeting. A most interesting point about the meeting is that the focus of discussion was not on the problem of water stealing—who had changed the gate, how he could be located, what punishment should be used. Rather, the discussion centered on the inappropriate behavior of the farmers who had not received their water in time. Most participants were concerned with the disrespectful behavior that farmers had shown toward government officials, including the behavior of waving guns. Most disturbing was the insulting behavior that had been used by the farmers in their complaints to the bureaucrats.

The group decided to take three actions, each reflecting an interesting perception of the problem. First, it was decided that the project administration would hold a meeting in the involved village to discuss with them the inappropriateness of their recent behavior. Second, it was decided that the police official would investigate to try to discover who had damaged the gate. Third, it was decided that the role of the water headman had not been fully utilized in avoiding such an incident. Thus, it was suggested that in the future the appropriate water headman should be on hand at the time water is rotated as a witness to the fact that the central administration was doing its part to operate the system as scheduled. If things went wrong after that, the water headman could assure the water users that the problem was caused by other agents than the water bureaucracy.

From this incident we can see the numerous difficulties that can arise when water users and bureaucrats attempt to interact, particularly in problem situations. The use of the irrigation headman as a link between these two groups is critical to project operation, as recognized in the staff meeting discussed above.

Clearly, this is a demanding responsibility and not everyone will be equally suited for the task. The following illustrates this point. During 1974, I met one water headman whose performance was considered so exceptional by local water users that he had been asked to serve five different small groups simultaneously. His success was apparently related to his ability to supervise and manage the behavior of water users. In his explanation, he indicated his special powers as a magician and his ability to prevent his skin from being pierced by any kind of object.[11] These special powers made others fearful of him and thus more willing to follow his orders, including orders about irrigation. In this case, the incorporation of a traditional role in the administration of the modern project also resulted in the mobilization of a special local resource: magical power!

For their activities and responsibilities, the water headmen are paid a small fee by each of the irrigators, reported to be about sixteen kilos of unmilled rice for each hectare of land served. In addition to this fee, each water user also pays a water fee to the project administration at the rate of 80 kilos of rice per hectare.[12]

Indigenous Social Organization and Irrigation Development

We began with a question regarding the utility of social organizational forms evolved in communal irrigation systems for the administration of modern irrigation projects. In view of this issue, the current use of the water headman role as an integral part of the irrigation administration of Nam Tan is suggestive.

Throughout Southeast Asia and elsewhere, those responsible for irrigation development are concerned with the issue of appropriate organizational forms for operating and maintaining costly irrigation systems. Most discussion of irrigation "organizations and institutions" deals with organization at the level of the water users and not at the level of the water bureaucracy. Consider the following excerpt from an Asian Development Bank publication (ADB, 1973: 50).

> The success of an irrigation project depends largely on the active participation and cooperation of individual farmers. Therefore, a group such as a farmer's association should be organized, preferably at the farmer's initiative or if necessary, with initial government assistance, to help in attaining the objectives of the irrigation project. Irrigation technicians alone cannot satisfactorily operate and maintain the project.

However, many nonlocal irrigation systems lack the formal type of water users' association prescribed above. Often, those who study such systems also

11. For an introductory discussion of the use of magic in Lao society, see Lebar and Suddard (1960).

12. To provide some perspective on these "costs," I have converted the payments into proportions of the harvested crop. On the assumption of an average yield of 1.5 tons per hectare, the payment to the water headman is just over 1 percent of the harvest; the payment to the system is about 5 percent.

It is also of interest to know something of the value of the payment to the water headman. I have converted the one-hectare payment into days of rice supply for one person. Using Gerhold's (1967) estimate of 0.85 kilos of rice as daily per capita consumption, the 16-kilo payment would allow the water headman to support one individual for approximately 19 days.

focus on the need for local organization. For example, Reidinger (1974) studied water distribution in a large system in northern India and concluded with a recommendation for local water associations as an alternative for improving system operation.

If it is accepted that an adequate strategy of irrigation development involves provision for local water-user organization, policy makers and implementers are still left with decisions regarding the organizational format to be followed. In many instances, though not all, irrigation development projects are intended to serve water users who have previous experience with indigenous forms of irrigated agriculture, including experience with local irrigation organizations, institutions, and leadership roles. Thus, there is the obvious possibility of using these previously evolved roles, institutions, and organizations, perhaps in a modified form, for the administration of the modern irrigation system.

Levine (1971: 29) has elaborated on this possibility:

> Almost all new irrigation projects recommend some form of public institution similar to that found in developed countries. There is very little attempt to work within existing institutions or to develop adaptations based upon indigenous social relationships. For example, the design for the Upper Pampanga River Project in the Philippines envisions either irrigation districts (U.S. style) or, more recently, irrigation associations (Taiwan style).

What is needed in irrigation development is recognition of the potentials of indigenous organizational arrangements for water management. This is not unlike Geertz's earlier recognition of the relevance of indigenous savings organizations for the development of savings behavior and credit institutions. In a well-known article, Geertz (1962) noted the prevalence of rotating credit associations in many parts of the world and explored their relevance for directed-savings programs. In anticipation of this subsequent discussion, he wrote:

> The rotating credit association, which assumes a remarkably similar form over a wide geographical area, although to be sure with local adaptations and variations, will be shown to be essentially a device by means of which traditionalistic forms of social relationship are mobilized so as to fulfill non-traditionalistic economic functions. It will be seen, in fact, to be an "intermediate" institution growing up within peasant social structure, to harmonize agrarian economic patterns with commercial ones, to act as a bridge between peasant and trader attitudes toward money and its uses. The rotating credit association is thus an institution of the sort Myrdal rightfully demands: one which fits into community patterns and yet aims at planned and "goal-directed" savings. (1962: 242)

Similarly, the role of the water headman in modern irrigation systems may be viewed as an intermediate pattern assisting with the transition from previous locally managed irrigation systems to present-day systems controlled and managed by field agents of the national government. To paraphrase Geertz, the contemporary water-headman role may act as a bridge between agency policies toward water and its uses and locally established patterns of water use and allocation. The case of irrigation development suggests another arena in which the study of "middle-rung" institutions is important.

Nevertheless, we know relatively little about the likelihood that indigenous organization can be articulated successfully with the formal administration of a modern irrigation project. The work of Ongkingco (1973) suggests that this has occurred in the Ilocos region of the Philippines, where the traditional irrigation associations (*zangjeras*) apparently continue parallel with the bureaucracy of the government's National Irrigation Administration.[13] I am not aware of other examples in the existing literature.

Some Generalizations

The role of water headman in Nam Tan, while significantly different from the traditional role in some aspects (particularly regarding duties and relationships with government agents) is highly analogous in others (particularly in the manner that incumbents are selected, rewarded, and judged).

It is important to note that the methods of recruitment and reward, and the normative expectations of the water headman, even in the modern context, makes the role incumbent accountable to the water users whom he serves. First, they are involved in his selection and have the ability to review his performance to consider his continuance or dismissal. In the example cited earlier, the water headman who was asked to serve five groups simultaneously was replacing other water headmen. Second, the incumbent is dependent upon the water users and not the bureaucracy for the payment of his fees. Reluctance to pay, or the actual nonpayment of irrigation fees to the water headman, provide a very direct form of job evaluation.

Both of these procedures reinforce the normative expectation that the water headman will act to meet the needs of the immediate water-user group that he serves, sometimes in lieu of the needs of the water bureaucracy. This is in considerable contrast to many of the roles used to link the water user with the larger system, which often do not provide for structural accountability to the water "clients" (see Reidinger, 1974, and Coward, 1973, for examples).

Having continued elements of the traditional role that maintain the incumbent's local accountability, it is possible to alter the traditional role with regard to its basic task activities and to expand the role to include relations with project officials. These alterations make the role compatible with selected administrative and management needs, especially the need to facili-

13. The *zangjera* irrigation associations have a long history in the Ilocos region of the Philippines. An early description is provided by Christie (1914). More recent discussions of the *zangjera* in both the traditional Ilocos area and new areas to which Ilocanos have migrated are found in Lewis (1971) and Hackenberg (1971).

tate the distribution of water within the areas of each of the small groups while also achieving an equitable distribution of water between these groups.

The importance of local accountability is further illustrated by the recent work of Gillespie (1975). In describing an irrigation system in northeast Thailand which is operated by the Royal Irrigation Department, an agency of the national government, he comments on the activities of the "common irrigator": a role with task activities very similar to the Nam Tan water headman. The following excerpts illustrate the cogent differences:

(1) ... the RID [Royal Irrigation Department] pays the common irrigator's stipend. [p. 15]
(2) RID personnel also take an active part in the selection of the common irrigator, although the farmers are supposed to independently choose him from among themselves. [p. 15]
(3) The project manager (RID) believed that the farmers did not have the experience to choose a competent common irrigator. [p. 15]
(4) According to the farmers... "they had no interest in who held the position." [p. 15]

In this case, a role with task activities and relations similar to the water headman has been created but with accountability oriented to the water bureaucracy and not to the water users. One important consequence is a lack of interest in the common irrigator's role on the part of the water users.[14]

From these observations, a hypothesis related to the utilization of indigenous organizational forms in developmental contexts may be derived. It builds on the simple observation that indigenous leadership roles in agrarian organizations are linked with various "follower" roles in ways that foster accountability. Often, in the process of linking agrarian organizations with central government activities, an attempt is made to use existing leadership roles while shifting leader accountability from the local community to the field agents of the central government. In so doing, the nature of the relationship between leader and constituents is significantly altered—sometimes precluding the effective performance of the leader.

Thus, a derivable hypothesis is as follows: Indigenous irrigation leadership roles can be mobilized to perform in modern irrigation projects even if task activities and relationships are altered (for example, shifting from maintenance duties to water distribution responsibilities and increasing the interaction with agency staff members) if accountability for job performance is linked to the local water users and not the water bureaucracy exclusively.

14. The Lam Pra Plerng irrigation project is a relatively new one, and a number of difficulties exist in these early stages. There are still important problems with the physical components of the system. At this time it would be difficult to distinguish between these effects and any organizational influences on system performance. Therefore, comments about the present accountability of the common irrigator should not be viewed as causative of system performance.

It is a hypothesis suggested by the discussion, clearly not tested by it, and it deserves increased research attention.

If supported by additional research, this hypothesis might have useful implications for irrigation development planning. As a part of such planning, attention would be given to the identification of irrigation leadership roles among those to be served by the modern system. Where such roles are present (or have recently been used), attention should be given to understanding the manner in which accountability between the irrigation leader and the water users was organized. Based on that critical information, organizational rules and procedures could be developed and tested to maintain this basic accountability to users while shaping the role to perform tasks and engage in relationships functional in the modern system. Ultimately, this concept of accountability might be extended to situations in which there are no indigenous institutions on which to build.

Summary

As irrigation development increasingly is implemented through projects constructed and operated by agencies of the central government, the problem of successful articulation between the project administrators and the water users is highlighted. In areas where the project serves local water users with previous irrigation experience, one solution to this problem is the adaptation of indigenous irrigation leadership roles to link bureaucracy and local users. It is hypothesized that these indigenous roles can be used for articulating these two groups if accountability for job performance remains largely with the local water-user groups. Identification of indigenous irrigation leadership patterns and the adaptation of these patterns to modern irrigation administration deserve increased attention by researchers and policy makers.

References

Asian Development Bank. 1973. *Regional Workshop on Irrigation Water Management*. Manila.

Bain, Joe S., Richard E. Caves, and Julius Margolis. 1966. *Northern California's Water Industry*. Baltimore: Johns Hopkins University Press.

Birkelbach, Aubrey W., Jr. 1973. "The Subak Association." *Indonesia* 16:153–69.

Bruneau, M. M. 1968. "Irrigation traditionelle dans de Nord de la Thailande: L'example du bassin de Chiengmai." *Bulletin de l'Association des Géographes Français* 362–63: 155–65.

Christie, Emerson B. 1914. "Notes on Irrigation and Cooperative Irrigation Societies in Ilocos Norte." *Philippine Journal of Science* 9:99–113.

Coward, E. Walter, Jr. 1971. "Agrarian Modernization and Village Leadership: Irrigation Leaders in Laos." *Asian Forum* 3:148–63.

______. 1972. "Irrigation and Incipient Commercialization in Western Laos." *Zeitschrift für Ausländische Landwirtschaft* 11:289-300.

______. 1973. "Institutional and Social Organizational Factors Affecting Irrigation: Their Application to a Specific Case." In International Rice Research Institute, *Water Management in Philippine Irrigation Systems: Research and Operations*. Los Baños, Philippines.

FRUTCHEY, ROSE HOUSE. 1969. *Socioeconomic Observation Study of Existing Irrigation Projects in Thailand*. Prepared for the Bureau of Reclamation, U.S. Department of the Interior.

GAITSKELL, ARTHUR. 1959. *Gezira: A Story of Development in the Sudan*. London: Faber & Faber.

GEERTZ, CLIFFORD. 1962. "The Rotating Credit Association: A 'Middle-Rung' in Development. *Economic Development and Cultural Change* 10:241-63.

______. 1967. "Tihingan: A Balinese Village." In Koentjaringrat, ed., *Villages in Indonesia*. Ithaca: Cornell University Press, 210-43. [Chapter 4 above.]

GERHOLD, CAROLINE. 1967. "Food Habits of the Valley People of Laos." *Journal of the American Dietetic Association* 50:493-97.

GILLESPIE, VICTOR A. 1975. *Farmer Irrigation Associations and Farmer Cooperation*. Papers of the East-West Food Institute, no. 3. Honolulu: Food Institute, East-West Center.

HACKENBERG, ROBERT A. 1971. "The Cybernetic Village." *Southeast Asian Journal of Sociology* 4:5-27.

HARRISS, J. C. 1974. "Problems of Water Management in Relation to Social Organization in Hambanota District." Paper presented at Cambridge, England, Workshop on Project on Agrarian Change in Rice-Growing Areas of Tamil Nadu and Sri Lanka.

JAY, ROBERT R. 1969. *Javanese Villagers: Social Relations in Rural Modjokuto*. Cambridge: M.I.T. Press.

KUNSTADTER, PETER. 1966. "Irrigation and Social Structure: Narrow Valleys and Individual Enterprise." Paper prepared for 11th Pacific Science Congress, Tokyo.LEBAR, FRANK M., and ADRIENNE SUDDARD, eds. 1960. *Laos: Its People, Its Society, Its Culture*. New Haven: HRAF Press.

LEVINE, GILBERT. 1971. "The Management Component in Irrigation System Design and Operation." Cornell University, mimeo. [Chapter 3 above, revised.]

LEWIS, HENRY T. 1971. Ilocano Rice Farmers: *A Comparative Study of Two Philippine Barrios*. Honolulu: University of Hawaii Press. [Chapter 7 above.]

MOERMAN, MICHAEL. 1968. *Agricultural Change and Peasant Choice in a Thai Village*. Berkeley: University of California Press.

______. 1969. "A Thai Village Headman as a Synoptic Leader." *Journal of Asian Studies* 28:535-49.

NASH, MANNING. 1965. *The Golden Road to Modernity*. New York: John Wiley.

ONKINGCO, P. S. 1973. "Case Studies of Laoag-Vintar and Nazareno-Gamutan Irrigation Systems." *Philippine Agriculturalist* 56:374-80.

REIDINGER, RICHARD B. 1974. "Institutional Rationing of Canal Water in Northern India: Conflict between Traditional Patterns and Modern Needs. *Economic Development and Cultural Change* 23:79-104. [Chapter 12 above.]

TAILLARD, CHRISTIAN. 1973. "Irrigation dans le Nord du Laos." *Bulletin des Amis du Royaume Lao*.

THORNTON, D. S. 1974. "The Organization of Irrigated Areas." Paper presented at Second International Seminar on Change in Agriculture, Reading, England.

VANDERMEER, CANUTE. 1968. "Changing Water Control in a Taiwanese Rice-Field Irrigation System." *Annals of the Association of American Geographers* 58:720–47. [Chapter 11 above.]

WIJEYEWARDENE, GEHAN. 1965. "A Note on Irrigation and Agriculture in a North Thai Village." In *Felicitation Volumes of Southeast-Asian Studies* 2:255–59. Bangkok: Siam Society.

———. 1973. "Hydraulic Society in Contemporary Thailand?" In *Studies of Contemporary Thailand,* ed. Robert Ho and E. C. Chapman, pp. 89–110. Canberra: Research School of Pacific Studies, Australian National University.

WITTFOGEL, KARL A. 1957. *Oriental Despotism*. New Haven: Yale University Press.

16

India's Changing Strategy of Irrigation Development

Robert Wade
UNIVERSITY OF SUSSEX

Irrigation Targets

It can be safely assumed that the next few decades will see a very large expansion of India's irrigated area. A doubling of the present acreage in the next 25 years would be well within currently estimated limits; and would be no greater than the rate of expansion in the past 25 years, during which time the area under irrigation has increased from 20m hectares in 1947 to 43m hectares in 1973 (which accounts for 30 percent of net sown area).[1] The likely need to provision twice the present population by the early years of the twenty-first century will encourage a high priority to be given to realizing the potential. In addition, it is now being seen that the margin of cultivation has recently been extended to land which should not be under crops at all, but rather put under pasture or trees so as to conserve a valuable and for all practical purposes nonrenewable resource, topsoil, and thereby to protect the (long-run) income of the local populations.[2] If this submarginal land is taken out of crops, a compensating increase in food-grain production elsewhere will be required.[3]

Previously published as "Water to the Fields: India's Changing Strategy," *South Asian Review* 8 (1975): 301–21. Reprinted by permission of the Trustees for The Royal Society for India, Pakistan, and Ceylon.

1. Draft Fifth Five-Year Plan, 1973, vol. 2, p. 105.
2. See Drought-Prone Areas Programme, in Draft Fifth Five-Year Plan, 1973, vol. 2, p. 91. Also Vohra, 1972, pp. 7, 10, 11.
3. Food imports and exports are ignored for present purposes. Even though the emphasis on self-reliance which has conditioned planning since 1968 is being softened, it seems likely that memories of the consequences of heavy dependence on food aid in the second and third plan periods will keep food aid at a low level. And it may be difficult to purchase grain against foreign exchange at prices India can afford to pay.

Nearly half the present irrigated area is supplied from government-financed canals, on the construction of which very large sums of money—amounting to nearly Rs. 30,000m (U.S. $4,000m)—have been spent since 1950.[4] It is likely that despite the rapid extension of tube wells, canals will continue to have a predominant role in future.[5] Of the 60 percent of total irrigation potential remaining unrealized at the end of the Fourth Five-Year Plan, it is estimated that most (65 percent) will be realized through major and medium surface schemes.[6] Current plans call for an additional *one million hectares* to be put under surface irrigation every year up to 1978/79.

Reasons for Underutilization

An expansion of government-financed canals of this magnitude poses staggering problems of cost, the need for trained personnel, and governmental organization. The problems are staggering not only because of the scale of the expansion but also because in post–Green Revolution conditions the new projects are much more complex than the easier alternatives of the past. Furthermore, serious concern has recently been voiced in a stream of government and nongovernment publications over the inefficiency of *existing* canals.[7] Their inefficiency is partly a matter of heavy water losses in transit from the head of the system to the field: in most Indian canal systems the losses are of the order of 70 to 80 percent.[8] The scope for improvement can be shown by a simple calculation: if losses could be reduced by 10 percent, the saving would amount to very roughly 20m acre-feet, which is the equivalent of about three additional Bhakra dams.[9]

Partly because of engineering inefficiency, partly because of unreliability in canal supplies, and partly because of poor on-farm use of water, average crop yields in canal-irrigated agriculture tend to be far below what one might reasonably expect.[10] Moreover, only about a fifth to a quarter of the irrigated

4. Vohra, 1973a, p. 1.

5. A reader has observed that tube wells too must have canals. When I refer to canals, I mean gravity-flow systems.

6. Hanumantha Rao, 1974.

7. See for example Draft Fifth Five-Year Plan, 1973: Irrigation Commission, 1973; Agricultural Prices Commission, 1975; National Commission on Agriculture, 1973: See also Vohra, 1972, 1973a, and 1973b. It is unofficially estimated that some 10m hectares of recently completed irrigation projects require an additional investment of U.S.$200–500m before they can become fully productive.

8. Chaturvedi, n.d., p. 75. The figure is a rough order of approximation only; little work is being done to measure water flows from head to field.

9. The Bhakra canal system, which waters large areas in Punjab and Haryana, is one of India's biggest. The dam's storage capacity is almost twice that of Shasta Dam, the keystone of California's Central Valley Project. See Reidinger, 1974. n. 3. The calculation in the text is of course highly simplified. It assumes that two-thirds of the total amount of water used for irrigation in India comes through canals. It ignores the fact that not all the water lost from a canal through seepage and percolation is lost to the crops. See Chaturvedi, n.d., pp. 60, 72.

10. This is widely believed, but I have not come across reliable figures. The irrigation

area is sown more than once.[11] If, to oversimplify, we assume that irrigated land has the capacity to produce two crops a year (to say nothing of higher levels of multiple cropping), the present underutilized capacity is of the order of 75 to 80 percent. By this standard the public-sector industries, such as fertilizers and steel, about whose underutilized capacity so much is heard, seem relatively efficient. The underutilization in the case of irrigated agriculture is especially serious in view of the fact that 80 percent of the irrigated land is under food crops.

Finally, the area which is actually irrigated is considerably less than the planned potential. It is estimated that only 80 percent of the canal potential created since 1950 is being used. But this figure may be misleadingly high, since the Irrigation Department counts as "utilized" any area which receives canal water, even if the land is waterlogged or even if the water arrives too late and too little.

Even worse, however, is the fact that the problem which plagued early canals in India—the deterioration of irrigated land due to waterlogging and salinity—has not been adequately dealt with even in some very recent schemes. In Chambal (Rajasthan), for example, the amount of formerly productive land taken out of production since the onset of irrigation in 1960–65 has averaged about 1 percent per year. Much of the waterlogging occurs because of very high seepage from unlined canals, and because no drains were built to carry water away. "It is a particularly distressing thought that lands which are sought to be benefited by being provided with irrigation at great public cost should run the risk of losing their productivity as a result of this very facility—merely because enough attention is not paid to the drainage requirements of the land when the distribution systems are being laid."[12] The lesson is now being drawn that in most soil and climatic conditions found in India, degeneration of unprepared land is inevitable in canal-irrigated areas; and that the cost of prevention is much less than the cost of reclamation.

A Conflict of Objectives

For these various reasons, then, existing canals are seen to be in need of substantial improvement. It should be noted, however, that concern over the inefficiency of existing canals is based, to some extent, on exaggerated expectations of what canals can do. It is often implicitly assumed that canals can eliminate the vagaries of the monsoon. But if rainfall is poor over one or more seasons there may be no water for the canal to supply; canals, too, depend on rainfall. Further, in large canal commands there are inherent conflicts between

---

Commission warns that the yield figures it gives for irrigated and unirrigated crops are extremely rough. See 1973, vol. 1, p. 204. See also Chaturvedi, n.d.

11. 1969/70 figures (Rao, 1974). The points in the remainder of the paragraph also come from this article.

12. Vohra, 1972, p. 13.

the objects of improving *predictability* of water deliveries (in timing and amount), *appropriateness* of water deliveries in relation to crop needs, and the *equality* of water deliveries per acre. When a reservoir-canal system is linked with a hydroelectric scheme, these objectives are still more difficult to attain, since water releases from the reservoir are determined partly by the demand for electricity rather than solely by the requirements of agriculture. Nevertheless, allowing for exaggerated expectations, one can still say that the potential for improvement is large. The government's attempt to realize this potential is known as the Command Area Development program.[13] Command Area Development (CAD) is now one of the major thrusts of current agricultural development policy.

The CAD program is directed primarily at land development at farm and outlet level, for it is here, in the opinion of the government and its advisers, that the potential for improvement is greatest. Traditionally, the government (that is, the Irrigation Department) was concerned with canal systems only down to the level of outlets serving areas of about 40 hectares, and development below that point (land shaping, construction of watercourses, drains, etc.) was left to the cultivators. But this development has lagged far behind canal construction and has been of generally poor quality. In rice areas of eastern and south India, the "field-to-field" method is common: the water is allowed to flood over the whole area below the outlet, passing directly from one field to the next instead of via watercourses along field boundaries. In wheat areas, water generally reaches the fields by a network of meandering, poorly maintained, and leaking watercourses constructed by the farmers through individual—sometimes informal cooperative—effort. Generally there are no (manmade) drains. Significantly, several IDA agricultural credit projects at present provide financing through commercial credit institutions for the construction of on-farm works by the farmers themselves, but uptake has been poor.

The CAD Program

Under CAD, the responsibility for on-farm and "on-outlet" land development is transferred to the government. The CAD program also embraces associated infrastructural improvements in canals (lining, gates, etc.), roads, and marketing facilities; and improvements in cropping patterns, livestock practices, input supply, and soil conservation.

The draft Fifth Five-Year Plan provides for as much as Rs. 3,000m (U.S. $400m) to be spent out of the Irrigation Department budgets alone on improving the use of water. In addition, provision is made for at least RS. 3,000m to

13. The "command area" of a canal is the area which it can (actually or potentially) water. For an overview of CAD, see Draft Fifth Five-Year Plan, 1973, vol 2, pp. 111–13.

be spent on these programs out of Agricultural Department budgets and cooperative and commercial bank loans.[14] The public-sector outlays will be divided roughly equally between states and the center. While it is true that the draft document which proposed these allocations is now regarded as obsolete, it is likely that the circumstances which made it obsolete—the sharp rise in oil, fertilizer, and food prices—will increase rather than diminish the priority attached to maximizing the benefits from existing irrigation projects rather than constructing new ones, because this provides the opportunity for the greatest improvement in agricultural output for each additional rupee of public expenditure.

Administrative Coordination

To plan and implement the various CAD components requires a high degree of coordination among several disciplines—in particular irrigation, agriculture, soil conservation, agricultural extension and cooperation—as well as support from credit and service organizations. In large commands covering one or more districts the scale of operations becomes huge, and, as the draft fifth plan points out, the problems are as complex and unwieldy as those faced by a large industrial enterprise. Hence the central government has recommended that the states should form unified agencies in each command area to bring the existing departments whose work is relevant to CAD into a single line of command. Administrative control over the staff of the various departments concerned, principally irrigation, agriculture, cooperation, soil conservation, and revenue, functioning within the command area would stand transferred to a Command Area Development Authority (CAA). The Authority would have as its head a full-time administrator vested with the powers of head of all constituent departments. There would be a separate allocation of funds for CAD schemes, independent of the budgets of the constituent departments. The administrator would have full powers over the budget and staff of the Authority. Each Authority would have a board of representatives of a wide range of official and nonofficial interests to supervise the programs and review progress. Finally, the Authority would take responsibility for the overall economic development of the command area, as well as for matters related directly to irrigation; it would be responsible for fixing and enforcing cropping patterns, strengthening extension training and demonstrations, planning and ensuring the supply of inputs such as credit, seeds, and fertilizers, ensuring the adequacy of marketing and communications throughout the command area, and planning local growth centers.

With varying degrees of mutation from this model some 32 Command Area Development Authorities have now been established, out of the 50

14. Figures from Vohra, 1973a.

projects designated for CAD treatment. Not all of them, however, as yet have more than a nominal existence.

A Trend towards Central Control

This administrative arrangement, at least as set out in the center's model, represents a new step in Indian agricultural administration. It is, to be sure, part of a trend in recent years toward special programs promoted from the center and administered at district level through specially created institutional arrangements which bring various departments into closer coordination. The Drought-Prone Areas program and the Small Farmers Development Agency program are examples, as is the older Intensive Agricultural Districts Program. Compared to these, the authority of the head of each CAD project over all heads of departments in the command area is—at least in the center's model—much greater and more clearly defined, for two main reasons:

1. Because a command area generally covers more than one district, and the project head must therefore be able to ensure collaboration and coordination across district borders (while the Drought-Prone Areas program, for example, can be administered by taking each district as a separate unit).
2. Because one of the departments most closely connected with CAD work, the Irrigation Department, is powerful, wealthy, and accustomed to going its own way; and on the whole has been slow to sympathize with CAD ideas.[15]

A further reason may be the center's concern to ensure that a program of such size and importance should not become a political pork barrel; by creating a special authority over the existing departments, the local political pressures which normally operate on district bureaucrats might be partially reduced.[16] Hence the CAD administrator is expected to coordinate not simply by prodding and exhortation, but by control over budgets and staff. The Command Area Development Authority thus represents a significant increase in administrative centralization at district and interdistrict levels.

But perhaps more important in the long run than the financial size of the program or the innovative aspects of its administration is the fact that it implies a very large expansion of governmental influence at *outlet* level, with the objective of bringing about not only physical but also institutional changes. In what follows I shall be concerned with three of the principal institutional components: consolidation of holdings, rotational irrigation, and water users' associations.

15. The Irrigation Department's attitude is changing. The Report of the Irrigation Commission, 1973, represented a victory for advocates of CAD.

16. In this sense, the CAA, the agency for the Drought-Prone Areas program, and the move towards reducing the role of cooperative banks in credit allocation and increasing the role of the nationalized commercial banks are all part of the same trend.

Land Development and Consolidation

The Kota Method

The problem facing the government is to find methods of land development which will improve the efficiency of water use, increase crop yields, lower costs of production, and maintain the fertility of the land; and be replicable on a large scale. Much attention has been given to the "Kota" method (Kota is the city in Rajasthan near which the technique was first implemented in a UNDP/FAO-assisted pilot project in the late 1960s). Briefly, this approach involves, to start with, a detailed study of the soils, topography, and layout of the area commanded by an outlet (or aggregates of outlet commands), and the leveling of land along the contours. Watercourses are constructed along natural ridges so as to irrigate strips lying on both sides, while drains are provided along natural depressions. In areas of high-seepage soils the watercourses may be lined, at least in their upper portions. Access-roads and paths are built along the drain bunds. This procedure reduces the length of watercourses and drains to a minimum; and the leveling of land in contour terraces involves the minimum movement of earth. Wider rather than narrower fields are preferred, as this facilitates mechanical cultivation and mechanical harvesting. However, to be successful this approach requires a redrawing of field boundaries, consolidation of fragments, and the reallocation of land to farmers in the contour terraces, so that each field can be connected to a watercourse at one end and a drain and access road at the other, and be rectangular in shape.

Backing from the World Bank. This is the method which until recently the central government has been urging the states to adopt; the World Bank, too, has backed it in making loans for land development. (What the Bank thinks about CAD is a very important influence on the direction as well as the pace of the program; CAD is the Bank's biggest program in India, and India is the Bank's biggest customer.) The Kota method has now been tried in many pilot projects throughout the country, but principally in Chambal (Rajasthan), the adjacent command of Chambal (Madhya Pradesh), and in Tawa, the other big new command in Madhya Pradesh. By the beginning of the coming monsoon about 15,000 acres in these three commands, and several hundred more in six others, will have been developed in this way.

Criticisms of Kota. But recently criticism of this method has been growing even with respect to its applicability to the conditions for which it was originally devised, those of Chambal (Rajasthan); and still more to conditions different from these. It is said that insistence on consolidation imposes enormous administrative strains on the Revenue Department, especially in states where land records are sometimes generations out of date; and this will constrain the implementation of the other components. It is also said that the cost per acre is far too high for many farmers to bear. Impressionistically, it appears that a small farmer would have to make a complete switch to higher yielding varieties *and* get the prescribed amounts of inputs in order to be able to repay. Even if a farmer is big enough for a Rs. 6,000 loan, say, to be within his repayment capacity, it is unrealistic to expect him to be happy to be forced to take out a loan of this amount when it is many times his annual income; and even if not forced, he may come to regret his voluntary decision later, and perhaps take action to have it rescinded.[17] The cost is high partly because of the prescription of long, wide fields, with an even slope from one end to the other, and access roads to each field. If the objective is rapid mechanization of agriculture, this may make sense. If a relatively labor-intensive agriculture is envisaged, and if attention is paid to the economic situation of small farmers as well as big "progressive" farmers, it does not.

Alternative Approaches

Opponents of the Kota method argue it is especially unsuitable in areas of shallow, variable soils, such as are found over much of the Deccan Plateau. Its field sizes and field slopes entail considerable disturbance of the topsoil, with drastic loss of fertility which cannot be recovered except by massive and costly fertilizer applications—and perhaps not even then. Moreover, it is in areas of shallow, variable soil that farmers' resistance to boundary realignment is greatest. It is here that farmers and their ancestors have had to work to build up soil fertility, and with good reason they fear that in the process of consolidation they will be given the land of someone who cared less about its fertility than they did about theirs. But farmer resistance is not confined to these areas. Even in those areas of deep, homogenous soils where the first few Kota-type pilot projects were initiated successfully, there is now beginning to be a reaction by farmers whose land has not been developed against allowing it to be done; and even on the pilot projects themselves there are some signs of discontent—except on those where the land was thoroughly waterlogged and virtually out of production.

Opponents also point to alternative techniques now being used in a number of commands which avoid the technical, economic, social, and administrative difficulties of Kota. In Ghataprabha (Karnataka), for example, about 50,000 acres have been irrigated by a method which involves government construction of primary watercourses and drains, located according to topographical criteria; government construction of field channels to each

17. One of the major issues preoccupying the CAD planners is what to do about the "ineligibles," the farmers who for one reason or another are ineligible to receive a loan for land development through commercial or cooperative bank channels. The issue is complex and far from resolved, but lies outside the scope of the present paper.

field, following property boundaries (thus reducing the need for boundary realignment, but at the cost of longer field channels); government design of land shaping; and government supervision of land-shaping work carried out by the farmers themselves. This type of development is less costly, arouses less resistance among farmers, imposes less administrative strain, and encounters fewer legal obstacles.

What opponents of the Kota method say, above all, is that it is suitable only for "pilot projects": it is too costly and too time-consuming to be done on a large scale. Given that something like 1m hectares of land will be brought under surface irrigation each year over the next few years, the choice is between applying the Kota package to a tiny part, in a pilot or demonstration role, leaving land development on the remainder to proceed mainly at a pace dictated by farmers' initiative as in the past, or adopting a less comprehensive, less elegant method which can be applied faster and at lower cost per acre. The choice of technique in land development, in other words, cannot be made without considering the time and resources available, and the area to be covered.

However, the Kota method still has its followers. They refute the arguments of its detractors point by point, and say, overall, that the benefits of the full package are so much larger than the sum of the benefits of the components treated individually, and the costs so much lower, that a vast potential would be missed if the full package were not applied—a potential which India can ill afford to forgo. On the question of farmer resistance, they say that if the farmers are approached the right way resistance is negligible. First, farmers within a single outlet are persuaded to talk about the problems they encounter in using water; then they are led on gently to ask themselves about possible solutions; they will then *request* consolidation. So far (say proponents), out of about a dozen attempts by way of this sort of approach there has been not a single case where the farmers refused.[18] On the other hand, several early attempts failed; and these were all organized and led by local officials who were convinced that the farmers in their area would never allow the boundaries of

18. According to an FAO expert, this approach is needed only for the first pilot project in an area; once the pilot is established, other farmers request the same thing to be done on their land. But I understand that in the only attempt at land development and consolidation in Kerala so far, the farmers strongly resisted consolidation and watercourse construction at first, for fear that they would lose even tiny amounts of their very small holdings. I understand, too, that on one command where officials are prone to tell enquirers that the farmers are entirely content with the post-land-development results, there is contrary evidence. For example, in 1974 the participants in a seminar on "command-area development visited this command to see the progress being made; and while they were inspecting one of the soil- and water-management projects, a number of farmers, hearing the officials telling how successful the whole thing had been, came up waving their fists and declaring that they had been swindled and were far from happy with the results. These are only two of several examples brought to my notice which question the argument of advocates of Kota that farmers, if approached in the right way, will not resist, and will be happy with the results.

their lands to be touched.[19] If resistance should not be negligible, compulsion must be used; the need for a restructuring of India's irrigated agriculture is so great that if farmers cannot be persuaded with carrots, they must, regrettably, be given the stick; laws must be changed to allow this, and the Command Area Authority must have all the powers necessary to see that farmers' objections and bureaucrats' resistance or inertia are overcome.

So the arguments go on. But there are signs that the balance of opinion is moving toward opponents of the Kota method, not only in the central government but also in the states; and also, very significantly, in the World Bank, which used to insist on "Kota or nothing." The next few years are likely to see a more pragmatic, less comprehensive, less dramatic approach to land development, with the Kota method confined to areas suffering from waterlogging or from very marked fragmentation of holdings. In the (very much larger) remainder, no more than very small boundary adjustments (shifts of up to 30 cms.) are envisaged. The question is what effects this change will have on output and distribution in both the short and the longer run.

Rotational Irrigation

Rotational irrigation, the second main institutional innovation, is an attempt to alter the principles on which water is allocated. Canals are devices for rationing a valuable input by control over, first, the size and position of channels and outlets, and second, the flow from the main canals into minor ones (the flow into some minors may be stopped for a period in order to distribute more of the total to farmers under other minors). But the most difficult type of control to implement is the third: control over who uses water in a given watercourse at any one time. In rotational irrigation, farmers on the same watercourse (or under the same outlet) follow a roster of turns to take water; normally the roster is arranged so that during their turn they are entitled to the whole supply available in the watercourses.

The Need for Certainty

The difficulties are obvious: if a farmer whose turn has passed believes his crop needs more water, he will be strongly tempted to take some out of turn, depriving another of his rightful share. If each farmer expects that others will

19. The argument of the pro-Kota people suggests that the resistance to consolidation comes more from officials and less from farmers than is often realized. If correct, it acquires considerable significance when put alongside the commonly expressed view that big institutional changes, such as organized systems of consolidation and drainage, are unlikely to be introduced in India within the present political order, and that, therefore, little increase in the rate of growth of food output can be expected. Of course, by comparison with, say, North Vietnam, the kind of consolidation talked about here, which in no way threatens individual ownership, represents a change of a minor order.

disregard the roster, none will be happy to wait in line. The essential condition for the roster to work is that the bulk of cultivators should be confident that if they do not take water whenever it is available, there will still be water left when their turn comes; or that if they do take water at the first available opportunity, the punishment will be too severe to make it worthwhile.

Rotational irrigation has been practiced from the beginning on the canals of Punjab, Haryana, and western Uttar Pradesh, where canal irrigation developed within the legal framework of the North India Canal and Drainage Act, 1873, which prescribed penalties for farmers who took water out of turn and defined the rotational procedures in detail.[20] Perhaps it is partly because there is a history of successful rotational irrigation in this part of India that there seems so little concern about how to get it to work elsewhere, or analysis of why attempts to introduce it into the new commands have met with small success, even in wheat areas. (One would expect wheat farmers to accept it more readily, because wheat does not require continuous irrigation. Nor does rice, but farmers generally believe it does.) It has not been adequately appreciated by irrigation and agriculture officials that the water requirements of the new seeds make the introduction of rotational irrigation today considerably more difficult than in the past. If the higher yielding varieties do not get water at certain critical periods, yields fall off dramatically; in contrast, the varieties which were prelevant in northwest India when rotational irrigation was institutionalized could withstand a wide range of variation in water supply without much effect on productivity.

Taiwan's Example

When drought conditions in Taiwan in the mid-1950s forced the introduction of rotational irrigation as a means of saving water, a large-scale national organization was set up to promote and enforce its adoption, and received strong backing from the top political leadership.[21] Water savings of 20–30 percent were achieved. There is no reason to think that the savings would be much lower in India. But so far there is no indication that the lesson of Taiwan's experience—that to introduce rotational irrigation is a major undertaking requiring high-level political commitment—has been taken to heart.

Water Users' Associations

Formal associations of water users exist in several East Asian countries (e.g., Japan, Taiwan, Bali) as well as in Spain and the United States. In South Asia,

20. For an account of rotational irrigation in India, see Reidinger, 1971 and 1974; Whitcombe, 1972.

21. See Levine, 1969, 1971, and Lee and Chin, 1961. For years Levine, an agricultural engineer, has been urging social scientists to concern themselves with water-use questions, and suggesting how they might go about doing so.

however, they are weak or nonexistent.[22] It is true that on at least some of the canals in northwest India, the Bhakra for example, there are village-based irrigation panchayats; but what little evidence is available suggests that in practice they are unimportant.[23]

Experience in Chambal

The CAD program, however, calls for the formation of such associations, and in several commands efforts are being made to set them up. They are expected to oversee the distribution of water, to help resolve disputes, to be a means of communication between users and irrigation officials; and also to collect water rates, and be responsible for the maintenance of watercourses and drains. It is particularly the need to delegate the last two functions which gives confidence that the introduction of the associations will be taken seriously. The government is anxious that, having built the watercourses and minor drains, it should not also have to maintain them, which would be a heavy and continuing burden on the public budget. And in at least one command, Chambal (Madhya Pradesh), it has been found that to delegate the collection of water charges to the head of a local irrigation association is a much more effective way of collecting revenue than the previous arrangement, in which low-level Revenue Department officials were responsible for collection. This experience is likely to encourage other commands to do the same.

Progress so far has been slow. In Chambal (Rajasthan), for example, efforts were made over a period of several years in the late 1960s and early 1970s to form associations. The UNDP/FAO project report (1974) summarizes the experience as follows:

> Farmers under one outlet, i.e., one sub-catchment of 20–100 ha. numbering from 10–15 owners, were expected to organize irrigation on their own. To this end water users' associations were to be formed and to be held responsible for the operation and maintenance of their system. During discussions at many levels of Government the project staff stressed the paramount importance of such associations for the lasting success of the scheme. Discussions with the farmers concerned had shown that 80–90 per cent would agree to cooperate and contribute their share. However, the resistance of the few objecting farmers often prevailed, owing to the lack of skill or patience of the officer leading the debate. The recommendation to put more socially experienced staff in charge of this important task still remains to be followed up.

Perhaps there is more to the problem than a "lack of skill or patience" on the part of the officer leading the debate.

22. They are known, however, on small tank systems. See Chambers, 1974. For a different sort of organization in small tank systems, see Leach, 1961.

23. Reidinger and Vander Velde, both of whom wrote theses on irrigation from the Bhakra canal, fail to mention irrigation panchayats, though the panchayats do exist.

In the adjacent command, Chambal (Madhya Pradesh), village-based irrigation panchayats have existed for the past five years, but until last year seem to have served little purpose. Last year, as noted previously, the head of each irrigation panchayat was given responsibility for collecting water charges in his village, in return for a commission, with the result that collections have been considerably higher than in the past. Casual observation suggests that this change has heightened cultivators' awareness of their collective dependence on water, and has raised their estimation of the importance of the irrigation panchayat.

Attempts are also being made in Chambal (Madhya Pradesh) to introduce organization on an outlet (rather than village) basis. Last year, for example, a young assistant engineer in charge of a subdivision attempted to introduce outlet associations over an area of about 500 acres. He found that farmers under the same outlet expressed strong reservations about being put in the same grouping as someone from a different village, caste, or, in some cases, political party. So he went ahead and encouraged them to form suboutlet associations, comprised of only five or six cultivators (having adjacent holdings on the same watercourse). So far nothing has come of them. As from next *rabi*, however, the engineers say they are planning to pay much more attention to the introduction of some kind of outlet or suboutlet organization.

The Need for Socioeconomic Research

The CAD program is costly, requires a vast expansion of governmental influence at outlet level, and takes a new step in Indian agricultural administration. One might expect a program of this magnitude and significance to attract the attention of social science researchers. Yet to the best of my knowledge very little research is being done (excluding work by engineers, agronomists, etc.). It is not that there is as yet nothing to study. The "Kota" method of land development (including consolidation) has now been tried in many pilot projects located in several command areas; a few of these have a history of three or four crop seasons. By the beginning of the coming monsoon, about 15,000 acres in three commands, and several hundred acres in six more, will have been given the full package. In Ghataprabha command (Karnataka), about 50,000 acres have been developed by an intermediate method which excludes consolidation, and in Jayakwadi (Maharashtra) about 12,000 acres have been irrigated by an even more limited approach. Attempts have already been made in at least one command to introduce rotational irrigation and to form water users' associations. Of the 50 projects designated for CAD treatment, over 30 now have Command Area Development Authorities, some of which are functioning on the ground as well as on paper; and one can already see substantial variation in their scope, powers, autonomy, and staffing patterns. There is certainly something to study.

Importance of the Early Stages

It might be argued that the program is too recent to warrant serious study at this stage. This I think is a mistake. It is true that the problems which are dominant in the first few years may not be those which will remain in the long term. But the difficulties of introducing changes, especially those of an institutional kind, are worth studying in themselves, in particular because given that much of the canal-irrigated area in India is and will be new, an important influence on the performance of irrigated agriculture (and therefore of total agricultural output) will be the performance of irrigation projects in the early operational stage, before some kind of equilibrium is reached.

Organizational Structures

However, the gap in CAD research is only part of a much larger gap in socioeconomic research on canals and canal-irrigated agriculture. In a country with a history of extensive canal irrigation going back before the beginning of this century it is a remarkable and unfortunate fact that there exists, to the best of my knowledge, not a single comprehensive account of how a canal is actually administered. There are not more than incidental accounts by sociologists and anthropologists of the relations among irrigators, between irrigators and Irrigation Department personnel, and among these personnel. Economists have done little better. There are a number of studies which use programming techniques to determine optimum water release rules, as well as studies of tube-well economics. But compared to the many detailed studies of the distribution of other inputs, such as fertilizers and credit, there is very little on the allocation of canal water.[24]

Few are likely to question the value of conventional economic studies of optimum water allocations and related matters. But what about institutional studies? Do the organizational structures of water use really matter, or are we

24. Exceptions are Minhas, 1972; Reidinger, 1971; Weaver, 1968; Vander Velde, 1971. See also Chambers, 1974; Hart, 1961; Hamilton, 1969. An outstanding exception from a historian is Whitcombe, 1972. Perhaps the clearest evidence for the argument that questions of water management have been overlooked comes from a number of case studies which focus directly on irrigation. Vander Velde, an economic geographer, reports (1971) on the distribution of benefits from canal irrigation in a Haryana village, but confines his occasional remarks about water organization at outlet level to footnotes. Weaver, an economist, has even less to say about water organization in his study area of Madhya Pradesh (1968), though his theme was optimal water allocation. The Sardar Patel Agro-economics Research Centre's study (Brahmbhatt, 1974) of the impact of canal irrigation in a Rajasthan village has no more than a few lines describing the organization through which water is allocated among cultivators. Epstein (1962) examines the social and economic effects of irrigation in a Mysore village, but tells us little about how the water—from a large-scale canal—was managed or what part it had in local politics. Hunter (1969) places questions of water availability at the very center of his chapter on "technical factors" in agricultural development: in subsequent chapters on "institutions" and "administration" he says not a word about water institutions or water administration.

just grinding sociological axes in saying that they do? Take water users' associations, for example: are they potentially important in the South Asian context and is it worth devoting public funds and the energies of public officials to the purpose of trying to initiate and sustain them? Or are they merely thrown into the CAD package in the wake of *garibi hatao*, or as part of a vision of what the good society should be like?

Managerial Control

The general answer is that these things do matter: because how water is distributed by surface flow over a command area depends on institutional as well as physical factors. The argument goes as follows: Institutional factors determine—or are a big influence on—the degree and location of managerial control. The degree of managerial control is inversely related to the water requirements of the system, such that the greater the degree of control, the greater the area that can be supplied with a given amount of water per acre from a given water source, or the greater the amount of water that can be applied to a given area (holding that seepage, percolation, evapotranspiration, and land preparation are constant). In other words, for a given water source, the design efficiency (in an engineering sense, i.e., the amount of water that has to be diverted into the canal system in order to supply a given amount of water to the fields) fixes the level of managerial control needed in order to irrigate a given area. The required level of control has important implications for relationships among farmers and between farmers and irrigation personnel: broadly speaking, increased cooperation and coordination are required as a given water source is used to irrigate a larger area.[25] The significance of this is clear: there are strong pressures from expectant farmers, politicians, and planners to extend canals to cover ever larger areas; but at the same time, the water requirements of the new varieties of seed are more exact than those of the older varieties. Improved managerial control is one means of reducing the conflict between these two objectives—with all that "improved managerial control" implies for relations among farmers and between them and irrigation officials. (Lining of channels, land preparation, better watercourse maintenance, and consolidation are other means.)

Southeast Asian Comparisons

Levine (1971) makes an admittedly crude analysis of differences in Southeast Asian canal systems to show that managerial control interacts with other factors to produce variations in engineering efficiency. Although the Philippines, Malaysia, and Taiwan have a similar monsoon climate, the efficiency

25. See Levine, 1971.

of their canals tends to vary widely: those in the Philippines tend to have efficiencies of less than 25 percent, those in Malaysia about 40 percent, and those in Taiwan 60 percent or more. The Taiwanese systems practice rotational irrigation within 50-hectare blocks; the lateral distribution channels are frequently lined; there are control gates and measuring devices at each 50-hectare outlet; the network of field channels is extensive; and irrigation goes on during the night as well as during the day. In the Malaysian systems, control up to the level of the primary and secondary canals is centralized and effective; water policy is specified by ordinance each year; but continuous (rather than rotational) irrigation is practiced; outlets serve relatively large farming areas; and distribution beyond the outlet is in the hands of the farmers, who have constructed only a thin network of field channels. The Philippines' systems practice continuous irrigation; there are few effective controls in conveyance channels and outlets; channel maintenance is poor; there are no measuring devices; irrigation is not done during the night; and farmer cooperation in water distribution is variable and frequently poor. As Levine points out, it is easy enough to list the differences; but we know little about how these factors interact or the reasons for their specific combinations. What seems clear is that farmers' behavior and the bureaucracy through which water is controlled are important variables.

Benefits for the Poor

Increased managerial control, then, can be used to save water and thereby increase output. But it is also important to ask about distributional issues: whose output? Water management is one of those happy areas of policy where there is scope for making many people better off and few people absolutely worse off. For example, a variety of measures can be used to reduce the uncertainty of water deliveries, including, of course, improved managerial control, but also including improvements in physical structures (canal lining, sluice gates, etc.). *A priori*, it is likely that reduced uncertainty of deliveries will tend to benefit small farmers relatively more than big farmers. First, tail-enders tend to suffer relatively more from uncertainty of deliveries than top-enders, who are likely to get proportionately more of what water is in the channel when water is scarce: partly for this reason, tail-enders tend to be the poorer.[26] Secondly, as between farm-price stabilization and farm-output stabilization, the latter is a more powerful means of stabilizing incomes; and since small farmers tend to be more adversely affected by income fluctuations than large farmers, reliable irrigation benefits them relatively more.[27] Thirdly,

26. See Reidinger, 1971, p. 109. In one case he mentions, the larger and more powerful farmers were originally placed toward the end of a watercourse, because their land was relatively low; they then got the position of the outlet changed so that their lands came to be near the new outlet, to the cost of the smaller farmers.

27. See Lipton, 1970.

and more speculatively, ties of personal dependence (as in the *jajmani* relationship) are a response to uncertainty; so a reduction in uncertainty is likely to facilitate the emergence of more single-interest relations between landlord and tenant or laborer; this, it may be argued, is a necessary first step to collective action by tenants or laborers.

The Effect on Output

Rotational irrigation is one way both of saving water and of reducing uncertainty.[28] In this case, however, direct redistribution is also likely to occur, and some will be made worse off. Daniel Thorner (1962) suggested that on the Sarda canal, in 1959, there were (crudely) two standards of water service: one for the strong and one for the weak. The strong grew profitable water-consumptive crops, in the expectation of water at the time and in the amounts required; the weak grew rain-fed crops and treated canal water as "an intermittent blessing, to be welcomed when it comes" but not to be relied upon. Something like this situation probably obtains on many canals where rotational irrigation is not enforced. To enforce a rotation would both reduce the uncertainty of water arrivals for the weak and also give them more water per acre, by reducing the water taken by the strong. The effect on aggregate output depends on the present position of the strong and the weak on their respective production functions; casual observation suggests it might well favor a redistributive strategy.

These are matters which require research. To what extent do large landowners get more water per acre than small landowners? What is the potential gain in aggregate output of a more equal distribution of water per acre? What would be required, of farmers and the public sector, to introduce rotational irrigation and enforce it effectively?

But perhaps the most important issue concerns the likely conflict between economic efficiency and equity objectives. A fixed rotation is inefficient in the sense that it allows no adjustment to need: officials are not meant to have the sort of discretion which allows them to say, "These farmers badly need water, these others do not; therefore the former shall be given water (although it is not their turn)." But once officials are allowed to allocate according to need, thus potentially increasing economic efficiency, it may well be difficult to prevent other criteria—political pressures, for example—from dominating their decisions. Although a fixed rotation, with the objective of equalizing the amount of water received by each acre of the command over the cropping season, is evidently unsuitable for meeting the exact water requirements of the

28. To say that rotational irrigation reduces uncertainty is too simple, for reasons analyzed in Reidinger, 1971. But where uncertainty refers to both timing and amount, rotational irrigation seems likely to be less uncertain than continuous irrigation, at least for those away from the outlet. There is the further problem of whether reduced uncertainty of delivery is obtained at the cost of increased divergence between expected deliveries and crop water needs.

new seeds, it may be the best compromise which can be reached between efficiency and equity objectives, given the capacity of the legal and administrative systems. The same point applies at the level of the outlet. There is talk of allowing elected leaders within each outlet command considerable discretion in altering the rotation within that outlet; and there is also talk of making water sales or trades legal. But with what effect on equity?

Adapting to Local Conditions

Water users' associations are another subject on which research might be expected to illuminate policy issues. Are *water* associations likely merely to reinforce the existing distribution of power, as is often the case with service cooperatives, or are there grounds for thinking that water associations might operate somewhat differently? Would it be better from the point of view of small farmers if the functions now to be performed by local users' associations were instead left with the bureaucracy? What types of associations—in terms of size, functions, operating principles, and principles of recruitment—are likely to operate best, in what socioeconomic conditions? Most new irrigation projects, not only in India, recommend some form of water users' organization based on models to be found in developed countries—particularly the United States and Taiwan. Little attempt is made to work within existing institutions or develop adaptations based on existing patterns of social relationships.[29] Little if any attempt is made to work out what must be done by the project authorities or the government sector generally to get organizations of a given type to work more or less as intended. While their importance may be recongized in a general kind of way, they are not, in sharp contrast to the engineering solutions, subjected to a detailed and site-specific analysis; nor, indeed, generally speaking, to any analysis at all. One reason, it may be suggested, is that we lack anything more than the rudiments of a methodology by which such analyses might be made; and that is because social scientists have not yet attempted to find one.

What Makes a CAA Work?

Take, finally, the CAAs. The type of organization recommended by the center breaks established lines of promotion, weakens the position of the Irrigation Department, and reduces the leverage of local politicians on the flow of governmental resources. Different states have created their own variants of the center's model: in some, control over funds and staffing lies at project level, in others at state level, and in still others the head of the authority is expected to achieve coordination using only his senior rank and informal pres-

29. See Levine, 1971.

sure, without real control over the resources or staff of constituent departments. In some cases the head of the authority can only be transferred with the approval of the chief minister, in others the procedures are less closed. Some projects are headed by IAS men, others by technical officers from Irrigation of Agriculture. Some have a special CAD cell, others rely on the staff of existing departments. Some authorities have power to raise loans on the commercial market, others do not. One may assume that the specific combination of these various features will influence how effectively the CAA is able to perform in its specific environment. But this, again, is a question for research.

References

BRAHMBHATT, D. M. 1974. *Impact of Irrigation on a Rajasthan Canal Village*. Sardar Patel Agro-economic Research Centre, Sardar Patel University, Vailabh Vidyanagar.

CHAMBERS, R. 1974. "The Organization and Operation of Irrigation: An Analysis of Evidence from South India and Sri Lanka." Paper given to seminar, Project on Agrarian Change in Rice-Growing Areas of Tamil Nadu and Sri Lanka, St. John's College, Cambridge.

CHATURVEDI, M. C. N.d. (1974). *Water in Second India*. Second India Studies, Ford Foundation, New Delhi.

EPSTEIN, T. S. 1962. *Economic Development and Social Change in South India*. Manchester: Manchester University Press.

HAMILTON, R. E. 1969. "Damodar Valley Corporation: India's Experiment with the TVA Model." *Indian Journal of Public Administration* 15, no. 1:86–109.

HANUMANTHA RAO, C. H. 1974. "Growth of Irrigation in India: An Outline of Performance and Prospects." Paper for seminar, The Role of Irrigation in the Development of India's Agriculture, Institute for Social and Economic Change, Bangalore, October 28–30.

HART, H. C. 1961. *Administrative Aspects of River Valley Development*. Indian Institute of Public Administration. Bombay: Asia Publishing House.

HUNTER, G. 1969. *Modernising Peasant Societies*. London: Oxford University Press.

India, Agricultural Prices Commission. 1975. *Report,* Rabi 1975.

______, Ministry of Irrigation and Power, Irrigation Commission. 1972. *Report*. 4 vols. New Delhi.

______, National Commission on Agriculture. 1973. *Interim Report on Modernising Irrigation Systems and Integrated Development of Commanded Areas*. New Delhi.

______, Planning Commission. 1973. *Draft Fifth Five-Year Plan, 1974–1979*. New Delhi.

LEACH, E. R. 1961. *Pul Eliya, a Village in Ceylon: A Study of Land Tenure and Kinship*. Cambridge: Cambridge University Press. [Chapter 5 above.]

LEE, T. S., and CHIN, L. T. 1961. *Development of Rotational Irrigation in Taiwan*. Far East Regional Irrigation Seminar, Taipei.

LEVINE, G. 1969. "Lowland Irrigation Requirements in the Humid Tropics, with Special Reference to the Philippines." Paper presented at seminar at International Rice Research Institute, March 13.

———. 1971. "The Management Component in Irrigation System Design and Operation." Department of Agricultural Engineering, Cornell University, September. Mimeo. [Chapter 3 above.]

LIPTON, M. 1970. "Farm Price Stabilisation in Under-developed Agricultures: Some Effects of Income Stability and Income Distribution." In *Unfashionable Economics: Essays in Honour of Lord Balogh,* ed. P. Streeten.

MINHAS, B. S., et al. 1971. *Scheduling the Operations of the Bhakra System: Studies in Technical and Economic Evaluation.* Indian Statistical Institute. Calcutta: Statistical Publishing Society.

RAO, V. K. R. V. 1974. "New Challenge before Indian Agriculture." Dr. Panse Memorial Lecture, Institute of Agricultural Research and Statistics, New Delhi, April 26.

REIDINGER, R. B. 1971. "Canal Irrigation and Institutions in North India: Microstudy and Evaluation." Ph.D. thesis in agricultural economics, Duke University. Ann Arbor: University Microfilms.

———. 1974. "Institutional Rationing of Canal Water in Northern India: Conflict between Traditional Patterns and Modern Need." *Economic Development and Cultural Change* 23, no. 1. [Chapter 12 above.]

THORNER, D. 1962. "The Weak and the Strong on the Sarda Canal." In *Land and Labor in India.* Bombay: Asia Publishing House.

UNDP/FAO. 1974. *Soil and Water Management Project, India: Project Findings and Recommendations.* Terminal Report. Rome.

VOHRA, B. B. 1971. *A Charter for the Land.* Ministry of Agriculture (Department of Agriculture). New Delhi.

———. 1973a. "Some Constraints in the Implementation of Water Management Programs in Canal Irrigated Areas." Paper presented at symposium on on-farm water management, Park City, Utah, October 1–8. Mimeo.

———. 1973b. "Implementation of Water Management Programs in Canal Irrigated Areas." Paper presented at symposium on on-farm water management, Park City, Utah, October 1–8. Mimeo.

WADE, R. H. 1975. "Mysteries of the Dhora." *Economic and Political Weekly* 10, no. 29.

WEAVER, T. F. 1968. Chapters 6–12 in *Developing Rural India: Plan and Practice,* by J. W. Mellor, T. F. Weaver, U. J. Lele, and S. R. Simon. Ithaca: Cornell University Press.

WHITCOMBE, E. 1972. *Agrarian Conditions in Nothern India,* vol. 1: *The United Provinces under British Rule, 1860–1900.* Berkeley: University of California Press.

Index

Irrigation and
Agricultural Development
in Asia

Designed by G. T. Whipple, Jr.
Composed by The Composing Room of Michigan, Inc.
in 10 point VIP Times Roman, 2 points leaded,
with display lines in Helvetica and Helvetica Bold.
Printed offset by Thomson/Shore, Inc.
on Warren's Olde Style Wove, 60 pound basis.

Library of Congress Cataloging in Publication Data
Main entry under title:

Irrigation and agricultural development in Asia.

Includes index.
1. Irrigation—Social aspects—Asia—Case studies. I. Coward, E. Walter.
HD1741.A78I77 301.18′095 79-24319
ISBN 0-8014-1132-7
ISBN 0-8014-9871-6 pbk.